T0256913

AN **EPISTEMOLOGY** OF THE **CONCRETE**

Experimental Futures:
Technological Lives,
Scientific Arts,
Anthropological Voices
A series edited by
Michael M. J. Fischer and
Joseph Dumit

AN **EPISTEMOLOGY**
OF THE **CONCRETE**

Twentieth-Century Histories of Life

Hans-Jörg Rheinberger

Foreword by Tim Lenoir

DUKE UNIVERSITY PRESS

Durham & London

2010

© 2010 Duke University Press

All rights reserved

Printed in the United States of America

on acid-free paper ∞

Designed by Amy Ruth Buchanan

Typeset in Cycles, Arepo, and Futura by

Tseng Information Systems, Inc.

Library of Congress Cataloging-in-

Publication Data appear on the last

printed page of this book.

Epistemologie des Konkreten was originally

published in German by Suhrkamp

Verlag Frankfurt am Main © 2006.

CONTENTS

Epistemology Historicized

Making Epistemic Things

TIM LENOIR, DUKE UNIVERSITY

Hans-Jörg Rheinberger emerged as the leading historian and philosopher of the biological sciences in the decade following the publication of his path-breaking book *Toward a History of Epistemic Things: Synthesizing Proteins in the Test Tube* (1997). Although conceived as a philosophy of experimental practice, this work has had a far-reaching impact beyond the history and philosophy of science, in fields ranging from anthropology, sociology, and economics, to literary studies. Rheinberger's book came out in a period characterized by the ascendancy of the biological sciences and a growing preference for practice-oriented as opposed to theory-dominated accounts of knowledge production. At the same time, critics of the monolithic discourses of identity and representation firmly embedded within many fields in the humanities and social sciences were gravitating toward epistemic models foregrounding repetition and difference; and all of this was occurring in the midst of a revolution in the cognitive neurosciences that was dissolving the traditional static epistemic framework of a solitary subject confronting a pre-given object in favor of distributed cognition, embodied reason, enaction and process.

Rheinberger's work resonated with these themes, and for good reason. He held degrees in both molecular biology and philosophy. For more than a decade prior to writing his now classic book, he was a researcher in one of the world's leading molecular biology laboratories—the Max-Planck Institute for Molecular Genetics in Berlin—developing models of the ribosome and its role in protein biosynthesis. While Rheinberger was steeped in every dimension of laboratory practice and experimental work, he also continued to engage deeply with traditions of continental philosophy, particularly the works of Edmund Husserl, Martin Heidegger, Jacques Derrida, Michel Foucault, Gilles Deleuze, Jacques Lacan, and others; indeed he was the translator of works of Derrida and Lacan into German. His background of experimental and theoretical work in molecular biology combined with his deep engagement in continental and postmodern (particularly French) philosophy has provided the foundations of a philosophical oeuvre with profound interdisciplinary significance.

Rheinberger's work is an exercise in historical epistemology, a distinctive Franco-German approach to the history and philosophy of science. In contrast to earlier traditions in the philosophy of science that treated truth as independent of the context of discovery and the history of scientific knowledge as a linear, progressive march in the elimination of error, asymptotically approaching nature, historical epistemology treats knowledge as historically contingent and focuses on uncovering the conditions of possibility and fundamental concepts that organize the knowledge of different historical periods. Rheinberger regards the aim of historical epistemology as to understand the conditions of possibility underlying the knowledge and practice of specific conjunctions of scientific and technical practice, the social, institutional, and cultural configurations in which they operate. Historical specificity is essential to this philosophical project.

Rheinberger traces the formative questions of an epistemology of the concrete to the work of Edmund Husserl, Ludwik Fleck, Gaston Bachelard, and Georges Canguilhem. Each of these epistemologists engaged in a concerted critique of the positivist movement in the natural sciences of the early twentieth century. Against the positivists, Fleck and Bachelard in particular show that the objects of scientific knowledge are not given ready made in nature. Indeed, according to Bachelard scientific objects do not even exist in nature; rather, they are technically produced in a continuous process of assemblage, rectification, and repetition—a process Rheinberger calls "recursion"; that is to say, the theory of knowledge considered classically as an existing structure of logic applied to fit lock-and-key to an externally existing, pre-given nature is replaced in Rheinberger's account by epistemology considered as a deeply historical process of constituting both the scientific object and our knowledge in a never-ending recursive process of reconfiguration and rectification.

Rheinberger's unique ability to engage molecular biology, continental philosophy, and the early architects of historical epistemology in a dialogue allows him to introduce a new conceptual repertoire to the analysis of knowledge production. Foremost within this new conceptual repertoire are the closely interrelated notions of "experimental systems," "epistemic things," and the "phenomenotechnique."

Bachelard's idea of the "phenomenotechnique" is crucial to Rheinberger's enterprise of historical epistemology. This is the notion that technology is not just an ancillary result of scientific investigation but also the very modus operandi of science. New phenomena, Bachelard tells us, are not simply discovered, but invented—constructed from scratch; and in another memorable formulation, Bachelard calls the phenomena that are the object of scientific investigation, such as the trajectories of isotopes in a mass spectrometer, "reified theorems." Accordingly the very concepts that science operates with are bound up inextricably with the instruments through which phenomena are produced and stabilized as objects of investigation. Rheinberger writes, "Phenomenon and instrument, object and experience, concept and method are all engaged in a running process of mutual instruction." The instrument represents the material existence of a body of knowledge—a reified theorem—while the emphasis on "instruction" underscores the point that the instrumentally mediated object becomes an agent too in the process of making knowledge.

Breaking with the notions of a pre-existent referent grounding scientific representations and theory-dominated accounts of knowledge construction, Rheinberger focuses on how research is enacted at the frontiers between the known and the unknown through the construction of experimental systems that give rise to epistemic things. Modern biological research, Rheinberger argues, is not theory-driven but rather dominated by the choice of experimental systems. The name of the game is constructing a robust experimental arrangement of instruments, chemical processes, physical structures, and biological materials capable of generating a network of experiments.

The experimental systems that Rheinberger investigates do not consist solely of scientific instruments but also include appropriately cultivated or even engineered bits of nature, such as a model organism. The many varieties of mutant Drosophila flies that fueled much of classical genetics, the nematode worm *C. elegans* central to modern molecular genetics and developmental biology, and the specially designed embryos of zebra fish used in contemporary environmental impact studies of nanotechnology are examples of model organisms. A model organism is a living thing from the plant, animal, or bacterial kingdom that has been tailored to experimental purposes; manipulating it can generate insights into the constitution, functioning, development, or evolution of an entire class of organisms. The operative criteria for selection of a model organism are the ease with which it can be maintained and handled, the quantity and quality of the knowledge already accumulated about it, and the relative accessibility of the phenomenon under investigation.

Just the right amount of vagueness rather than theoretical precision is the key to a powerful experimental system, because the scientific object, the research object, or "epistemic thing" is latent within and emerges out of the technical conditions embedded within the experimental system. As Rheinberger explains, "epistemic things" are what one does not yet know, things contained within the arrangements of technical conditions in the experimental system. Experimental systems are thus the material, functional units of knowledge production; they co-generate experimental phenomena and the corresponding concepts embodied in those phenomena. In this sense, experimental systems are techno-epistemic processes that bring conceptual and phenomenal entities—epistemic things—into being. Epistemic things themselves are situated at the interface, as it were, between the material and conceptual aspects of science.

Once revealed, epistemic things become materialized interpretations that form the components of models. According to Rheinberger, the referent of scientific work is the model, comparison never being made to nature but always to other models, a process that Rheinberger analogizes to the operation of Derrida's supplement. In Rheinberger's study of the molecular biology of protein synthesis, for example, different cellular components are defined by centrifugation, sedimentation properties, radioactive tracers, and chromatograms using a DNA sequencing gel. The scientific object is gradually configured from the juxtaposition, displacement, and layering of these traces. The experimental systems molecular biologists design are "future generating machines," configurations of experimental apparatus, techniques, layers of tacit knowledge, and inscription devices for creating semi-stable environments—little pockets of controlled chaos—just sufficient to engender unprecedented, surprising events. When an experimental system is working, it operates as a difference-generating system governed by an oscillatory movement of stabilization-destabilization-re-stabilization—what the molecular biologist François Jacob, echoing a similar statement of Derrida, called the "jeu des possibles." At the heart of the laboratory/labyrinth are experimental arrangements for transforming one form of matter into another and inscription devices for transforming matter into written traces. The products of this complex of experimental arrangements and inscription devices are trace-articulations, which Rheinberger calls "graphemes." They represent certain aspects of the scientific object in a form that is manipulable in the laboratory. Graphemes in turn are the elements for constructing models. They are the manipulable signs scientists use in "writing" their models.

Also contributing to the graphematic traces of a well-oiled lab and experimental system is what Rheinberger refers to as an "economy of the scribble": the notes and scribbling of the lab members, from excerpts of literature to notes on basic concepts and fragmentary ideas, striking correlations, sketches of experimental setups, data from single sets of experiments, tentative interpretations of experimental results, corrections, provisional calculations, calibrations of instruments, designs for new equipment, and so on. Through the economy of the scribble, the organization of an experiment in space and time is projected onto a two-dimensional surface, which facilitates exploration of new ways of ordering and arranging data. More than serving a merely reductive function, laboratory notes, protocols, and the economy of the scribble are themselves new resources and materials that give research its distinctive contours and prevent it from being prematurely closed off.

For Rheinberger and the architects of the epistemology of the concrete the production of knowledge is not only a deeply historical process. It is also a product of culture. Husserl emphasized that scientific knowledge is the product of scientists working together, and Ludwik Fleck famously argued in the *Genesis and Development of a Scientific Fact* (1935) that the historical is deep at work in the research process shaping scientific facts from local interactions among communities of scientists he called "thought collectives." For Bachelard as well, scientific knowledge always marks an ongoing technical and instrumental evolution emerging from processes situated within cultural configurations. But while the construction of knowledge is a deeply historical and cultural process, none of the historical epistemologists of interest to Rheinberger's account regard science as an arbitrary or purely social construct. The centrality of the "phenomenotechnique" provides the primary counterweight to a purely social constructivist account. For Rheinberger, nature appears simultaneously as agent and epistemological obstacle in the construction of both knowledge and society. Like Bruno Latour, another close student of Bachelard's work, Rheinberger describes the entities constructed in the labs as hybrids of nature and culture. The scientific phenomenon, technically produced, is also discursively constructed, so that phenomena are always material and discursive entities; and just as the objects contemplated by science are not immediately given but always constructed in a continuous material-discursive dialectic, so too, the scientific mind is not immediately given. Discussing Bachelard's elaboration of this important notion, Rheinberger writes that "The knowing mind has to externalize itself and become instrumental; for it is itself technically mediated, as are all its concepts, the categories of scientific knowledge not excepted. The consequence is that the scientific mind and scientific object enter into a symbiotic relationship of reciprocal externalization, and simultaneously, interiorization." The centrality of the phenomenotechnique to the production of knowledge has the consequence that the external world is deeply intertwined with the production of the scientific mind and particular scientific communities.

In learning about how epistemic things arise from the hybrid assemblages of machines, practices, concepts, and humans inhabiting laboratory spaces, the reader is struck by the similarity of Rheinberger's experimental systems to Andy Clark's or Edwin Hutchins's accounts of distributed systems of cognition in which cognition resides as much in distributed external objects as in the human cognitive agents themselves. Thus, Rheinberger talks about the extended tangle of networks of experimental systems and technical objects and how researchers are embedded within them as symbiotic elements of the machinic assemblage. Elaboration of the network, extracting the signals that carve out the epistemic things, requires the researcher's intimate familiarity with the system. This process takes time, which explains why, as Rheinberger argues, once experimenters have established their experimental network, they often stick to it in an almost symbiotic fashion. But once the system has become familiar to those who inhabit it, its own momentum may take over. The more the experimenter learns to manipulate the system, the better the system comes to realize its intrinsic capacities: in Rheinberger's terms the experimental system starts to manipulate the researcher and to lead him or her in unforeseen directions.

This radically historicized account has two important implications for the shape of historical epistemology. On the one hand the obstacles at the heart of research activity are never fully containable or predictable. Empirical thought is clear only in retrospect, Rheinberger tells us. While knowledge gained through this agonistic encounter with nature is confirmable through acts of recursion, there is no pre-specifiable golden path or universal method that can provide a blueprint for knowledge production. Research is inherently untidy and requires an appropriate cultural setting that sustains innovation by allowing the new and unpredictable to emerge. Environments that sustain controlled chaos are the heartland of innovation.

The deeply historical and constantly evolving character of epistemic events has as a further consequence that the epistemology of the sciences is best investigated through detailed historical case studies. This view contrasts sharply with earlier philosophers of science, such as Karl Popper and Imre Lakatos, who proclaimed that the history of science belonged in the footnotes of epistemologies treating the logic and methodology of the growth of knowledge. Science, according to Rheinberger, might be described as the most profoundly historical of cultural enterprises, since scientific activity is defined precisely by its need to render itself obsolete in order to remain scientific. As we have seen, scientific knowledge is constructed in highly localized and constantly evolving contexts. Knowledge emerging from such localized processes is disunified, fragmented, and always driving toward specialization. The history of epistemic events that Rheinberger maps is a rhizomatic structure. Accordingly, epistemic history should not be a history of grand unified theories but of conjunctions that give rise to new concepts and instrumental configurations. In this view the history of science is focused on dense material-discursive configurations, conjunctures, or nodal points in which something comes about that has its origins in minor decisions constrained by local, particular conditions.

Rheinberger argues that molecular biology, for instance, did not arise from the articulation of a preexisting theoretical paradigm. What was involved was an "assemblage" arising from work on bacteriophages and the tobacco mosaic virus. This work crystallized in the early 1950s with the biophysical characterization of the double helix structure of DNA and peaked with the deciphering of gene expression and the genetic code between 1960 and 1965. The research of this period was conducted with instruments and techniques such as the ultracentrifuge, liquid scintillation counters, electron microscopy, chromatography, and electrophoresis. These instruments had their origins in contexts having nothing to do with molecular biology or classical genetics. Molecular biology took over various biophysical and chemical analytical techniques without prior theoretical coordination. In the midst of this assemblage, new, biologically specific concepts emerged shaped by the language of information, giving molecular genetics its distinctive profile by the end of the 1960s.

The 1970s brought a second assemblage fashioned from the development of genetic engineering techniques, such as cloning, recombinant DNA, the polymerase chain reaction (PCR), as well as new molecular biology techniques involving nucleic acids and enzymes. Unlike the first phase of molecular biology, these new techniques were not imported from other research fields but were born from the first phase of molecular biology itself, and they became instruments for manipulation of materials at the intracellular level. These new developments enabled increasingly complex conceptions of the expression and transmission of genetic information, and, indeed, a new and richer set of conceptions of genes.

The lesson Rheinberger wants to impress upon us with this history of the two phases of molecular biology is that epistemic history should be focused on examining the configurations of instruments, techniques, and the conditions that have brought about productive experimental settings and the operational concepts materially tied to them, rather than on the history of theory. This position is forcefully represented in this book by an examination of the multiple transformations the concept of the gene has undergone over the period of more than fifty years since the discovery of DNA. The "gene" in Rheinberger's account is an example of an "epistemic thing." From the outset, the gene was a fuzzy concept, an imprecise boundary object with no exact, systematically fixed meaning. The fertility of imprecise concepts like it depends on their ability to generate productive tension—what Rheinberger calls "contained excess"—an operational potential for being integrated into different experimental contexts in accordance with changing needs. Over the course of its history the gene has not been a unified concept, and molecular biologists, in Rheinberger's view, have not needed a unified, unambiguous, rigorously determined notion of the gene to make major advances in their field. On the contrary, molecular biology, despite its claim to general validity, has been a mosaic of many contexts, consisting in contributions from different disciplines, experimental systems, and views of the genome. Molecular biology is a hybrid science combining experimental systems from biophysics, biochemistry, and genetics. It makes use of a great diversity of model organisms in its search for biological functions at the molecular level. It is no wonder, writes Rheinberger, that its concepts are hybrid. Indeed the expansive potential of its discourse resides precisely in the hybrid nature of its concepts.

Hans-Jörg Rheinberger has given us a profound treatment of the philosophy of experiment and a powerful set of instruments for understanding the construction of knowledge. Intertwined among the carefully drawn case studies in the book is one penetrating analysis after another in which Rheinberger's proposals for an epistemology of the concrete are shown to resolve the key debates that have shaped science studies and the philosophy of science over the past three decades—issues such as constructivism versus realism, theory domination versus practice and experiment, and more—all discussed beautifully in an organically interconnected exploration of experimental systems in classical and molecular genetics, the close mutual dependence of model organisms and key instruments in the formation of concepts that have shaped molecular biology, and the role of writing and graphemes in shaping the hybrid nature/culture entities that become epistemic things. *An Epistemology of the Concrete* offers a methodological framework and a set of research exemplars that will shape science studies for years to come.

This book brings together case studies and reflections on the history and epistemology of the life sciences. As building blocks of an epistemology of the concrete, they constitute a sequel to earlier studies in which I defined the basic features of the relationship, in experimental systems, between epistemic objects and the technical conditions of their production.[1] A number of the texts included here have seen publication elsewhere; these have undergone extensive revision for the present volume. All the chapters in the book represent new attempts, each undertaken from a different angle and adopting a slightly different historiographic attitude, to communicate something of the fascination of *scientific things*—those organisms, spaces, apparatuses, and techniques that have been colonized and transformed by research, which they in turn have transformed and diversified. Communicating this fascination is one of the ambitions of a history of the material culture of the sciences.

The book is divided into four parts, each with a different focus. Part One is about the ways of historicizing scientific knowledge as encountered in the work of four twentieth-century epistemologists: Ludwik Fleck, Edmund Husserl, Gaston Bachelard, and Georges Canguilhem. Four self-contained case studies in the history of genetics and reproductive biology make up Part Two. All four are accounts of work pursued by scientists active in the Kaiser Wilhelm Institute for Biology in Berlin-Dahlem. Presented here are various configurations of biological and, in particular, genetic experimentation in the first half of the twentieth century; the case material is drawn from the experiments of Carl Correns, Max Hartmann, Alfred Kühn, and the Dahlem "Virus Research Workshop" (including Gernot Bergold, Rolf Danneel, Georg Melchers, and Gerhard Schramm, among others). These case studies focus on the function of model organisms in the life sciences. Part Three concentrates on the interaction between apparatuses and concept formation in molecular biology. At the center of the first study in Part Three is the gene, the pivotal concept in both classical and molecular genetics. The second looks at an apparatus for measuring radioactive traces that has become indispensable to molecular biology: the liquid scintillation counter. The third examines the language of information typical of the new biology. In the fourth and last part of the book, I consider a number of fundamental epistemological issues bearing on the relationship between instruments and objects of knowledge; laboratory preparations as a special class of epistemic objects; and the note-taking and write-up techniques utilized in research labs.

: : :

The early decades of the twentieth century witnessed a deep crisis of positivistic thinking in every field of knowledge. No quick and easy solution appeared on the horizon. In the nineteenth century, the history of science had been dominated by descriptive models that told a story of progressive discovery. Only gradually did historians of the sciences learn to take their distance from the notion of the linear forward march of the disciplines and bring the historicity of scientific knowledge itself into their field of vision; only gradually did they begin to comprehend science by setting out from both its conditions of production in the narrower—material and institutional—sense of the term, and, simultaneously, its broader cultural contexts. For the philosophy of science, this held out, not least, the prospect of again bringing together the long sharply separated contexts of "justification" and "discovery," of the validity of knowledge on the one hand and the historical conditions of its production on the other. This development in philosophy of science was by no means a process internal to philosophy alone; rather, it came about in a reciprocal exchange with the concrete scientific revolutions of the twentieth century, with their premises and consequences. The process of historicizing epistemology was perhaps the most important twentieth-century contribution to the philosophy of the sciences.

Historicism in the human sciences, the counterpart to positivism in the natural sciences, gave way to a historical epistemology that found its first consistent formulation in the work of Ludwik Fleck and Gaston Bachelard, who contend that the sciences do not find their objects ready-made but have to constitute them using specific epistemic settings. Thus the French historian and philosopher of science Bachelard declares that these objects "do not *exist* in nature; they have to be technically produced."[2] The implication is that the objects of research all have their particular, longer or shorter, histories not just in the humanities but also in the natural sciences; that they are operationalized in different ways, at different levels, and for different lengths of time in different scientific fields; and, finally, that they have effects of differing intensity in these different fields.³ The scientific disciplines, which had emerged roughly in their present form by the end of the nineteenth century, were in their turn ultimately only institutional expressions of this fundamental process of the production of different scientific objects.

The primordial dichotomy of nature and culture inscribed in the division between the natural sciences and the humanities has been challenged ever since. Is there any reason to hold fast to this stubborn, yet central difference? The sociologist and philosopher of science Bruno Latour has urged us to reconsider the advantages of a kind of pre-modern relationship to the world and simply make the best of the fact that we have always already inhabited a universe of hybrids between nature and culture; he has even advocated convening a "parliament of things."⁴ This provoking metaphor indicates, at the very least, the next step to take: precisely because no scientific fact is ever simply given, but is, literally, *made* by the joint efforts of a community, we should cultivate a certain readiness to work with "mixed" explanations.

For Bachelard, it is not the only defining feature of modern knowledge formation that it constructs its objects or tailors them to specific scientific ends. Another characteristic of the modern sciences that he singles out and casts in a positive light is its drive toward subdivision and diversification. The consequence of this drive is that the conceptual dynamic of a particular scientific field is no longer separable from the specific objects or phenomena in and by which they are constituted. According to Bachelard, historians and philosophers of science must therefore analyze the sciences in all the detail of their regional manifestations. In much the same spirit, Ernst Cassirer, in his late work *The Problem of Knowledge: Philosophy, Science and History since Hegel*, regarded the diversification of the sciences as a hallmark of modernity. In his view there were, and quite rightly, as many different epistemologies as there were scientific fields and particular interests; "the real, inner, moving forces" responsible for this diversity "often lie, deeply hidden, within the sciences" themselves. Coming to terms with them, as Cassirer saw it, called for "a persistent, patient steeping of oneself in the work of the separate sciences."⁵ Bachelard, for his part, went still a step further in advocating a "dispersed"⁶ philosophy of science of particular scientific acts. For under twentieth-century conditions, historical epistemology can presuppose neither a fundamental unity nor a fundamental disunity of the sciences; it must rather come to grips with a displacement of borders that occurs time and again as an effect of scientific thought and action itself, whose shape has become as malleable as the border between the natural and the human sciences. To the extent that our present-day societies can be described as knowledge societies, Karin Knorr Cetina argues that the individual "knowledge cultures and epistemic cultures merit a good deal more consideration—together with the symbolic constitution of scientific-instrumental rationality, that is to say, our Western rationality." If, Knorr Cetina goes on, "there is still the tendency in sociology to draw a sharp distinction between symbolic, 'cultural' beliefs and orientations, on the one hand, and technical activities, demands for efficiency, etc. on the other,"[7] then history of science has all the more reason to present scientific work as one innately pluralistic form of cultural expression among others; and *no* single form of science has, on principle, a privileged relationship to nature.

The debate about the relationship between the natural sciences and the humanities has gained new actuality today. The humanists find themselves confronted with the demand that they take issue with the newest findings of molecular genetics, the neurosciences, and the nanosciences. The debate is marred by the tendency of natural scientists to argue on the basis of a naive conception of their objects, such as, for instance, the "genes" in the life sciences. Both parties to this quarrel would do well to keep their sights trained on the moving battle-lines being drawn at the frontiers of research, for the boundary between nature and culture is constantly in flux there, if it is not, indeed, disappearing altogether.[8] Following such epistemic metamorphoses and their scientific dynamics up close—as I do below in the case of the history of genetics and molecular biology—will perhaps also help us to arrive at a more accurate picture of current developments in the natural sciences.

∶ ∶ ∶

The molecular biologist and historian of science Michel Morange once described molecular biology as "all those techniques and discoveries that make it possible to carry out molecular analyses of the most fundamental biological processes—those involved in the stability, survival, and reproduction of organisms."[9] He hastened to add that it was by no means easy to decide just what was and was not to be ranged among "all those techniques and discoveries"—a formulation obviously designed to sidestep all binding disciplinary or theoretical choices. Yet at the very least, Morange noted, it was possible to assign to the "molecular revolution" fairly sharp temporal boundaries. "The new conceptual tools for analyzing biological phenomena were forged between 1940 and 1965. The consequent operational control was acquired between 1972 and 1980."[10] There are indeed good reasons to assume that molecular biology does not owe its rise to any prior theoretical paradigm. Nor was it an explicitly organized endeavor, as was, say, the subsequent human genome project, although a number of historians have highlighted the early promotion of molecular biology by the Rockefeller Foundation. In view of what we know today about its history, I am inclined to argue that what was at stake was rather the emergence of a new, active "assemblage," to cite the anthropologist of science Paul Rabinow. The history of the sciences, says Rabinow, is marked by dense configurations or nodal points at which something comes about that "has its origins in nothing but minor decisions; decisions that are, to be sure, constrained by conditions, yet not entirely predetermined."[11] "From time to time," he says elsewhere, "new forms emerge that have something significant about them, something that catalyzes previously present actors, things, institutions into a new mode of existence, a new assemblage, an assemblage" that not only puts things in a different light, but makes them "work in a different manner."[12]

The history of molecular biology has witnessed two such conjunctures, as I have called them, or assemblages. The first came about in the two and a half decades referred to above. It had its origins in research on bacteriophages and the tobacco mosaic virus; it crystallized in the early 1950s with the biophysical characterization of the double helix structure of deoxyribonucleic acid (DNA); and it peaked between 1960 and 1965 with the biochemical deciphering of gene expression and the genetic code. This period was marked, first of all, by the invention of a series of apparatuses and procedures designed specifically for use in research.[13] Among them were ultracentrifugation, electron microscopy, X-ray structure analysis, liquid scintillation counting, chromatography, and electrophoresis. The corresponding instruments had their origins in different contexts, which often had nothing to do with either biology or, in particular, contemporaneous classical genetics. Also characteristic of this period was a transition from traditional biological models to new, simple model organisms such as lower molds, bacteria, and viruses. Third, these years saw the emergence of hitherto unusual forms of disciplinary cooperation among biologists, physicists, chemists, and engineers. It follows that molecular biology emerged not as a straightforward continuation of classical genetics but from a parallel assemblage *sui generis*. At the methodological level, molecular biology took over various biophysical and biochemical analytic techniques without any prior theoretical co-ordination. Ultimately these techniques generated a new set of concepts. At the heart of this conceptual shift stood a new conception of biological specificity shaped by the language of information; it gave the new field of molecular genetics its distinctive profile.

The 1970s brought a second assemblage of equally far-reaching importance. It resembled its predecessor in that the first generation of molecular biologists had not intentionally prepared the way for it. It was bound up with the establishment of gene technology and went hand in hand with the introduction of *molecular* techniques of a new kind. Classical molecular biology borrowed methods from biophysics and biochemistry; the methods of genetic engineering, in contrast, evolved out of those established during the first phase of molecular biology, which is to say that molecular tools now operated within the same horizon of meaning within which they had been elaborated. In these molecular procedures, biological macromolecules—especially enzymes and nucleic acids—themselves acquired instrumental functions and what is more, intracellular ones. This shift found its concrete expression in an increasing complexification of the expression and transmission of genetic information and, consequently, of the concept of the gene as well.

∴ ∴ ∴

As we shall see, organisms used as models began to play a central role in the life sciences at the beginning of the twentieth century—that is, relatively late in history, given that modeling and the construction of models compose an essential dimension of the experimental practices of the sciences. That model organisms and the concept of model organism could emerge at all in this period presupposed the idea of a *general biology*, the notion that certain attributes of life were common to all living things and could consequently be experimentally investigated using *particular* organisms that were representative of all others. In previous centuries, it was the *differences* between various living creatures that had commanded the interest of scientists, who in the natural history tradition had sought to account for life forms in all their diversity. With the goal that biological knowledge now set itself, these differences were invested with an altered, instrumental meaning: they could be used to arrive, by way of such particularities, at general characteristics of life. If biology had asked, upon entering the ranks of the sciences around 1800, what distinguished living from nonliving things, it tended to ask, around 1900, what constituted life as such and what pertained to all living beings as compared to each other. The concept and present-day meaning of the expression "model organism" arose under these epistemological conditions. There is good reason to affirm that general biology had no choice but to create model organisms for itself, if it was to become, like the specialized disciplines, an experimental science, and, as such, undergo subsequent internal diversification.

How are model organisms defined? A model organism is a living thing from the plant, animal, or bacterial kingdom that has been tailored to experimental purposes; manipulating it can generate insights into the constitution, functioning, development, or evolution of an entire class of organisms. The operative criteria for selection of a model organism are the ease with which it can be maintained and handled, the quantity and quality of the knowledge already accumulated about it, and the ease of access to the phenomenon under investigation. The pertinence of these criteria will become evident when we look, in the two central parts of this book, at the series of model organisms that were used by classical and molecular genetics in the course of the twentieth century and exerted a decisive influence on various stages of its development. Here we can identify certain trends which, while telling relatively little about the central issues in genetics, have all the more clearly facilitated the laboratory process: in particular, the dwindling size and increasing reproduction rate of laboratory organisms. In other cases, the choice of a model organism was more closely geared to a particular research program. The fruit fly *Drosophila melanogaster*, for example, was very well suited to Thomas Hunt Morgan's gene-mapping project because it had only a few chromosomes. In contrast, the thirty or so chromosome pairs of Alfred Kühn's flour moth *Ephestia kühniella* were of no significance in the first phase of the attempt to trace gene action chains; but in the course of the experiments another feature of the animal became decisive: the fact that it lent itself to tissue transplantation.

Model organisms have their conjunctures as well, which depend largely on the way the relevant research field evolves. To stick with our example, *Drosophila* played a decisive role in the establishment of classical formal genetics and the mapping and determination of relative gene positions, but was replaced by bacteria and viruses in the next high-water phase of molecular genetics. *Drosophila* has been making a comeback, however, for about the last two decades, in the context of developmental molecular biology. *Ephestia*, in contrast, turned out to be an ideal object with which to inaugurate analysis of the basic relationship of particular genes to a cascade of metabolic processes, but was eclipsed when George Beadle and Edward Tatum introduced the filamentous fungus *Neurospora crassa*. *Neurospora*'s mutants defective in the synthesis of particular amino acids made it possible to draw up a complete map of the synthetic pathways leading to these building blocks of proteins. The general trend in twentieth-century genetics was to make use of ever simpler organisms. In the virus and phage research of the 1940s and 1950s, the steady miniaturization of life systems culminated in the physical and chemical characterization of genes themselves. This extraordinarily successful condensation had a long-lasting traumatic effect on modern biology: genes took over. Here Claude Lévi-Strauss's remark that in models "knowledge of the whole precedes knowledge of the parts" has proved valid in a peculiar fashion.[14] In the case at hand, the part *became* the whole—a process that is rendered problematic whenever this whole is projected from the easily understood experimental world back onto the world of complex systems from which it was earlier extracted as one tiny part.

I would like to mention one last peculiarity of certain model organisms. As material supports for scientific work, they can survive the disappearance of entire research programs, which is a consequence of the continuous accumulation of knowledge embodied and actualized in the preparation of the models themselves. An especially interesting example is provided by tobacco mosaic virus research, which has been under way for more than a century now. As documented by a number of path-breaking original papers in the field,[15] the TMV model went through four conjunctures that succeeded one another at irregular intervals. In the opening decades of the twentieth century, TMV was a model for ultrabacterial infectious plant diseases, thus securing itself a firm place in phytopathology. In the 1930s the virus, as a crystallizable, physically and chemically definable protein, came to embody the frontier between the realms of the animate and the inanimate. In the 1950s it was promoted to the rank of model object for the analysis of the relations between nucleic acids and proteins, the molecular self-assembly of organic complexes, and the genetic code. Finally, it has in the past two decades emerged from another latency period to become a model in the process of making plants resistant by genetic engineering, thus returning as a technical object to the field from which it came. By definition, every model stands for something that it represents. Work on the model turns on this ever precarious relationship. Georges Canguilhem once quipped that models are distinguished precisely by a certain lack of knowledge: they are relevant to research only as long as they leave something to be desired.[16] We can extend his idea: from the standpoint of the research process, models maintain their function for only as long as this representational relation remains somewhat hazy, only as long as we cannot say exactly what a particular model ultimately represents. The emergence of certainty about a particular question abolishes the need for models altogether.

: : :

Let us turn back to Cassirer. His *Philosophy of Symbolic Forms* has been de-scribed by the historian of philosophy Michael Friedman as an attempt to understand the sciences of nature and culture as *a single* historical con-figuration of expressive, representative, and symbolic knowledge in which each pole remains dependent on the other.[17] Cassirer did not wish to bind philosophical reflection to the exact sciences as Carnap did, thus subordi-nating it to their logic, nor did he want, as Heidegger, to oppose it to the mode of reasoning of the exact sciences. Cassirer held fast to the notion of a basic complementarity of the natural and cultural sciences, each of which was to be understood, "in its uniqueness," as making a different contribution "to the building up of a 'common world.'" When Cassirer writes that the emergence of a "'meaning,' which is not absorbed by what is merely physical, but is embodied upon and within it . . . is the factor common to all that content which we designate as 'culture,'"[18] he does not counterpoise the knowledge of nature to cultural content as such, but rather effectively treats it too as a central component of culture. He is clearly aware that "the era of the great constructive programs, in which philosophy might hope to systematize and organize all knowledge, is past and gone." Nevertheless, he contends, "the demand for synthesis and syn-opsis, for survey and comprehensive view, continues" to exist "as before," since "only by this sort of systematic review" can "a true historical under-standing of the individual developments of knowledge be obtained."[19]

: : :

The "practical turn" in history of science of the last three decades has, by its nature, privileged micro-histories. Yet the shorter the time-spans studied, the greater the risk of systematically obscuring the historical *longue durée*. Precisely in order to forestall a return to the grand narratives of progress, the time has come to examine such local histories with an eye to their implications for the historical long term, without sacrificing the richness of detail that is their strength.

In this spirit, Michel Morange, in the conclusion of his history of mo-lecular biology, suggests that the ideas of the great historian of the Medi-terranean basin Fernand Braudel about differentiating historical tempo-ralities should be exploited by the history of science as well. Following his suggestion, Morange distinguishes three strands of differing duration, or three time-frames, in the history of molecular biology, to which three distinct traditions or currents can be assigned. In the "reductionist time-frame," the longest, the new biology takes its place alongside long-standing attempts to reduce the phenomena of the material world to their physical and chemical foundations: attempts at making such reductions have characterized the modern sciences in the West for a good three hundred years. Once we have placed molecular biology in this "long time-frame," we can go on to ask whether the molecular understanding of life that took shape in the mid-twentieth century will also prove constitutive of all other fields of the life sciences. The intermediate time-frame might be summarily defined as the history of the disciplines: what influence has the molecularization of life had on the general configuration of the biological disciplines? A first answer might be that the boundaries between medicine and biology have been blurred and, simultaneously, redrawn. But it might also prove worthwhile to investigate the stronger suspicion that the molecularization of life represents only one aspect—of crucial importance to the life sciences, to be sure—of a general process of dissolution that affected all the classical nineteenth-century scientific disciplines in the twentieth century. This would suggest, however, that it will become increasingly difficult and even impossible to analyze the dynamics of recent science in terms of the history of the disciplines—with all the attendant consequences for traditional history of science. As Paul Forman points out, the "devaluation of disciplines" can be traced back to the reorganization of the internal relationships of the sciences, but also to the increasing commercialization of scientific procedures and products.[20] The market orientation of research alters not just the way knowledge is produced but also the way it is managed. Morange finally identifies the third and shortest time-frame as the history of events, the category into which most case studies fall, those presented in this book included: this is the history of experiments, concepts, model organisms, instruments, and the whole gamut of institutional, political, and social factors that determine the actual course of the development of knowledge. Here assemblages—historical conjunctures—set the conditions for the emergence of epistemic novelty.

: : :

I would like to thank Leona Geisler, Antje Radeck, and Jan Bovelet for valuable help in preparing the manuscript. I am grateful to G. M. Goshgarian for his excellent translation. Finally, I would like to acknowledge Ken Wissoker and Tim Elfenbein of Duke University Press for their help throughout the production of the book.

PART I

Historical Epistemology

The chapters in this part give voice to four historians and philosophers of science, Edmund Husserl (1859–1938), Gaston Bachelard (1884–1962), Ludwik Fleck (1896–1961), and Georges Canguilhem (1904–95), who played a crucial role in setting the frame for what has become known, particularly in the French context, under the concept of "historical epistemology" (see also Rheinberger 2010). Each of the three short chapters sheds light on a particular aspect of their efforts to understand the sciences and their development as a fundamentally historical and cultural phenomenon. My historical thinking has been shaped and sharpened by analysis of their work, and traces of this analysis will be recognized throughout the subsequent case studies. However, none of these chapters is meant to be an exhaustive assessment of the authors. Instead, each chapter highlights one particular aspect of each author by a close reading of a number of central texts.

Husserl sets the stage for this historical and cultural analysis of the sciences with his insistence in principle that a philosophical understanding of the sciences can be attained only from a historical perspective. Fleck focuses on the idea of science as a cultural practice. In the case of Bachelard, the notion of phenomeno-technique helps to expose his view of science as a process. And, finally, Canguilhem's consideration of the object of the history of science and what it means to write the history of science as a history of concepts (and objects accordingly) gives us an example of the kind of multi-layered reflexivity that must be involved in dealing with the sciences as historical and cultural phenomena.

1. Ludwik Fleck, Edmund Husserl

On the Historicity of Scientific Knowledge

At the beginning of the twentieth century, the "crisis of historicism" was much debated, particularly in Germany. Interestingly, the crisis of historicism soon came to be seen as a "crisis of reality." This twofold crisis—of historiographical thinking and ways of experiencing reality—had repercussions on the internal and external perception of the history of science. Between 1880 and 1930, the history of science ceased to be an ancillary endeavor of mainly pedagogical interest, becoming a field of research with genuine historical and epistemological pretensions. As such, the emerging discipline of the history of science began to contest the dominion of the philosophy of science over the space of epistemology. In what follows, I would like to show that this deep-reaching change was bound up closely with the crisis affecting the sciences early in the twentieth century.

In 1872, the German physiologist Emil Du Bois-Reymond delivered a lecture entitled "On the History of Science" at the Leibniz Session of the Academy of Sciences in Berlin. Speaking entirely in the spirit of a history of science conceived as a pedagogical auxiliary to the established disciplines, he emphatically affirmed that "whether what is involved is an organism, state, language, or scientific doctrine, it is always developmental history that best reveals the significance and internal connections of things." "In contrast to the not infrequently dogmatic presentation of the textbooks," Du Bois-Reymond went on, "I hold an inductive approach to be the right way to teach physiology, be it in textbooks or the lecture hall." "When we adopt such an approach," he concluded, "we teach a science and its history at the same time."[1] Du Bois-Reymond's reasoning exemplifies the unproblematic and, as it were, "natural" assessment of the history of the sciences prevailing *before* the critical turn. This kind of historical perspective pervaded both the humanities and the life sciences of the late nineteenth century. But how was history in general conceived at the time?

Even physicists such as Ludwig Boltzmann unhesitatingly proclaimed the nineteenth century to be the "century of the mechanical conception of nature, Darwin's century,"[2] not, as might be expected, that of the steam engine or electricity. This association of mechanics with natural history shows that for Boltzmann, Darwin's theory of evolution by no means amounted to a comprehensive historicizing of nature. It was just the opposite: the theory of evolution was, in his eyes, living proof of the possibility of mechanizing history—or, at any rate, natural history. The German philosopher of science Gregor Schiemann is thus quite right to affirm that "the temporalization of explanations of nature [implies] a deteleologization of the knowledge of nature that goes even further than the early modern causal conception of science. Therefore it considerably sharpens the contrast between the knowledge of nature, on the one hand, and, on the other, historical knowledge based on understanding [*Verstehen*]."[3] Max Hartmann, a protozoologist with an interest in philosophy who had in 1914 become a director of the Kaiser Wilhelm Institute for Biology in Berlin-Dahlem, took a similar view. In *Allgemeine Biologie* (1927), he observes about "the question of the logical and methodological foundations of historicism in biology" that

> the vexed philosophical question as to whether history, insofar as it is intellectual or cultural history, rests, from the standpoint of the theory of knowledge, on different epistemological bases than the natural sciences— whether it can be grounded only through an appeal to notions of value, in Windelband's and Rickert's sense—need not be discussed here, because the history of biology is assuredly not a historical science in that sense. . . . It can hardly be doubted that, although historical knowledge is unquestionably more closely geared to the individual and to that which occurs only once, scientific biology does not and cannot content itself with ascertaining a historical state of affairs, but aspires above all to arrive at knowledge of the laws that condition such a state of affairs, seeking explanations for historical developments like the one that Darwin provided with his magnificent essay on the origin of species.[4]

In Hartmann's estimation, biology had not yet gone very far down this path. It was, "notwithstanding Darwin's grand, fruitful attempt, still miles away from rational, nomothetic comprehension of the history of the biological system."[5]

It is plain that in the closing decades of the nineteenth and the first decades of the twentieth century, an encompassing historicizing of nature that simultaneously put forward mechanistic claims stood in curious contrast to the "waning interest natural scientists took in the historicity of their own knowledge."[6] The positive sciences of the late nineteenth century defined themselves, including their evolutionary-biology sub-disciplines, analytically and mechanistically.

This situation was bound to change with the subsequent development of the sciences themselves. Basically, two phenomena would lead, between 1900 and 1930, to a growing resistance to mechanistic thinking: the revolutionary developments in physics, and the debate about the unity of the sciences. Both contributed decisively in the 1920s to an intellectual mood that I shall describe in greater detail below with reference to the early work of Ludwik Fleck and the late work of Edmund Husserl. It should be noted that the radical changes in the theoretical edifice of physics gave rise to two insights that have proved relevant to thinking about the history of science. Hartmann formulated them with exemplary clarity. The first of his conclusions was that one must "keep one's eyes and mind open for the discovery of new conceptual systems, when called for by experience." According to the second, we have to take seriously the admonition "that it is dangerous for thought and science to transpose, with dogmatic overconfidence, the results of a period such as that of the nineteenth-century heyday of classical physics to other fields in the natural sciences and humanities or even the intellectual realm as a whole. For to do so is to exceed the competency and limits of one's own scientific discipline."[7]

A new awareness of the historicity of all scientific knowledge, his own included, makes itself felt in Hartmann's remarks, if only in the form of a presentiment. And it does so with reference to the very field that has been emblematic of scientific knowledge since the beginning of the early modern period: physics. Recent developments had not made classical physics either obsolete or superfluous. Yet it had to be qualified in light of its own time-bound character, situated as one historical stage in the age-old development of knowledge about the physical world, and seen as limited by its reference to particular phenomena. Scientific knowledge could no longer be perceived as something inherently destined for closure and perfection; it was now apprehended as "endless progress," its horizon no more predetermined than its direction. Hartmann's categorical summary of the matter is quite radical: "Every statement about the way the world really is [*die Beschaffenheit der Weltwirklichkeit*] is a time-bound statement."[8]

At the same time, in France, Gaston Bachelard[9] emphatically declared, in his *Essai sur la connaissance approchée* (1928): "As we see it, the nature of . . . reality, given the inexhaustibility of the unknown, is eminently such as to invite us to pursue research without end. Its very being resides in its resistance to knowledge. We shall therefore take it as an epistemological axiom that our knowledge is inherently open-ended." Pursuing this idea,

Bachelard emphasized the moment of surprise and unpredictability in the process of knowledge acquisition: "As the history of the sciences teaches us, every big step towards an ultimate reality has shown that it was discovered in an altogether unexpected quarter."[10]

Science's unpredictability led to the second phenomenon, the debate over the problem of the unity of the sciences. Even biologists as committed to the principle of causality as Hartmann recognized a qualitative difference between physics and the life sciences, without having recourse to neovitalistic premises for that marker. "The task of causal research in biology," Hartmann maintained, "is not to trace biological events back to physical and chemical reactions, but, rather, to elucidate the specific laws of complexification that determine the essence of particular, individualized natural bodies."[11] This fundamental claim to a terrain in which other things were investigated and explained than in physics went hand in hand with a de facto fragmentation of that very terrain itself. The life sciences had broken down into different disciplines with extremely variegated knowledge horizons, from physiology through developmental mechanics to genetics and morphological phylogenetics. Such fragmentation provided good reason to suspect that it was not just biology that was irreducibly many-sided and multiform, but the sciences in general. At the very moment when a movement in the philosophy of science, the Vienna Circle, was engaged in yet another bid to save the unity of the sciences, the actual developmental dynamics of the sciences themselves pointed in a different direction. In a paper entitled "Die Krise der Wirklichkeit" that appeared in 1928 in a journal published by the Kaiser Wilhelm Society, *Die Naturwissenschaften* (a competitor, at the time, of the British *Nature* and the American *Science*), the Frankfurt philosopher and cultural politician Kurt Riezler summed the situation up: "To begin with, some of what we had thought of as the laws of nature were revealed to be mere statistical regularities. . . . To this moment, the divergent paths of development taken by the individual sciences added another. . . . [The individual sciences] were not converging, but diverging; they were developing their conceptual systems in different directions."[12]

Both these moments, the demarcation of biology from physics and the internal differentiation of the life sciences into disciplines that could be neither reduced to each other nor derived from each other, preoccupied philosophically minded biologists from the end of the nineteenth century. As a rule, these biologists presented their reflections on the fundamental relationship between physics and biology under the rubric "theoretical biology," whereas the internal differentiation of biology itself generated diverse attempts at drafting a comprehensive "general biology." Accounts that lump these efforts together as so many subordinate expressions of a holistic Zeitgeist have a point, but the situation is more complex. Manfred Laubichler remarks, "While it is certainly true that there are connections between particular trends in German culture and history on the one hand and certain features of holism on the other, holistic science can by no means be traced back to this cultural context alone." Indeed, the "grand historiographic challenge" is precisely to understand, Laublicher goes on, "the many different intellectual and scientific developments in their own right, rather than constantly seeking the explanation for them in the shadow following them,"[13] that is, in the political developments in their wake. The various attempts to formulate a theoretical and general biology in Germany in the 1920s certainly involved wholeness and synthesis; yet as a rule they proceeded exclusively on the basis of experiences and experiments specific to a particular field. The Gestalt psychologist Wolfgang Köhler thought, for example, that it was impossible to lay claim to the holistic or synthetic as something specifically biological. Köhler's school held that holistic phenomena occurred in all natural domains, which—as attributes of material systems—could be investigated experimentally; they were by no means entities that could be conceived only in transcendental, let alone vitalistic, terms, as Anne Harrington has shown in *Reenchanted Science*.[14] Again let us attend to what Hartmann had to say on the subject: "Vitalism basically puts the frontier of knowledge in the wrong place, namely, between the organic and the inorganic, when it really runs between the rational and the irrational."[15]

Ludwik Fleck

A year after Riezler's essay in *Naturwissenschaften* (1928), there appeared a reply by a young, unknown, philosophically minded microbiologist from Lemberg with ten years of laboratory experience as an immunologist. His name was Ludwik Fleck. *Naturwissenschaften* had also made room for his contribution, which bore the evocative title "Zur Krise der 'Wirklichkeit.'" Riezler in his essay had distinguished between the historically produced common sense of three realities: the reality of the continuous stream of our external and internal perceptions; the reality of our objectivizing knowledge of the world; and the absolute reality underlying our historically changing knowledge of the world. For Riezler, the development of the sciences in the opening decades of the twentieth century, beginning with the progress made by physics, had profoundly shaken the belief that our second reality was steadily approximating the third, absolute reality and would ultimately coincide with it. With that he effectively identified the crisis he had diagnosed as an epistemological one. Fleck makes reference to Riezler's paper only in passing, but with a surprising shift. In one passage, he notes, in lapidary fashion, that "no progress is to be expected from . . . trying to solve the problem of the origin of knowledge, in traditional fashion, as the individual concern of a symbolic 'man.'" "Hence I do not know," he went on, "why and to what end I should distinguish between a first and a second reality as depicted by, among others, Riezler."[16]

This remark announces a reorientation that was to have fundamental consequences. In a first step, Fleck redefines the epistemological "crisis" affecting the relation between the second and third reality as a crisis affecting the conception of realities one and two. The passage just cited goes on, "one should precisely not neglect the social moment in the emergence of knowledge." Scientific knowledge becomes, in consequence, something that is in principle relational: "As a member of some society, every thinking individual has his own reality, in which and according to which he lives. Indeed, every human being possesses many, partially conflicting realities: the reality of everyday life, a professional, religious, and political reality, and even a small scientific reality. Hidden from others' eyes, he also has a superstitious reality that is governed by fate, a personal reality that makes an exception of his own ego."[17] Yet Fleck does not leave matters at what might at first glance appear to be a general knowledge relativism. In a second step, the problem of reality, initially redefined in sociological terms, is transformed into a *historical* problem: "Every epistemology must be brought into relation with the social," he says, "and, further, with the history of culture, if it is not . . . to come into sharp contradiction with the history of knowledge and everyday experience."[18] Thus Fleck no longer conceives of knowledge, in the Cartesian tradition, as the relationship of a knowing *ego* or "symbolic 'man'" to his object; nor does he simply reformulate this relation as a multiple social relationship with the environment, in which the individual belongs to different social groups and consequently moves in different "worlds." Rather, Fleck historicizes the problem of knowledge. Knowledge acquisition—situated between everyday experience on the one hand and a cultural history of knowing on the other—is an inherently supra-individual process: "For knowing is neither passive contemplation nor acquisition of the only possible insight into something that is already finished and given in advance. It is an active, living entering-into-relation with, a transformation and being transformed, in short, creation [*ein Schaffen*]."[19] Referring explicitly to Niels Bohr's quantum postulate and the non-negligible interference between atomic phenomena and the devices for measuring them, Fleck affirms that "observation, knowing [Erkennen], is always a feeling-out-of, and therefore, literally, a transformation of the object of knowledge."[20] There is a striking similarity here to Bachelard, who, like Fleck, was a self-taught epistemologist assignable to no school or tradition; one cannot miss the reference to Bohr in Bachelard either. Bachelard, however, goes even further in the direction of an irreducibly historical epistemology: "Since the phenomenon is absolutely inseparable from the conditions of its detection, we must also characterize it in terms of those conditions."[21] We are accordingly "justified in thinking knowledge in its unfolding, not its sensory origin."[22] In question here is, as Bachelard puts it, a new "realism without substance,"[23] a realism of knowledge as process—and, let us add, the transition from the classical theory of knowledge to historical epistemology: that is, from the contemplating subject's relation to the world it contemplates to a conception of knowledge as an always technically and culturally implemented process.

In his essay, Fleck repeatedly defines and describes this process as a "democratic" one that is shaped by the community of "experts" (Fachleute) and shapes them in its turn: "For natural science is the art of forming a democratic reality and orienting oneself toward it—which means being transformed by it. It is eternal work, rather more synthetic than analytic, that can never be completed—eternal like the work of a river that forms its own bed. Such is true, living natural science. Its creative-synthetic and social-historical aspects must not be forgotten."[24]

Fleck, a scientist with quite a bit of laboratory experience, insists that an end be put to the "paper form" and philosophical elevation of the natural sciences. "The natural sciences as they are," he maintains, "are confused with the natural sciences as they are supposed to be or, in fact, as one would like them to be."[25] Divested of their ideal and ideologized philosophical form and brought down to the level of other real, worldly phenomena open to empirical investigation, they become an object of historical research, a multiply articulated cultural phenomenon which, in Fleck's estimation, embodies a democratic attitude in its internal structure and constitution. For Fleck, science is collective work sustained and driven by its necessarily provisional character, always ready to distance itself from its past: either to surpass it or to pass it by. Fleck's aim is to show that the natural sciences, bereft of their power of fascination and often criticized for being "mechanistic," do not, in actual practice, operate mechanistically, and have by no means been disenchanted. This brings us back to the broader context of the crisis of the mechanistic worldview. Fleck's reasoning represents an attempt not to diabolize the natural sciences, but to free them from the image of rigidity (generally also their self-image) that they had in the surrounding culture.

Neither Fleck's essay nor the book he published in Basel in 1935, *Genesis and Development of a Scientific Fact*,[26] enjoyed immediate success. His ideas and suggestions began to exert a real influence only some forty years after he had published them, in the wake of the debate about the structure of scientific revolutions touched off by Thomas Kuhn's concept of the scientific paradigm. The reasons are plain. Fleck's writings had no time to make an impact; Nazi Germany shrouded them in silence. Fleck himself, a Jewish doctor, was deported in 1941 to the ghetto of his home town of Lemberg and later to the concentration camps in Auschwitz and Buchenwald. He survived, resuming his work on immunobiology at the Polish Academy of Sciences after the war. The intensive, ongoing discussion of his ideas on the sociology and history of science, his "knowledge view" [*Wissensanschauung*],[27] commenced only in the late 1970s, thanks to the republication and English translation of his book.[28] His message is perhaps best summed up in a play on words that we owe, on Bruno Latour's witness, to Bachelard: *un fait est fait*—a fact is fabricated. A scientific fact has to be made.[29]

Fleck condensed his laboratory experience as an immunologist into a compact, plastic image of the way local communities of scientists form, produce, discard, and elaborate their scientific objects in the day-to-day "fabrication process," gradually hardening them into facts that are themselves subject to ongoing, and, on principle, interminable modification. Individual experiments that demonstrate the validity of hypotheses are atypical of laboratory science and, consequently, of modern natural science as a whole. What are more typical are series of exploratory experiments that communicate among themselves with varying intensity, constituting an experimental texture in which equally unexpected condensations and eradications can occur at any time. Fleck repeatedly compared the development of science to a meandering river that determines its course by its own movement through a contingent terrain. But that is by no means to say that the river is exempt from the laws of gravity. Fleck denies that science is arbitrary. The capacity for resistance that phenomena possess, and thus also their reality, remain constitutive. Gravity, however, does not determine the actual course of the river of knowledge. Rather, the path of knowledge is fundamentally conditioned by what Fleck calls the "sequence of the discoveries" themselves.[30]

Fleck here brings an intrinsic historicity into play that is neither an iron-clad, if perhaps always merely asymptotic approximation of absolute reality—Riezler's third reality—nor a mere succession of events, that is, a random stream of discoveries or a series of constructions dominated by particular interests. The historical is now seen to be at work at the core of the research process, at the heart of the epistemological, inasmuch as research is a cultural configuration endowed with its own temporality. In the sciences as well—indeed, in the sciences above all—the acquisition of knowledge becomes for Fleck an iterative procedure, in which the possibilities for taking the next step depend at any given moment on the current state of epistemic affairs. Bachelard, whom I am here putting side by side with Fleck to suggest the European dimension of this movement of historicizing epistemology, also talks about a "rectifying movement" and "thought in action" that should be regarded as the "true epistemological reality."[31] The reality of epistemology is accordingly that of perpetual crisis.

Bachelard never tires of pointing out that the modern natural sciences produce their phenomena in laboratories by technical means and consequently bring forth appearances that as a rule are purely and simply inaccessible to immediate, everyday experience.[32] For Bachelard, the simplest, most basic things with which the natural sciences have to do today are, in a kind of inversion of the primal Cartesian obviousness of clear and distinct ideas, among the most thoroughly derivative, because they are the most thoroughly subjected to the phenomenotechnical work of purification. Bachelard attended closely to this work, investigating it in detailed historical studies. Characteristically, his epistemology of the detail also emphasizes the constitutive entanglement of the subjective aspect of knowledge production: in an act of phenomenotechnical externalization and self-surpassing, the knowing subject must, as it were, outwit itself to arrive at knowledge that is inaccessible to the anticipating imagination because it lies beyond the bounds of the actual given experience that thus acts as an "epistemological obstacle."[33]

Edmund Husserl

On the eve of the First World War, the German theologian Ernst Troeltsch diagnosed a "powerful longing" on the part of the postwar generation to see historical life "again knitted together around unified forces and goals," another forceful evocation of a crisis in the foundations of historicism.[34] However, this time it did not in his view primarily affect historiography as a specialized discipline. It was the general philosophical foundations of historical thinking that had been profoundly shaken: the pillars around which the nexus of history was to be "contemplated and construed." Late in life Edmund Husserl developed a similar argument from a philosophical standpoint in *The Crisis of the European Sciences and Transcendental Phenomenology*, the book project on which he was working from 1934 to 1937, that is, until shortly before his death. The triumphal march of the positive sciences, with their uncontested theoretical and practical successes and orientation toward "prosperity," had, Husserl affirmed, provided "the basis for the support and widespread acceptance" of a pervasive "philosophical and ideological positivism." This kind of science had nothing more to say to the younger generation in particular; indeed, as he saw it, the young were overtly hostile to science. This development had led not just to a crisis of philosophy and the modern sciences in their philosophical universality, but to a "crisis of European humanity itself in respect to the total meaningfulness of its cultural life, its total '*Existenz*.'"[35]

The occasion for this late text was a lecture that Husserl, then seventy-five, delivered in Vienna at the invitation of the Wiener Kulturbund in May 1935. The manuscript of his lecture, entitled "Philosophy and the Crisis of European Humanity," has been preserved in the philosopher's archives. In it he outlined the program that he went on to develop more systematically in his last major work. My concern here, however, is not with transcendental phenomenology and its philosophical ramifications. I wish instead to point out the surprising parallels between Husserl and Fleck, which come to the fore with particular clarity in their programmatic considerations. They are all the more surprising in that the academic background and disciplinary experiences of the physician and microbiologist Fleck and the mathematically trained philosopher Husserl could hardly have been further apart.

Husserl too is aware that what is ultimately at stake is a new self-conception of the natural sciences. For him it is clear that there is no arriving at this new conception by way of a rationalistic epistemology. It is a question of fundamental epistemological decisions that cannot be had from a theory of science rooted in the prevailing forms of methodological reflection on knowledge acquisition in the natural sciences. Fleck had therefore rejected Riezler's distinction between a first reality based on living experience and a second based on scientific experience. Against the confiscation of all thinking about culture by objectivism, naturalism, and positivism, Husserl suggests that the academic disciplines be brought back within the general horizon of the history of culture. At the very beginning of his exposition, he asks: "Is it not absurd and circular to want to explain the historical event 'natural science' in a natural-scientific way, to explain it by bringing in 'natural science' and its natural laws, which, as an intellectual accomplishment, themselves belong to the problem?"[36] Such "fatal naturalism," he argues, can be overcome only by a sweeping historical gesture capable of revealing that the "intellectual accomplishment" of the knowledge of nature results from a long historical chain of efforts of "natural scientists working together." This is Husserl's philosophical version of Fleck's "thought collective." Husserl's hope is to retrieve knowledge of nature from the residual existence and "residual concept" assigned to it by positivism and to relocate it within the ambit of the philosophically motivated knowledge of the world characteristic of European humanity.

For Husserl, this does not imply a blunt relativization of scientific knowledge. Rather—and here too the similarities to Fleck as well as Bachelard leap to the eye—science designates "the idea of an infinity of tasks, of which at any time a finite number have been disposed of and are retained as persisting validities."[37] Thus scientific questioning becomes a process that is in principle interminable: a "mobile forward process from one set of acquisitions to another" effected by an "open chain of generations of those who work for and with one another."[38] It is crucial to this process that those who work for and with one another bear constantly in mind that their questions ultimately originate in their life-world, and furthermore that even the oldest sedimentations of the long scientific tradition still have a presence in this life-world, notwithstanding the countless transformations that they have undergone. "Where is that huge piece of method subjected to critique and clarification—[that method] that leads from the intuitively given surrounding world to the idealizations of mathematics and to the interpretation of these idealizations as objective being?"[39] The question contains a programmatic demand. The crisis that Husserl had diagnosed was in his view due to the failure of those working in science to engage in this epistemological task, in self-criticism and self-clarification, the "*functionaries of mankind*" included—his name for philosophers in this period of intensifying nationalism.[40]

Husserl left this "huge piece of method" to posterity's—historical —"critique and clarification." In the short treatise known as the *Origin of Geometry*, posthumously published by Eugen Fink and translated into French by Jacques Derrida about twenty years later,[41] he sketches by reference to the crisis the contours of a historical and critical epistemology. Like Bachelard, he rejects any dogmatic distinction between "epistemological *elucidation*" and "historical *explanation*." His thesis is that both should coincide in one and the same originary founding act. "Certainly theory of knowledge [*Erkenntnistheorie*] has never been seen as a peculiarly historical task. But this is precisely what we object to in the past. . . . Fundamentally mistaken is the limitation through which precisely the deepest and most genuine problems of history are concealed."[42] Husserl's struggle bears on the "deepest and most genuine problems of history." A "genuine history . . . of the particular sciences" cannot be written as a merely retrospective account of their development, cast in the form of a list of contingent historical events. On the contrary, a genuine history of the individual scientific disciplines is for Husserl "nothing other than the tracing of the historical meaning-structures given in the present, or their self-evidences, along the documented chain of historical back-references into the hidden dimension of the primal self-evidences which underlie them."[43] For the phenomenologist Husserl, what is "historically primary in itself is our present."[44] Consequently, from a Husserlian standpoint "the past of a present-day science is not the same thing as that science in its past," as Georges Canguilhem once put it.[45] It is precisely the past of a present-day science that interests historical epistemologists—in contrast to historians of science in the strict sense, for whom this science in its past may well be of greater interest.

Unfortunately, Husserl did not follow the path he sketched to the end. His attempt to define basic forms of knowing led him to make a programmatic sketch of a historical epistemology. Ultimately, however, he continued to seek originary certitudes, which for him meant the "historical a priori[s]" of a "universal teleology of reason."[46] He held fast to reason as the telos of history—which assuredly distinguishes him from Fleck. Yet his work has played a key role in later studies of the history of the sciences and media concerned with the peculiar historicity and materiality of even the most elusive objects of the sciences. "The important function of written, documenting linguistic expression," we read in the *Origin of Geometry*, "is that it makes communications possible without immediate or mediate personal address; it is, so to speak, communication become virtual."[47] To his enduring credit, Jacques Derrida seized on this idea and developed the potential of Husserl's late writings for an epistemology of particular forms of iteration.[48]

2. Gaston Bachelard

The Concept of "Phenomenotechnique"

As David Hyder has shown in an essay on the historical epistemology of the sciences in Michel Foucault, Jean Cavaillès, and Edmund
Husserl,[1] we can distinguish two divergent tendencies in history and philosophy of science in France in the mid-twentieth century, put simply:
subjectivism and conceptualism. Both critically engage Husserl's phenomenology. The subjectivist tendency, represented by Maurice Merleau-
Ponty, among others, follows in the tradition of the philosophy of consciousness. The conceptualist tendency is represented by philosophers
and historians of science ranging from Gaston Bachelard and Cavaillès
through Georges Canguilhem to Foucault and constitutes a tradition of
philosophy of knowledge. In this chapter I consider Bachelard and, more
particularly, his concept of the "phenomenotechnique" (*phénoménotech-
nique*),[2] in his work on "historical epistemology."[3] Interestingly, Bachelard's philosophical and historical work came in for vehement criticism at
the very moment in the late 1970s when students of science such as Bruno
Latour, mobilizing the notion of "technosciences,"[4] began to take an interest in the "making" of science. They took issue with a precisely identifiable aspect of Bachelard's work: the stress put on the idea of an epistemological break between pre-scientific thought anchored in the immediacy
of everyday experience and scientific thought, which according to Bachelard can develop only within the confines of particular scientific cultures.
This distinction dominated the reception of Bachelard in the 1960s. It was
criticized on the grounds that it privileged scientific knowledge and that
it brought in its wake the whole set of values of classical philosophy of science with its emphasis on truth. If science were to be analyzed in a strictly
social-constructivistic fashion and understood as a genuinely social phenomenon, then philosophers of science would have to dispense with epistemologists such as Bachelard. Perhaps it is time for a reassessment.

Undeniably, Bachelard's writings testify throughout to "scientific
progress," in the sense of a permanent break with established experience. At the 1949 International Congress of Philosophy of Science in Paris,
Bachelard went so far as to say that "it seems to me that the very *existence*

of science is defined as a *progress* of knowledge."[5] But it is crucial to understand that "progress" in Bachelard's sense is less a process of perfection or a gradual approach to a predetermined end than it is a steady movement, a continuous, differential production of new knowledge. Cristina Chimisso has recently remarked that Bachelard's writings are in many respects easier to grasp in their critical impetus if we bear in mind that he was deeply involved in the French educational system of the early twentieth century, with its ideals of the "scientific spirit," and that he taught physics and chemistry for many years at the *collège* in his home town, Bar-sur-Aube.[6] It should also be recalled that Bachelard considered becoming an engineer before setting out on a career as a secondary-school teacher and ultimately rising to the Chair of History and Philosophy of Science at the Sorbonne.[7] Against this backdrop, we can better understand a basic ambivalence in his epistemology. From the perspective of the scientist *as an individual*, his epistemology appears as a psychoanalysis of the scientific mind; yet from the perspective of science as a *process* it appears as a praxeology of scientific *work*.[8]

Premises

Although Bachelard radicalized his thinking about the relationship between science and technology in his second series of epistemological writings, on the "rational application" of knowledge,[9] there is also a remarkable continuity between these later texts, produced in the late 1940s and early 1950s, and his epistemological work of the late 1920s and early 1930s.[10] Key to this overarching concern is the notion of "phenomenotechnique." According to Jean Gayon, Bachelard sought above all "to define the precise nature of the technical aspect of science. The first step in this undertaking was his elaboration of the concept of 'phenomenotechnique'. . . . which is without a doubt at the core of his philosophy of applied rationalism."[11] Teresa Castelão-Lawless comes to a similar conclusion: "Phenomenotechnique is potentially one of the richest concepts that Bachelard has to offer to contemporary philosophy of science and science studies in general."[12] According to this concept, technology is not just a possible consequence or fortuitous by-product of scientific activity and one of the forms in which it is manifested in society, but rather an integral part of the modus operandi of contemporary science. This conception of technology as a central component of science also brings its epistemic function into the foreground. The message of the notion of phenomenotechnique is that science does not merely describe the world, but actually produces new phenomena in the course of its work. In an essay on microphysics from 1931, Bachelard argues, "We might therefore say that mathematical physics corresponds to a noumenology very different from the phenomenography to which scientific empiricism claims to restrict itself. This noumenology sheds light on a phenomenotechnique by means of which new phenomena are not simply discovered, but invented—constructed from scratch."[13] Some twenty years later, invoking the example of the separation of isotopes in the mass spectrometer, Bachelard reaffirms that contemporary scientific activity consists in the "noumenal preparation of phenomena" that are themselves "technically constituted." "The trajectories that enable us to separate isotopes in the mass spectrometer do not *exist* in nature; they have to be technically produced . . . they are reified theorems."[14] What is commonly perceived as something simply given in the real world is in fact something made, the product of a material and discursive circuit—part thing, part theorem—which philosophers of science must take seriously and analyze if they are to come to terms with the immense productivity of contemporary science and technology. This simultaneously material and discursive circuit is for Bachelard dialectically constituted. It has no clearly definable point of origin, whether as noumenon or phenomenon. Hence we have to go beyond both an empiricism privileging phenomena and a rationalism privileging noumena to arrive at a proper philosophical understanding of what goes on in science. Such understanding presupposes that philosophy of science immerse itself in the specifics of each of the sciences taken individually—in the concrete detail of their phenomenotechnical activity.

Thus Bachelard adopts a productivist stance on the objects of knowledge: they are techno-epistemic products cast in the special form of embodied theorems. The conceptual dynamics of the sciences must accordingly be deemed inseparable from the production of the specific phenomena in which they are manifested. Bachelard further assumes that the sciences necessarily undergo increasing fragmentation in consequence of the ever tighter interplay between increasingly specific theories and the corresponding phenomena. The resulting patches of scientific activity must be studied in all the details of their complex development if the sciences are to be properly understood. Philosophy of science therefore in Bachelard's view has to position itself beyond the usual dichotomies of positivism and formalism, empiricism and conventionalism, realism and idealism, because these traditional positions all derive from illegitimate abstractions. To bring out the dialectical tension within his own philosophy of science, he calls it "applied rationalism" or "technical materialism."[15] In *The Philosophy of No*, Bachelard argues for a "dispersed" or "distributed philosophy"[16] that will engage the processes by which the sciences emerge and develop, instead of merely analyzing codified knowledge. Such a philosophy of science orients itself essentially toward a specific problem; each significant cluster of questions, each experiment, each equation even calls for specific philosophical reflection. As Gayon puts it: "Bachelard the epistemologist reaffirms time and again that his concern is with a rationality of detail that pluralizes the object in precisely the same measure as it *rectifies* and *realizes* it."[17] Bachelard is convinced that it is not the epistemologist's task to lay down timeless, universally valid norms for the acquisition of scientific knowledge. The sciences alone have the right to define their constantly shifting boundaries. For philosophers to think adequately about the practice of contemporary sciences, they must commit themselves to observing concrete laboratory practice and scientific activity. Philosophy of science must necessarily take issue with scientific practice. Philosophers of science, according to Bachelard, must ask scientists to "tell us what [they] are thinking, not as [they] *leave* the laboratory, but during those hours when [they] quit ordinary life to *enter* scientific life. Give us, not the empiricism of [their] evenings, but the vigorous rationalism of [their] mornings."[18]

To trace the dynamics of knowledge acquisition back to the permanent fragmentation of the sciences is to postulate methodological pluralism and epistemological openness. Bachelard outlines a program for an epistemology of process, an epistemology of emergence and innovation. He speaks of a "philosophie au travail."[19] To realize his program, he deems it necessary that philosophers of science closely follow and reflect upon the sciences as they develop. "We have to attain a concrete rationalism, one that is of a piece with experiments that are always particular and precise. This rationalism must also be *open* enough to take on new determinations from experiments."[20] An epistemology that sets out to grasp the dynamic nature of scientific thinking should accordingly be as plastic, mobile, fluid, and open to risk as scientific thinking itself. Two requirements for the philosophy of science follow from this.

The first is horizontal mobility. According to Bachelard, modern science branches out into transient kernels, defined as islands of scientific culture possessing specific cultural codes and characteristic forms of emergence accessible only to those intimately familiar with them. These kernels can be understood best from the inside—that is, according to their particular codes and forms of emergence. Bachelard often likens these kernels to cantons, regions, or domains of knowledge in the "city of science": he speaks, for example, of the "relativistic canton" in the "city of mechanics."[21] This means that epistemology must engage new, specific complexities with the appearance of each new, insular formation.

The second requirement is vertical mobility. One of Bachelard's central claims is that "the scientific spirit is essentially a rectification of knowledge. . . . It judges its historic past by rejecting it. Its structure is its awareness of its historical errors. Scientifically, one thinks truth as the historical rectification of a persistent error, and experiments as correctives for an initial, common illusion."[22] This passage helps explain Bachelard's concept of "recursion," a permanent process of rectification and reorientation that has to do less with epistemological breaks between scientific thinking and "initial illusions" than with a process unfolding within scientific activity itself once that activity has been initiated, ceaselessly transforming the truths of the present into the errors of the past. However, since the scientific truth of the present is produced solely by way of recursion, it also acquires its peculiar ambivalent character from it: judge of the past, it is simultaneously the accused in an ongoing tribunal of rectification. A scientific truth of the present must always be ready to find itself in turn as an error of the past. Therein lies the essence of the historicity of the sciences; it is this that stamps them as special cultures of veridiction and verdict. "The scientific spirit has a variable structure from the moment that knowledge has a history."[23] Thus knowledge *is* evolution of the scientific mind, not something that can be attained once and for all. It is grounded, not in the enduring unity of the thinking ego, but in its ongoing transformation and genealogical fungibility, in the cunning of reason and its self-critical powers. Methods are temporary strategies for the acquisition of knowledge employed by a science that, "made confident by the acknowledgement that it has paid its debts, risks what it has already acquired to acquire more."[24] Methods are not mere catalysts that remain external to their object; they modify and also exhaust themselves in their own application.

Two Basic Features of Bachelard's Epistemology

We shall now take a somewhat closer look at two basic features of Bachelard's epistemology. One bears on the object of science, the subject of science, and objectivity; the other has to do with the social character of science.

The Scientific Mind and Its Objects

To begin with, let us see what consequences Bachelard's epistemology of process has for the concept of the object of science and the notion of scientific objectivity.

First, the objects with which the sciences operate and the phenomena they treat are not immediately given to the senses. Built into Bachelard's concept of science is, rather, the idea that the objects of science are entities that cannot be grasped in unmediated fashion. They are always already elaborated products of earlier scientific work. Hence, "scientific objectivity is possible only if one has first broken with the immediate object, refused to yield to the seduction of one's first choice, and curbed and contradicted the thoughts that arise from one's first observation."[25] Elsewhere, Bachelard notes that "it must therefore be accepted that there is a very real break between sensory knowledge and scientific knowledge."[26] This break (*rupture*) must be thought of as a kind of originary "epistemological act"[27] by which the perception of a phenomenon is transformed into a scientific object.

With this act, something that is a given for the senses is transformed into a *problem*. A scientific object is a phenomenon that has been drawn into a process of rectification; it is not, however, constituted once and for all, but remains a scientific object only by virtue of its constant reconfiguration and rectification. The epistemological break marks the transition from everyday experience to the acts of scientific knowledge, while at the same time inscribing itself in these continuing acts and thereby becoming one essential hallmark of a persistent scientific engagement with a piece of the world.

The second aspect of Bachelard's concept of scientific objectivity consists in the thesis that just as the objects of science are mediated, so the mind that has knowledge of them is not immediately given. "Movement towards an object is not initially objective."[28] The knowing mind has to externalize itself and become "instrumental,"[29] for it is itself technically mediated, as are all its concepts, the categories of scientific knowledge not excepted. The consequence is that scientific mind and object enter into a relationship of reciprocal externalization and interiorization. Instruments stand at the heart of this epistemic ensemble in modern science. On the one hand the instrument embodies an already acquired knowledge; on the other, it helps produce the object as technophenomenon. In modern science, Bachelard contends, "the instrument is a necessary intermediary in the study of a phenomenon that has been truly instrumented or designated as the object of a phenomenotechnique."[30] Thus the instrument is not a passive apparatus positioned between a Cartesian mind and the world in order to extend the observer's capacities of discernment; it is "truly a reified theorem."[31] The instrument always already represents the material existence of a determinate body of knowledge. Phenomena are called up in the form of problems within the field it defines; new concepts may then be required to accommodate and assimilate them. Phenomenon and instrument, object and experience, concept and method, are all engaged in an ongoing process of mutual instruction. Bachelard occasionally speaks of "construction" in this context. In *The Formation of the Scientific Mind* he writes that "Nothing is given. Everything is constructed."[32] As a general rule, however, he uses the term "realization." "Science *realizes* its objects without ever just finding them ready-made. Phenomenotechnique *extends* phenomenology. A concept becomes scientific in so far as it becomes a technique, in so far as it is accompanied by a technique that realizes."[33] His highest preference, however, is for the term "instruction," in so far as it accentuates the reciprocity of the process in which the object itself becomes an agent of the process of knowledge. The object enters into an epistemic "engagement": "The position of the scientific object, or, more exactly, of the object serving as instructor at a given moment, is much more complicated and *engaged*. It calls for solidarity between method and experience. One must therefore know the *method of knowing* in order to grasp the *object to be known*, which is to say—in the realm of methodologically valorized knowledge—an object that is itself capable of transforming the method of knowing."[34] The knowledge relation is thus by its nature counter-intuitive and unnatural. The scientific mind has to form itself in opposition to nature and the force of its own habits. By no means, however, can it instruct itself by simply "educating itself"; its only valid option is to come to grips with the objects of the world, a process in the course of which it "purif[ies] natural substances and . . . bring[s] order to a jumble of phenomena."[35] Thus neither the scientific mind nor the objects of science are immediately given or apprehensible; they take shape only in a punctuated historical process of purification and ordering.

Bachelard describes the subjective dimension of this engagement in what he calls the *psychoanalysis* of objective knowledge. It is one of the hallmarks of his historical epistemology that it does not postulate a preexistent structure or "logic" of the scientific mind. Quite the contrary: Bachelard expressly conceives of scientific thinking as a *psycho-epistemic* activity. As he sees it, the acquisition of knowledge is an act in which the scientist's whole being is involved: it is *labor*. The "epistemological obstacle" stands at the center of what he describes as a "phenomenology of labor."[36] It is not a difficulty that imposes itself from the outside, such as the complexity of the world or the physical limitations of our senses. "At the very heart of the act of cognition . . . by a kind of functional necessity, sluggishness and disturbances arise."[37] Epistemological obstacles are not simply to be overcome, for they are what sustains the process of acquiring knowledge. In the act of knowing, no transparency reigns: "Reality is never 'what we might believe it to be': it is always what we ought to have thought. Empirical thought is clear only *in retrospect*, when the apparatus of reasons has been developed."[38] While knowledge can be confirmed in an act of recursion, no set of rules can ensure its production. Knowledge production continuously occurs in untidy places: in places of confusion and stubborn, ready-made opinions; in places where new things are tried out. That is why there will always be a need for "epistemological acts . . . that generate unexpected impulses in the course of scientific development."[39] Unlike many of his contemporaries, Bachelard does not want to wall this untidy space off from epistemology; he proclaims it epistemology's heartland.

Science as Social Process

Bachelard proceeds on the assumption that scientific activity as described has an essentially social or communitarian dimension. Modern knowledge acquisition can no longer be pictured in traditional fashion as a relation between an object and a solitary subject: knowledge acquisition is set in motion and carried out as a *collective* enterprise, organized and sustained by a scientific community through the "real interpsychological labor" of language and experimentation.[40] On Bachelard's view, this labor takes place in the scientific "cultures" of cantons or neighborhoods within the "scientific city." He defines a scientific culture as "an accession to an emergence,"[41] that is, an environment allowing for innovation and unpredictable events. While for Bachelard acts of artistic creation bear the mark of the individual, in the sciences, "these emergences are, emphatically, socially constituted."[42] Even the epistemological acts of scientific geniuses are deeply embedded in such cultures as a rule.

Scientific cultures are group enterprises directed by the pursuit of a common epistemological project. "Individuals hesitate," he says, "but the school—in the sciences—does not. The school—in the sciences—drives research forward."[43] Scientific schools are accordingly social expressions of the process of recursion characteristic of scientific knowledge acquisition. They forge the relationships required for the "dialectic of attachment and engagement" to attain its potential.[44]

Scientific cultures are characterized by specialization, which Bachelard, as we have seen, regards in a positive light. Specialization, he says, "actualizes a generality."[45] "It dynamizes the mind. It works. It works unceasingly," with the result that it is precisely "the most specialized cultures that are the most open to substitutions."[46] But such cultures also unleash the "power of fixation," the counterweight to the openness and flexibility of the modern scientific mind.[47] Together specialization and fixation form the basic condition for the play of the constant replacement of methods, categories, and phenomena. The more narrowly defined an area, the more readily conventions, measurements, descriptions, and classifications can be modified or transformed, and under certain circumstances carried over into other fields of research. Specialization engenders epistemic flexibility: whence the idea of the patchwork structure of knowledge production. Finally, the *theoretical* and the *technical* societies cooperate in the scientific city. Their "intimate, active interpenetration" constitutes in Bachelard's estimation the genuinely "new philosophical reality" of the modern sciences. The convergence of precision (theory) and power (technology) is a thoroughly *social* accomplishment, not by any means a *"natural* necessity." In the final analysis, this convergence is responsible for the "phenomenon factory" of the contemporary scientific world.[48] This account shows that there is an irreducible (albeit not always conspicuous) social and cultural dimension to Bachelard's epistemology of process. The dialectic of the technical and the noumenal at the heart of Bachelard's conception of the research process translates in his thinking into a close social relationship between science and technology. His late epistemological work, beginning with *Le rationalisme appliqué* (1949), particularly emphasizes the importance of this relationship. But this relationship is also fraught with tension, which Bachelard notes as early as 1928 in his *La connaissance approchée.* Gayon describes the tension as follows: "Science constantly runs up against 'the fundamental irrationality of the given.' Industry, in contrast, materially *realizes* a rationality that is plainly recognized because it is deliberately sought out. 'The one seeks the rational, the other imposes it.' Technology *fully realizes its object,* and, consequently, avoids the spontaneous skepticism that is one of the most typical cognitive features of scientific thinking."[49]

Technoscientific Productivity

The two-way relationship between the technical and the scientific recalls the "phenomenotechnique." To stress that each of its two poles conditions the other, Bachelard calls his own philosophical enterprise "applied rationalism" or "technical materialism." Materialism here refers to a "reality" "transformed" by rationalism and thus always already branded with "the mark of rationalism."[50] Bachelard's concept of "rational application" operates at another level than the traditional dichotomy between basic research and applied science, commonly formulated as follows: where basic research is conducted, experiments are carried out and hypotheses are formed, with great freedom in both instances; eventually, the solutions to fundamental problems discovered by basic research find technological applications in society. This way of looking at things usually goes hand-in-hand with the idea that basic science is largely value-free and that the results it obtains become socially and ethically charged only in the wake of their translation, dissemination, and mass production—in a word, as a result of their industrial application to good or bad ends.

Bachelard in contrast holds that the epistemic and the technical stand in a much less easily dissolved relation to each other, a relation of co-evolution grounding the productivity of modern science. "Application" is not extrinsic to modern knowledge; it is not the mere shell of an epistemic kernel. Rather, it produces effects at the level of concept formation itself. The technical is an integral part of the theoretical essence of the modern sciences. In *The Formation of the Scientific Mind*, Bachelard unmistakably affirms that "in order to accommodate new experimental proofs, we must . . . *deform* our initial concepts, examine these concepts' conditions of application, and, above all, incorporate *a concept's conditions of application into its very meaning*."[51] Applicability is thus inscribed at the heart of concept formation in the modern sciences. Stephen Gaukroger identifies this assumption as the "essence of [Bachelard's] phenomenotechnics."[52] Applied rationalism is accordingly technically implemented materialism. Science may search for possible applications, but it is recognized as such precisely because it has always already evolved *within* the applied realm. The technical comprises an integral part of its epistemological structure; application informs not just the meaning of concepts, but the very rules of concept formation. The technical is internal to experimental phenomena. Indeed, noumena are embodied in instruments and always already possess an instrumental form that grants the development of the phenomeno-technical apparatus as a whole. Bachelard generally eschews descriptions of the situation in terms of "construction" or "artifice." "The factitious," he contends in a discussion of Hans Vaihinger's *Philosophy of the As If*, may well provide a "metaphor" for the experimental situation, "but, unlike the technical, it cannot furnish us with a syntax capable of linking arguments and intuitions."[53] In his estimation, this criticism applies not only to Vaihinger's fictionalism, but also to utilitarian and pragmatist interpretations of science, whose "scattered pluralism" Bachelard likewise rejects in the name of a "coherent pluralism."[54] Against the superficial methodological flexibility of a "fragmented pluralism," he insists on the value of the claim to truth, the source of the "power of permanent integration of modern scientific knowledge":[55] the claim to truth counterbalances the productive regionalization of knowledge, which is, however, no less crucial, since "there can be no objectivity without a proliferation of viewpoints."[56] On the one hand, "the language of science is . . . in a state of permanent semantic revolution."[57] On the other, the interconnection of the sciences in a material network mediated by instruments and spreading in the form of standardized industrial products endows them with the necessary minimum of coherence. In the final analysis, such coherence is owing to the structurally grounded applicability of modern science. "In the new rationalism, the key concept is technique. Modern science is, intrinsically, a 'technically' constituted science."[58]

Historical Epistemology

In closing, I would like to return to Bachelard's notion of a historical epistemology. Its two main problems concern technoscientific productivity and the relation of present to past knowledge. The first problem is related to the second, in that scientific objects—technophenomena—are the vectors of a history that is inseparable from them: for, once they have been torn from the grip of immediate experience, they have always already been invested with experimental work. For Bachelard, the vertical coherence of science takes a special form because it comes into being on the serpentine path of science's incessant self-rectification; this path is open-ended, pointing simultaneously beyond itself back to itself. In this respect, "the history of the sciences appears . . . to be the most irreversible of all histories."[59] Scientific objects are always transformations of earlier scientific objects and thus intrinsically historical entities. Thus "the nineteenth century's 'electrical reality' is quite different from the eighteenth century's,"[60] and yet both are perceived as transformations of the same range of phenomena. By the close coupling of the noumenal and the technical in the scientific realm and of science and technology in society, applied rationalism appears as part of the material culture of humanity. However, despite science's irreversible historicity, its emergences are neither historically, nor socially, bound to take a necessary direction. No "historical Reason" in the strict Hegelian sense is at work in the history of the sciences.[61] The synthetic achievements of the sciences are profoundly emergent phenomena, and their emergence is by no means teleologically programmed. To strengthen his claim, Bachelard cites the French Nobel laureate Louis de Broglie in this connection: "Many of today's scientific ideas would be different from what they are if the paths that the human spirit took in pursuit of them had been different."[62]

The second, related problem in Bachelard's historical epistemology concerns the requirement that the present state of knowledge serve as a foil for the evaluation of past science, despite the theoretically grounded impossibility of anticipating scientific "progress" and the non-existence of historical Reason. Historical judgment is a recursive activity always accompanied by a teleology of hindsight; history appears in the light of a "finality of the present"[63] which both illuminates it and partitions the history of the sciences into the sanctioned and the superseded. *Science itself makes this distinction*, since it is in essence an unending process of self-limitation, self-distinction, internal polemic, and self-negation. The historical epistemologist who wishes to follow the movement of his objects from the vantage point of his own time is accordingly obligated to respect the same rules in his own work.

Proper handling of these recursions calls for "a sure touch," since it involves a "self-destructive element." "The philosophical position I take here," Bachelard warns, "is assuredly not just difficult and dangerous, but even harbors a self-destructive element rooted in the ephemeral character of the modernity of science. If one adopts the ideal of the modernist tension which I suggest that history of science should, then that history must be often rewritten and reassessed."[64] If the sciences themselves constantly revise their judgment of the past, epistemology's task is to accompany them in a kind of co-transformation. Historical epistemology then becomes a historically variable enterprise in its turn. If the scientist's culture is a "history of permanent reforms,"[65] the epistemologist's cannot be much different. Epistemology, too, in its tight coupling to the advancement of science advocated by Bachelard, must be prepared to modify its past judgments in the light of new scientific developments.

3. Georges Canguilhem

Epistemological History

History of science as understood and practiced by Georges Canguilhem plainly displays, in its conceptual orientation, features of an epistemology.[1] But what kind of epistemology? Following Dominique Lecourt, Jean Gayon maintains that the term "historical epistemology" is more appropriate for the work of Canguilhem's predecessor Georges Bachelard than for Canguilhem's own. Gayon argues that we have to reverse the terms to characterize Canguilhem's work correctly as "epistemological history."[2] The main subject of this chapter is the relationship between history and epistemology in Canguilhem. I concentrate on his post-1960 writings, ignoring his medical thesis *Essai sur quelques problèmes concernant le normal et le pathologique* (1943), the *Connaissance de la vie* (1952), and *La formation du concept de réflexe aux XVIIᵉ et XVIIIᵉ siècles*, his doctoral dissertation in philosophy (1955).[3] In recent years, the characterization of Canguilhem, based on *The Normal and the Pathological*, as a "vital rationalist"[4] has attracted considerable attention in the United States and France. I would simply like to note here, in passing, that there is a certain affinity between Canguilhem's historical vision of health, disease, and life and his conception of science, if only to the extent that he regards both life and the sciences as genuinely self-corrective, fragile processes. Although to this point there has been no serious attempt to periodize Canguilhem's thought, commentators and students such as Etienne Balibar have suggested that there occurred, "if not a break, then at least a new departure" in his work around 1960.[5] By that time, Canguilhem had succeeded Bachelard as holder of the Chair in History and Philosophy of Science at the Sorbonne, where he met Michel Foucault. Foucault asked him to prepare the academic jury report on his dissertation in philosophy, *Folie et déraison: Histoire de la folie à l'âge classique* (*Madness and Civilization: A History of Insanity in the Age of Reason*).[6] Foucault's next book, *Naissance de la clinique: Une archéologie du regard médical* (*The Birth of the Clinic: An Archeology of Medical Perception*),[7] appeared shortly thereafter in a series called *Galien*, under Canguilhem's general editorship, on the history and philosophy of medicine. We owe to Foucault one of the most penetrating characterizations of Canguilhem, the teacher who became his student's student: "This man, whose work is austere, deliberately delimited and carefully devoted to one particular domain in a history of the sciences that in any case is not regarded as a spectacular discipline, was in some way present in the debates in which he took care never to figure himself. . . . In the whole debate of ideas that preceded or followed the movement of 1968, it is easy to find the place of those who were shaped in one way or another by Canguilhem."[8] In an essay on "science and truth" in Canguilhem's philosophy, Balibar cites a televised interview with Canguilhem conducted by Alain Badiou in the mid-1960s:

> *Question*: Do you mean to say that the term "scientific knowledge" is a pleonasm?
>
> *Answer*: You have understood me very well. That is just what I am trying to say. Knowledge that is not scientific is not knowledge. I maintain that "true knowledge" is a pleonasm, as are "scientific knowledge" and "science and truth," and that they all come to the same thing. This does not mean that the human mind has no aims or values outside the truth. It does mean, however, that we cannot call that which is not knowledge knowledge, and cannot give that name to a way of life that has nothing to do with truth, that is to say, with rigor.[9]

A few years later, in February 1968, Canguilhem referred to this interview in a talk at the Sorbonne. Before a student audience mobilized by the *Union rationaliste* and the editorial board of the journal *Raison présente*, he declared:

> One day, it seems, I shocked all the philosophy students taking part in a television program—the students, and many of their professors as well. The reason was that I said that there is no truth other than scientific truth, that there is no philosophical truth. I am perfectly willing to repeat here what I said then. But to affirm that there is no objectivity outside scientific knowledge is not to affirm that philosophy has no object. . . . There is not an object of philosophy, however, in the sense in which there is an object of science, which is precisely what science constitutes, theoretically and experimentally . . . but I don't mean to say that philosophy has no object.[10]

At first sight, both declarations would seem to say the exact opposite of what would normally be expected from a historically oriented epistemology. Moreover, the second remark displaces the problem, shifting the focus from the question of knowledge or truth to that of the object of science. Indeed, the question of the object of science—like the object of the history of the sciences and the object of philosophy pointed to by Canguilhem in the passage above—can help us arrive at a more accurate under-standing of his epistemology. I shall therefore set out from an analysis of "L'objet de l'histoire des sciences," a paper he delivered in October 1966 at the invitation of the Canadian Society for History and Philosophy of Science in Montreal. The overriding importance that Canguilhem attached to the object of the history of science is suggested by his inclusion of this Montreal lecture at the head of an essay collection published two years later under the title *Etudes d'histoire et de philosophie des sciences*.[11] Canguilhem had already discussed the subject with his students at the Institut d'Histoire des Sciences et des Techniques in 1964–66, some of whom wrote seminar papers on it.[12] In what follows, I shall also refer to Canguilhem's introduction to the essay collection *Ideology and Rationality in the History of the Life Sciences*, which he published about a decade later.[13]

Canguilhem opens his talk on the object of the history of the sciences with a critique. The relation between history of science and science, he says, is not at all as unambiguous and unproblematic as is often assumed: the relation of the sciences to their objects can by no means be projected onto the relation of history of science to *its* object, the sciences. In this way, Canguilhem emphatically keeps his distance from straightforward historical positivism or "pure history," as he calls it later in the Introduction to *Ideology and Rationality*. "A pure history of eighteenth-century botany would consider 'botanical' nothing but what botanists of the period took to be within their field of inquiry. Pure history reduces the science it studies to the field to which the scientists of the day assigned it and the gaze that they turned on this field. But does this science of the past constitute a past for the science of today?"[14] The history of science, understood as the past of a present-day science, is by no means a blue-print, read backwards, of the chronological unfolding of a given science in a given period. "Between the chemistry of oxidation and the biochemistry of enzymatic reductions, plant physiology first had to become cellular physiology (and cellular theory of course met with tremendous opposition) and had then to rid itself of its early concepts of the cell and proto-plasm in order to study metabolism at the molecular level. . . . It should . . . be clear why the past of a present-day science is not the same thing as that science in its past."[15] Canguilhem credits Bachelard with having first insisted on this distinction, which is central to Canguilhem's think-ing about the sciences. Indeed, he derives the need for an epistemological foundation for the history of the sciences from it. "Without epistemology," he contends, "it would be impossible to distinguish between two kinds of . . . history of science: the history of obsolete knowledge and the history of sanctioned knowledge, by which I mean knowledge that continues to play an active [*agissant*] role in the present."[16] It might seem that Canguilhem here throws the door wide open to what had long since been criticized by historians of science in particular and historians in general as "whiggish history,"[17] and has more recently been rejected as illegitimate "normative epistemology" by representatives of sociological and anthropological science studies.[18] Why should it be the task of historians of science to distinguish between still sanctioned knowledge and obsolete knowledge? Why should historians of science have to familiarize themselves with the sciences of the present and their epistemological conditions of existence?

The implications of the distinction between sanctioned and obsolete knowledge go much further than is generally assumed. This distinction does not just carry the whole burden of the notion of "historical recursion" developed by Bachelard in his late epistemological writings. Historical recursion constitutes a means of thinking about the history of the sciences in non-teleological fashion, while simultaneously upholding the idea that it is "the most irreversible of all histories."[19] Here Canguilhem takes the concept of historical recursion a step further, making it a central element of historical methodology itself. With Alexandre Koyré and Bachelard and against Emile Meyerson, he declares:

> The history of science . . . is not an inverted history of scientific progress, not a picture of discredited stages in perspective, with today's truth at the "vanishing point." It is an effort to discover and explain the extent to which discredited notions, attitudes or methods themselves discredited others in their day—and therefore an effort to determine in what sense the discredited past remains the past of an activity that should still be called scientific. It is as important to understand why something once counted as instruction as to find out why it later merited destruction.[20]

The objects of the sciences are thus, according to Canguilhem, permanently changing entities. "Today's truth" is not eternal. It is quite emphatically "of the present," not the static, abiding truth of a unique "Nature." For the "natural object, considered independently of any discourse about it, is not . . . the scientific object. Nature is not in and of itself divided up and classified into scientific objects and phenomena. It is science which constitutes its object from the moment it devises a method for forming, from propositions that can all be coherently combined, a theory controlled in its turn by the effort to catch it making a mistake."[21] Science might even be described as the most profoundly "historical" cultural enterprise, since scientific acheivements are defined precisely by the possibility in principle to become superseded. Science *is* science only as a constant process of *becoming*. "In other words," Gilles Renard observes, "[Canguilhem] does not refer the progress [of science] to an ontology, knowledge of which would lie outside that progress, or to a reality accessible without recourse to it, but to science's technical and theoretical *self-rectification*."[22] Thus Canguilhem attributes a fundamental temporality to scientific activity. Yet this temporality is not homogeneous; it unfolds at a variable rate in the different fields in which knowledge establishes itself as scientific knowledge: "the time of the advent of scientific truth, the time of veri-fication, possesses a liquidity or viscosity that is different for different disciplines existing in the same period of general history."[23]

What are the implications for the object of the history of the sciences? First, the history of the objects of science cannot be written as if it were a "natural history," to use Canguilhem's words.[24] Another level of object-formation must be introduced, one corresponding to the epistemological dimension of the historian's practice:

> The object of the history of science has nothing in common with the object of science. The scientific object, constituted by methodical discourse, is secondary to, although not derived from, the initial natural object, which might well be called (in a deliberate play on words) the pre-text. The history of science bears on these secondary, nonnatural, cultural objects, but it is not derived from them any more than they are derived from natural objects. The object of historical discourse is, in effect, the historicity of scientific discourse, to the extent that this historicity represents the realization of a project with its own internal norms. But it is a historicity shot through with accidents, delayed or diverted by obstacles, and interrupted by crises, that is, by moments of judgment and truth.[25]

If science projects its objects into the future, history of science draws *its* objects from the past. As Helge Kragh once observed, the position of the historian of science inevitably involves a certain "anachronism."[26]

The historiography of science must consequently obey its own temporality, establishing a temporal order of its own in which to represent scientific cultural objects. One resultant problem lies in the relationship between continuities and breaks in the development of the sciences. Canguilhem is aware of the theoretical dangers involved in Bachelard's concept of the "epistemological break" between scientific thought, which can flourish only in the environment of a research culture and a prescientific mode of thought entrenched in the immediacy of everyday life and its lived illusions. The question is: how do we reflect our own historical periodizations? Canguilhem cannot answer by mobilizing Thomas Kuhn's essentially socio-psychological construct of alternating phases of normal and revolutionary science. Science, according to Canguilhem, is inherently revolutionary; yet the pace of scientific substitutions can vary sharply as a function of cultural circumstances and the status of the scientific object itself. I shall return to this problem in discussing Canguilhem's concept of scientific ideology.

The search for precursors in the history of the sciences is a typical historical manifestation of the unclarified relationship between continuity and break. Canguilhem rejects any such effort. In his view, the notion of the precursor epitomizes a history of science that confuses its object with the object of the sciences themselves. "When we substitute the logical time of truth relations for the historical time of their invention, we align the history of science with science and its object with the object of science. We thereby create an artifact, a counterfeit historical object—the precursor."[27] According to Canguilhem, to proceed in this fashion is to rob science of its essentially historical dimension, reducing it to a logically sequenced quest for the Holy Grail: "Strictly speaking, if precursors existed, the history of science would lose all meaning, since science itself would then merely *appear* to have a historical dimension."[28] The precursor, the veritable counter-image of the concept of historical recursion, appears as a virtual caricature of the history of science. Canguilhem brings all the intellectual power and sophistication of his epistemology to bear on his critique of this notion, notably in his early case study of the development of cell theory[29] or his late study of the history of the life sciences since Darwin, where he considers the work of (among others) Gregor Mendel, whose name has come to be synonymous with—or even to stand as a "historical pleonasm" for—the idea of the precursor.[30]

Let us now examine how epistemological reflection within the temporal order of the history of the sciences is connected to the ongoing development of the sciences. The progress of the sciences implies that the historian of science and his way of reconstructing and presenting things must change: "The history of the sciences, a history of the progressive relation of intelligence to truth, secretes its own temporality. The way it does so varies, depending on the moment of progress from which it sets out with a view to reviving, in previous theoretical discourses, what contemporary language still allows us to understand."[31] Thus history of science is also a fundamentally historical enterprise, one whose changing cultural context helps determine how the cultural contexts of nascent objects of knowledge can be represented. Just as science produces scientific objects that are constantly rendered obsolete, so the historian in accounting for this productive process of obsolescence produces constructs that are rendered obsolete in their turn—albeit not by the same kind of self-rectification characteristic of the sciences. What motivates the changes of the discourse of the history of the sciences is not primarily its tendency to surpass already achieved standards; its discourse is epistemologically linked to the dynamics of scientific development itself:

> It doubtless goes without saying that the progress of the sciences resulting from epistemological breaks requires that the history of a discipline be frequently rewritten—a discipline that one cannot quite call the same discipline, since what is involved is a new object designated by a name that has remained the same thanks to the force of linguistic habit. [Thus] François Jacob's *Logic of Life* (1970)[32] differs from the second edition of Charles Singer's *History of Biology* (1950),[33] the personalities of the two authors aside, not just because the volume of knowledge increased in the interim, but because the structure of DNA was discovered in 1953 and because new concepts were introduced into biology—some, such as organization, adaptation, and heredity, bearing old names, others, such as message, program, or teleonomy, bearing new ones.[34]

The historian of the sciences can acquire a "feel" for historical breaks and filiations only through contact—even if it is a contact mediated by epistemology—with the science of his day, whose breaks punctuate his own discourse. Such contact enriches epistemology, "on condition that, as Gaston Bachelard taught, it remain vigilant. History of science, so understood, can only be uncertain and is bound to engage in self-correction."[35]

One rather trivial aspect of the cultural embeddedness of history of science is the influence that particular historical conjunctures have on historians of science—and historians in general—in choosing their objects and preferred connections. But for all that, historians appear swayed less by contingent circumstances than by a kind of felt necessity that mandates them to periodically rewrite the history of the disciplines and the transdisciplinary dynamics of the history of scientific objects. This necessity is governed by the historicity of the scientific objects themselves, their development and displacements. "A science is a discourse whose norms stem from its critical rectification. If this discourse has a history whose course the historian believes he can reconstruct, it is because it *is* a history whose meaning the epistemologist must reactivate."[36] A simple thought experiment may illustrate the point. What would the history of evolution look like if new findings either corroborated the view that acquired characteristics are not hereditarily transmitted or else pinpointed that they are, to one degree or another? In the first case as in the second, the historical place and significance of Lamarck and Darwin would change dramatically.

The decisive point remains that the object of the historian of the sciences differs from the scientist's. The historian's object is the scientific object in context, in the framework of a specific historical and epistemological discourse. It is an object doubly bearing the stamp of a culture, not a universal object:

> If the history of the sciences . . . bears on an object delimited in this fashion, it has to do not only with a group of sciences lacking any intrinsic relation to each other, but also with "non-science," that is, ideology and political and social praxis. This object accordingly has no natural theoretical locus in one or another science, where history might find it ready-made, any more than it has such a locus in politics or pedagogy. Its theoretical locus can be sought only in the history of the sciences itself, for it is this history, and this history alone, which constitutes the specific domain in which the theoretical issues posed by the development of scientific practice find their place.[37]

Since the object of history of science is not identical with the scientific object, it follows for Canguilhem that the history of science "is not a science,"[38] but belongs to a different epistemological order. It is not easy to draw an accurate picture of an activity that constantly traces and retraces on its own terrain historical connections whose possibility and plausibility depend, in the last instance, on the state of affairs currently prevailing in a specific scientific field. History of science accordingly presupposes a never-ending constructive effort: it cannot be written once and for all, but must constantly be *made* and *remade*. In this way, the Canguilhem of the 1960s and 1970s reflected and rationalized his own learning process.

Canguilhem fills the space thus marked out for an epistemological historiography with diverse examples drawn from the history of the life sciences. His epistemology is inseparable from its concrete historical material. His primary objective is to produce a history of concepts,[39] understood as a history of the different problems that arise and recur in the historical space of the sciences. In other words, his objective is the "restitution," as Renard says, "of a conceptual path"[40] or conceptual trajectory. Where he says "scientific objects," Canguilhem means concepts. In this respect, his work is an integral part of the reasoned history of ideas represented in the mid-twentieth century by such authors as Alexandre Koyré, Arthur Lovejoy, Alistair C. Crombie, and Karl Rothschuh. His emphasis on concepts has, however, still another dimension. Balibar, analyzing Canguilhem's conception of science, draws attention to "the fundamental idea according to which the typical units of knowledge are not 'theories' but 'concepts.'" He goes on to note that "it is in the circulation—that is, the translation, transposition, and generalization—of concepts that they get applied or that their 'work' gets performed; here it is that they are put to the test."[41] François Jacob, among others, took up this idea, observing that the life sciences in particular seem to be organized less around theories than around concepts that, depending on the period and field, have a broader or narrower scope.[42]

Let us look at one example more closely. On several occasions, Canguilhem points out the crucial function that Claude Bernard's conception of the *internal environment* had in developing an autonomous order of physiological thinking in the latter half of the nineteenth century. In Canguilhem's view, this concept marked a turning point in the history of the life sciences that went hand in hand with the opening up of new realms of experimentation. He suggests that this concept was elicited by the new experimental culture and its conceptual circumstantiality, as Bachelard discussed it in *The New Scientific Spirit*. According to Bachelard, "concepts and methods alike depend on the experimental realm. A new experiment may lead to a change in scientific thinking as a whole. In science, any 'discourse on method' can only be provisional; it can never hope to describe the definitive constitution of the scientific spirit."[43] In the case at hand, Canguilhem argues that neither the anatomy nor the chemistry of the first half of the nineteenth century was capable of resolving genuinely physiological questions; what was required was a new kind of experimentation on animals that, by making it possible to observe the mechanism of a function in a living organism, opened the door to the discovery of hitherto unknowable phenomena in a way that could not have been foreseen by anyone. Bernard's technique of vivisection, that new mode of experimenting with living animals, was, however, indissolubly bound up with and in a sense even enabled by a fundamental *conceptual* transformation. "It must be emphasized," Canguilhem says, "that the *concept* of the internal environment [provides] . . . the theoretical foundation for the *technique* of physiological experimentation."[44] The in vitro experiments of twentieth-century biochemistry and enzymology would make the *milieu intérieur* an object of intensive investigation, in the course of which the concept would be completely reconfigured. Thus while Canguilhem recognizes that "the history of science can . . . distinguish and accommodate objects at various levels within the specific theoretical domain it comprises: there are always documents to be classified, instruments and techniques to be described, methods and questions to be interpreted, and concepts to be analyzed and criticized," the fact remains that only conceptual labor of that kind "confers the dignity of history of science" upon all the other activities of the historian.[45]

In the case of Claude Bernard, it is possible to speak of a localized epistemological break. In other instances, however, the movements that occur in a conceptual space appear much more widely scattered, if not pervasive. Canguilhem examines one such case in a study of the concept of biological regulation in the eighteenth and nineteenth centuries.[46] Here he shows that "we cannot undertake to write the history of 'regulation' unless we start with the history of 'regulators,' which is a question involving theology, astronomy, technology, medicine, and even nascent sociology."[47] Taking the broad approach that this calls for, he carves out a path through nineteenth-century science by way of Leibniz's idea of a pre-established harmony, Linné's natural economy, Buffon's spontaneous organization of living matter, Watt's steam engine regulator, and Malthus's principle of regulation of the human population. He focuses on the opposition (virulent in France) between Auguste Comte's regulation from the outside and Claude Bernard's regulation from within, a controversy in which the concept of the *internal environment* again plays a crucial role. This trajectory culminates in early twentieth-century *Entwicklungsmechanik*. It is only with Hans Driesch's 1901 *Die organischen Regulationen*, Canguilhem contends, that the decisive turn appears, thanks to which a self-critical discourse of biological regulation can at last emerge.[48] For Canguilhem, the plural in the title of Driesch's book is key: from this moment on, it is assumed that regulatory networks are essential to understanding biological functions. They are introduced as scientific objects capable of self-correction, as a "particular problematique that informs [a science] and defines a homogeneous field of conceptual filiations."[49]

Another important concept elaborated in Canguilhem's later writings comes into play at this point: the concept of "scientific ideology" that we mentioned in setting out. Canguilhem acknowledges that the contemporaneous work of Louis Althusser encouraged him to introduce and develop this concept.[50] The basic idea behind it is threefold. First, scientific ideologies should be characterized as systems of thought whose objects are hyperbolic as measured by the standards of the systems to which they refer; that is, they are not or not yet in the systemic field and under the system's control. Second, a science never arises in unmediated fashion from lived experience; it never springs up on virgin soil, but always emerges on a terrain already prepared by a scientific ideology. Conversely, a scientific ideology generally has its origins in the extension of an existing science into a neighboring field. But, third, a scientific ideology must not be confused with "false science." A scientific ideology derives its power and prestige from an already established science whose principles it proposes to put to work in another field.[51] A suggestive formula that I borrow from one of Canguilhem's essays on the history of physiology encapsulates the point: "problems . . . do not necessarily emerge on the terrain on which they find their solution."[52] Whether the term "scientific ideology" is a particularly happy one may be doubted; the choice has to be seen against the backdrop of the debates of the day. One might also speak of the guiding function of powerful generalizations.

Balibar concludes that Canguilhem's work on the concept of scientific ideology is the vanishing point on which a great many indications scattered throughout his work converge; hence, "there can be no history of truth that is *exclusively* a history of truth, nor a history of science that is *exclusively* a history of science."[53] There is "a dialectic of scientificity and ideologization always already [at work] in history," or, "better still, a dialectic, constitutive of knowledge, *of the ideologization and de-ideologization of concepts.*"[54] Canguilhem's own formulation runs: "to try to write the history of the truth exclusively is to end up writing an imaginary history."[55]

Reading between the lines, we might argue that for Canguilhem the micro-breaks generated in the process of scientific self-rectification—or the ongoing replacement of the truths "of the present"—are themselves always simultaneously embedded in macro-breaks and displacements in which the dynamic of whole fields of knowledge is reoriented and reconfigured. If Canguilhem's famous remark—taken up by Foucault in "The Discourse on Language"[56]—about Galileo's "being in the truth"[57] implies that "the truthfulness [*veridicité*] or truth-saying of science does not consist in the faithful reproduction of a truth eternally inscribed in objects or the intellect,"[58] then it is all the more imperative to interpret these broad discursive shifts as far-reaching cultural transformations. After all, to cite Renard's quasi-Husserlian, yet apposite summary of Canguilhem's ultimate message, "it is humans who produce science and not the other way round."[59] The sciences accordingly remain anchored in the social and technological concerns from which they arise. As Canguilhem observes about one of his favorite objects of analysis, Claude Bernard's experimental medicine, "experimental medicine is thus just one of the figures of the demiurgic dream dreamed by all industrial societies in the mid-nineteenth century, the period when the sciences, thanks to the applications of them, became a social force."[60] History of science à la Canguilhem thus can be seen as paying tribute to an affirmation of Bernard's: "with the help of these *active experimental sciences*," we read in Bernard's *An Introduction to the Study of Experimental Medicine*, "man becomes an inventor of phenomena, a real foreman of creation."[61]

PART II

Model Organisms

STUDIES IN THE HISTORY OF HEREDITY

AND REPRODUCTION

This section assembles four case studies spanning the time from the end of the nineteenth century to the end of the Second World War. They reach from the establishment of classical genetics to the dawn of molecular biology in Germany. The scientists involved, from Carl Correns and Max Hartmann, to Alfred Kühn, and to Georg Melchers and his colleagues, were all housed at one time or another at the Kaiser Wilhelm Institute for Biology in Berlin-Dahlem, founded in 1911. This institute was one of the few and prominent places in which genetics took shape in the early decades of the twentieth century in Germany.

The four case studies describe four experimental systems, each with a different model organism. In the case of Carl Correns, it was a hybridization system based on the garden pea *Pisum sativum*; in the case of Max Hartmann, a cellular propagation system based on the unicellular green alga *Eudorina elegans*; the flour moth *Ephestia kühniella* served as the experimental model for Alfred Kühn and his group as they developed a system for studying gene actions, in particular in eye pigment formation; Georg Melchers and the virus group concentrated their efforts on tobacco mosaic virus. Each group used its model organisms in different ways, testifying to the role their experimental systems played in generating unprecedented knowledge. But the focus of each group was also different. The case study on Correns examines the micro-dynamics of a particular experimental re-orientation. The case study on Hartmann looks at the experimental narrowing of a biological function. The chapter on Kühn focuses on the productive interaction of model organism and experimental system. And the study on the Dahlem virus group fore-grounds the institutional dynamics of research. Together with my case study on the history of protein synthesis research in the test tube after the Second World War (Rheinberger 1997), the studies of this volume amply demonstrate the role of experimental systems as historical actors for the life sciences in the first half of the past century.

4. Pisum

Carl Correns's Experiments on Xenia, 1896–99

Classical genetics as a discipline was born around the turn of the twentieth century. The standard history of genetics has long included the particular circumstances of its birth. Yet despite considerable historiographical effort the question why Gregor Mendel's experiments took more than thirty years to find a wide audience remains unanswered.[1] It is still debated how the so-called "rediscovery" of Mendel's laws came about around 1900.[2] As is well known, the conventions of scientific writing in general and priority disputes in particular can obscure the way certain findings were historically obtained and certain conclusions were historically drawn. In Mendel's case, his "rediscovery" occurred at least three times, making far more complicated the usual story of belated scientific recognition.[3] In what follows, I reconstruct the story by referring to Carl Correns's unpublished research notes on his hybridization experiments with *Pisum sativum* and *Zea mays*. The resulting picture is of a scientist who initially seeks to solve a particular problem, finds himself struggling with his experimental material, and finally arrives in the course of his experiments at an epistemic regimen in which previously incidental observations and considerations suddenly acquire new significance. The whole account is a textbook example of the impossibility of rationally reconstructing the birth of a science, since neither its origins nor its critical turning points, the hallmarks of scientific success, are easy to discern. The magnifying glass of Correns's research protocols sharpens our view of the hesitations and delays constitutive of empirical scientific work. His research notes also give us an idea of the intricacies and material peculiarities of his hybridization experiments and their effect on his conclusions.

This case study is based on protocols housed in Berlin-Dahlem,[4] which means that they must have accompanied Correns on his itinerary from Tübingen by way of Leipzig and Münster to Berlin, where they survived the Second World War. This was lucky, since as Correns's biographer Emmy Stein informs us most of Correns's archive was destroyed early in 1945.[5] What remains are primarily observations and lists of findings noted on loose sheets roughly arranged according to experimental object and order of experiment. Unfortunately, Correns did not date all of his notes and Stein, who worked on the archive during the last years of the war, did not always keep the files in the original condition.

With Hugo de Vries[6] and Erich von Tschermak-Seysenegg,[7] Carl Correns is one of the three botanists credited with the rediscovery of Mendel's laws. Robert Olby remarks that of the so-called three rediscoverers and the several other plant breeders who published Mendelian ratios in and around 1900, "Correns showed the deepest understanding," and assumes that he arrived at the "correct explanation" without knowing about Mendel's paper.[8] But the story is more complicated.[9] In the following pages, I retrace Correns's path to the segregation law. His first publication on peas (1900) offers this formulation of it: "The hybrid produces sexual nuclei that bring together the dispositions [*die Anlagen*] of the parents in all possible combinations, with the exception of those of the same character pair. Every combination occurs with roughly the same frequency."[10]

A Short Biographical Note

Carl Erich Correns (1864–1933) was born in Munich, where he studied with Carl Wilhelm von Nägeli toward the end of the latter's life. He completed his dissertation, entitled "Über Dickenwachstum durch Intussusception bei einigen Algenmembranen" (On Thickening by Intussusception in the Membranes of Several Algae) in 1889, two years before Nägeli died.[11] From 1889 to 1892, he worked as an assistant to Gottlieb Haberlandt in Graz, Simon Schwendener in Berlin, and Wilhelm Pfeffer in Leipzig. He obtained his *Venia legendi* (authorization to teach) at the University of Tübingen in 1892, when Hermann Vöchting granted him his habilitation for the work "Über die Abhängigkeit der Reizerscheinungen höherer Pflanzen von der Gegenwart freien Sauerstoffs" (On the Dependence of Excitation Phenomena in Higher Plants on the Presence of Free Oxygen).[12] Correns spent the next ten years in Tübingen, where he performed his first hybridization experiments (with peas and corn, among other plants) in the small botanical garden of that South German university town. He also cultivated experimental plants on the truck farms of commercial gardeners in the Tübingen area.[13] He continued to carry out extensive breeding experiments to the end of his career, which led him to Leipzig as an associate professor in 1902 and to Münster seven years later as a full professor and director of the botanical garden. In 1913, he was appointed director of the newly founded Kaiser Wilhelm Institute (KWI) for Biology in Berlin-Dahlem, where he pursued his research until his death in 1933.

The Early Tübingen Years

Correns investigated *Pisum* in his years as a *Privatdozent* (untenured lecturer) in Tübingen (1892–1902). A glance at his extant notes of the 1890s shows that he cultivated a wide range of interests in this period. He pursued his work on the morphology and growth of the cell membrane, subjects already taken up in his doctoral dissertation;[14] explored, in the wake of his *Habilitationsarbeit*, the physiology of movement in plants, especially the physiology of tendrils;[15] carried out extensive research on the vegetative reproduction of mosses, publishing the results in a book in 1899;[16] undertook a systematic study of algae and cyanobacteria (the *Oscillatoria*);[17] and collected and published his findings about alpine flora.[18]

Correns's academic training had been centered on physiology, morphology, and systematics. He started to do breeding experiments only around 1894. In a short, undated autobiographical sketch, he says: "In my ten years as a *Privatdozent* in Tübingen, my works and publications tended to follow beaten paths. . . . However, as soon as I obtained the habilitation and gained access to a botanical garden, I started to conduct, in addition, a wide diversity of experiments that we would today call 'genetic,' experiments that I had long had in view."[19] At first, Correns's crossing experiments seem to have had no real impact on his research questions, which were still physiologically oriented: these experiments were in any case limited in scope and rather unsystematic. They were manifestly carried out as assays designed to elucidate problems of developmental physiology, such as the formation of adventitious embryos in *Hosta* (Funkia), a problem Correns vainly tried to solve through hybridization. In his autobiographical sketch, he himself characterizes his early breeding experiments as *"fooling around."*[20] However, as we shall see, they moved ever closer to the center of his concerns as the years went by. Yet it was not until he had finished his book on the vegetative reproduction of mosses in fall 1899—a book with no direct bearing on questions of heredity—that problems of breeding assumed a commanding position in Correns's work, occupying him for the rest of his scientific life.[21]

Xenia

Correns's interest in maize was aroused when he looked into Darwin's observations on xenia. Xenia are characters of a pollen-giving plant that appear directly on the mother plant, especially its seeds and fruits, and not just in individuals of the next generation. When Correns began his investigation of xenia in 1894, there was no valid explanation for the phenomenon. By the time he completed it six years later, Sergej Navashin in Kiev and Léon Guignard in Paris had discovered and published the basic explanation for it: a consequence of double fertilization in angiosperms, leading to the embryo on the one hand and the endosperm, the early nourishing tissue of the embryo, on the other.[22] Correns carried out his first crosses with *Zea mays* in 1894, and maize remained his most important experimental object for studying xenia. But he also combed the literature on other plants for references to xenia. Over the next few years, he added *Lilium*, *Matthiola*, and *Pisum* to the list of his experimental plants.

In a folder containing Correns's notes taken between 1896 and 1900, there is a page dated "15/IV," entitled "Gaertner's experiments with *Pisum*."[23] The reference is to Carl Friedrich Gärtner's *Versuche und Beobachtungen über die Bastarderzeugung im Pflanzenreich* (Experiments and Observations on the Creation of Hybrids in the Plant Kingdom).[24] There is every reason to suppose that this page of notes dates from the early period when Correns had just started experimenting with peas—it was probably written in 1896—since the remarks in which he summarizes his observations indicate that he was working on the question of xenia in peas at the time of writing: "Thus—as G. himself also emphasizes, a very striking influence on the fertilized seed (not the fertilized fruit!). Very striking that *all* the seeds in the same pod, or perhaps even all the pods of a single cross (c. 5) usually attested this influence! Does the coloration of the seed depend on that of the cotyledons? This would make it possible to explain the result, understood as an *intermediary formation*!"[25] Another undated note on *Pisum* summarizes a paper of Wilhelm Rimpau's, singling out Rimpau's observations of "reversions": "Rimpau easily achieved several crosses, all showed a great inclination to reversions, several not constant even after 6 years."[26]

The biggest surprise in this folder is a note titled "Mendel (66)" dated "16. IV. 96" (fig. 1).[27] The note is transcribed in Table 1, below. We must conclude that Correns read Mendel's paper in April 1896, his later recollections notwithstanding. In the autobiographical sketch already mentioned, he says:

It was in a sleepless November night, near daybreak, that I suddenly hit upon the explanation for the observations of *Pisum* and *Zea*. But it was only when I sat down to prepare the publication that I systematically reviewed the literature. It was then that I realized, thanks to Focke's Pflanzenmischlinge [1881], that Mendel had already discovered and published all this thirty-five years earlier. . . . If I had found the explanation earlier, I would

Table 1

---

16. IV. 96

Mendel (66) distinguishes:
dominant and recessive characters. For our

cases is	dominant:	recessive:
—form of seed	round	angular
—seed coat: ("albumen")	grey to brown	*white.*
—cotyledons:	yellow	pale yellow, *green*
—pod:	smoothly rounded	wrinkled
— :	green (unripe)	yellow (unripe)

The dominant and recessive characters are expressed from the first generation on in such a way that the former are present in 3 individuals, the latter in 1.

> The hybrid form of seed shape and cotyledons develops immediately, directly through fertilization

thus

Cot. yellow ♀ + green ♂ = yellow ♂ + green ♀ = ¾ yellow + ¼ green
Form. round ♀ + angular ♂ = angular ♀ + round ♂ = ¾ round + ¼ angular

The seed coat, the form of the pod, and the color of the pod do not *change.*

Later, however, *Mendel* says, e.g., that A (seed round, cot. (p. 19) yellow) pollinated with B (seed angular cot. green) yielded *nothing but yellow* seeds that were *round.*

---

have published it in a preliminary note, although I was at work on my book on mosses. For the significance of the results was pretty much clear to me from the first.[28]

Correns gives a similar version of events in a letter he wrote on January 23, 1925, in response to a query from Herbert F. Roberts. The indignant undertone is unmistakable: "The date of the day upon which, in the autumn (October) of 1899, I found the explanation, I no longer know; I do not make note of such matters. I only know that it came to me at once 'like a flash,' as I lay toward morning awake in bed, and let the results again run

16. IV. 96.

Mendel (66) unterscheidet:

dominirende u. recessive Merkmale. Für unsere
Faelle ist dominirend: recessiv:
– Samenform   rund   kantig
– Samenschale : grau bisbraun   weiss.
  ("Albumen")
– Kotyledonen : gelb
                                             blassgelb, grün
– Frucht : einfach gewölbt
                                             runzlig
–   : grün (unreif)   gelb (unreif)

Die dominirenden und recessiven Merkmale treten
gleich bei der ersten Generation so hervor, dass die ersteren
je 3, den letzteren je 1 Individuum aufweist.

| Die Hybridform von Samengestalt und Kotyledonen
entwickelt sich unmittelbar direct dch die Befruchtg |
       also

Cot.   gelb ♀ + grün ♂ = gelb ♂ + grün ♀ = 3/4 gelb + 1/4 grün.
Form.   rund ♀ + kantig ♂ = kantig ♀ + rund ♂ = 3/4 rund + 1/4 kantig

Nicht verändert wird die Samenschale, die Fruchtform
u. die Fruchtfarbe.

Spaeter aber giebt Mendel z. B. an dass A (Samenrund, Cot.
(S. 19) gelb) mit B (Samen kantig Cot. grün) bestäubt,
lauter gelbe Samen gab; die rund waren.

1. Correns's excerpt from Mendel's paper. MPG Archive, Section 3,
Folder 17, No. 115, File "Pisum-Kreuzungen 1896–1900."

through my head. Even as little do I know now the date upon which I read Mendel's memoir for the first time; it was at all events a few weeks later."[29]

This version of events has gone virtually unchallenged since. In recent years, to the best of my knowledge, only Onno Meijer has aired the suspicion "that Correns read Mendel's paper before that inspired night, that he did not realize Mendel's importance on his first reading of it, and that the paper may have made its way through his unconscious, only to re-emerge 'in a flash' later."[30] Let us see what the research protocols can tell us about this.

To begin with, we should take a more careful look at the excerpts Correns made from Mendel's paper the first time he read it. It is certain that he did not rely on the account of the paper contained in Wilhelm Olbers Focke's book on *Pflanzen-Mischlinge* (plant hybrids),[31] since he refers in his notes to the pagination of the original publication. There are indications that the offprint which Mendel had sent a correspondent of his in Munich, von Nägeli, who later became Correns's teacher and mentor, was in Correns's possession; it is probably the copy held in the offprint collection of the Kaiser Wilhelm Institute for Biology.[32] Correns's papers in Dahlem in addition contain various notes and drawings by von Nägeli.

Closer examination of the excerpt reveals that while Correns took note of Mendel's offspring ratio of 3:1 with respect to dominant and recessive characters, something else seems to have captured his immediate attention: all those seed characters that might have something to do with the xenia phenomenon he was then studying in maize. This is suggested by the fact that he framed the following sentence with little vertical lines: "the hybrid form of seed shape and cotyledons develops immediately, directly through fertilization." Correns doubtless considered this an observation relevant to the xenia problem.

It should also be mentioned that Correns was in all probability misled by Mendel's terminology. Mendel calls the second generation of crosses the "first generation of hybrids." Correns's summary reads: "The dominant and recessive characters are expressed from the first generation on in such a way that the former are present in 3 individuals, the latter in 1." At the end of his notes, however, he points—erroneously—to a "contradiction" in Mendel: "Later, however, *Mendel* says, e.g., that A (seed round, cot. (p. 19) yellow) pollinated with B (seed angular cot. green) yielded *nothing but yellow* seeds that were *round*." We can take this as evidence of a quite cursory reading.

One last observation points in the same direction. In the sentence just quoted, Correns does not use the symbols A and B the way Mendel does in his paper. Mendel uses the capital letters (A and B) exclusively for *dominant characters*, whereas, in his excerpts, Correns calls the two *plant varieties* to be crossed A and B.

To judge by the notes that Correns left behind, he had very probably decided to carry out experiments with peas even before reading Gärtner, Rimpau, and Mendel, since he had already selected and purchased in April 1896 the six varieties on which he wanted to begin experimenting. This explains why he refers to "our cases" in his note on Mendel and proceeds to list five of Mendel's seven paired characters. The three characters he chose to follow were the form of the seed, the coloration of the seed coat, and the coloration of the germs (the cotyledons). A fourth character, the coloration of the seed, meant a combination of the coloration of the seed coat and that of the germs or cotyledons. If his aim was to solve the xenia mystery, all this is just as one would expect. On April 23, 1896, a week after reading Mendel's essay, Correns sowed six varieties of peas in six pots, planting three peas of each variety in each pot (fig. 2).[33]

The hard part of our reconstruction begins here. There is no reason to assume a priori that Correns was simply lying when twenty-five years later he wrote the lines to Roberts cited above or when he noted in his autobiographical sketch (undated, but also written in the 1920s) that he first read Mendel late in 1899 (that is, after he had completed the fourth generation of his pea experiments and *after* he had "arrived at the explanation" for his observations). Those who accept Meijer's conjecture and our own analysis of the note on Mendel will be more inclined to conclude that Correns had regarded Mendel's experiments from a very different angle when he first read his paper in spring 1896. He may have written his note on Mendel and given it no further thought until it resurfaced in his memory a few years later. At that point the note took its place in the experimental context that he himself had created in the meantime.

The 1896 *Pisum* Protocols

What do the protocols housed in Dahlem tell us about all this? Correns did not simply fasten on an especially clear-cut pair of characters, which is what one would expect of a "demonstrative" experiment designed to confirm Mendel's results. What he did was to cross five of the six varieties sown in April by means of reciprocal pollination. The first artificial fertilizations took place on June 19, 1896, as the plants Correns had sown began to bloom. He harvested the first seeds in late July-early August. The results are listed in the table entitled "Pollination and Yield 96" (fig. 3).[34] The

2³/IV. 96.

Erbsenaussaat.

6 Sorten, je 6 Töpfe mit 3 gequollenen Erbsen
   beschickt

Haage u. Schmidt. Name. Zeichen. Oberfläche, Schale, Cotyled.
   Nº Sorte.

Nº	Sorte	Name	Zeichen	Oberfläche	Schale	Cotyled.
1250	K.	Bohnenerbse	P.B.E.	grubig	graubräunlich	gelb
1255	K.	grüne Folger	Pgr	glatt	farblos	grün
1258		Purpurschote	Pp	etwas faltig	rothbraun	hellgelb
1265	M.	Pride of th. Market	Ppm	fast glatt	farblos	grün
1275	z.M.	Alliance	Pw	faltig	farblos	weisslich
1300		Golderbse	Pgo	glatt	farblos	gelborange

Am 19. Juni 96:

1250 P.B.E.

1255 P.gr.

1258 P.p.

1265 Ppm.

1275 P.w.

1300 Pgo.

K = Kneifelerbse
M = Markerbse
z. M. = runzlige Markerbse.

2. Page of Correns's research protocol entitled "Erbsenaussaat [peas sown]."
MPG Archive, Section 3, Folder 17, No. 115, File "Pisum-Kreuzungen, Resultate 96."

3. Page of Correns's research protocol entitled "Bestäub. u. Ertrag 96 [pollinization and yield, 1896]." MPG Archive, Section 3, Folder 17, No. 115, File "Pisum-Kreuzungen, Resultate 96."

first number at the bottom and to the right of each square represents the number of pods; the second gives the total number of seeds in each pod; the number in the center of the squares represents the number of artificial pollinations. For example, the reciprocal crosses of P_{gr} ("grüne, späte Erfurter Folgererbse," abbreviated "Grüne Folger") with P_p ("purpurviolettschotige Kneifelerbse," abbreviated "Purpurschote") yielded four and ten seeds, respectively, for a total of fourteen seeds resulting from twenty-six (7 + 19) artificial pollinations. These numbers make it abundantly clear that the crosses were not realized with an eye to the kind of statistical evaluations that Correns would surely have made had he been following in Mendel's footsteps at this point.

This non-statistical approach stands in striking contrast with Correns's statistical treatment the same year (1896) of the question of the number of pollen grains required to achieve optimal fertilization of *Mirabilis* plants.[35] In the context of these pollination experiments with *Mirabilis Jalapa* and *Mirabilis longifolia*, he conducted extensive statistical thought experiments on paper, which is to say that he was plainly already familiar with statistical methods of treating experimental data where he thought them appropriate.[36] The unpublished 1896–97 notes on *Mirabilis* also contain indications as to Correns's reading at the time, which included Joseph Gottlieb Kölreuter, Gärtner, and Charles Naudin.

In the case of the peas, Correns was obviously not planning large-scale

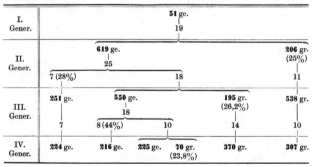

4. "Versuch I [Experiment I]." A cross between a "late green Erfurt Folgererbse" with a green germ and a "purpurviolettschotigen Kneifelerbse" with a yellow germ. Figure from Correns 1900a, 162.

experiments, nor did he initially set out to calculate average values. Instead, he carefully described the coloration of each harvested seed. He also compared artificial pollinations with self-pollinated samples in extensive control experiments, noted all the striking phenomena he observed, and in a few cases even measured the diameters and weight of the seeds. The last-named properties of the seeds clearly had nothing to do with the characters chosen for the crossing experiments. About the hybrids P_{pm+gr} ("Pride of the Market" fertilized with pollen from "Grüne Folger"), Correns noted laconically—as if to remind us of the question at the origin of his experiments—"all Xen [xenia] have failed."[37]

My objective here is not to provide a detailed account of all of Correns's *Pisum* crosses through the various breeding generations, let alone the parallel crosses that he carried out with maize. I shall, rather, restrict my attention to the pair already mentioned, Grüne Folger (P_{gr}) and Purpurschote (P_p), while adding pertinent information from other protocols where appropriate. The result obtained with this pair—namely, that yellow as the color of the embryo is dominant over green—is also described in "Experiment I" in Correns's first 1900 paper on *Pisum* (fig. 4);[38] we can therefore take this experiment as a reference point for the rest of our analysis. Let us retain for the time being that in describing his first-generation crosses, Correns calls the color of the seeds of the hybrid P_{gr+p} (Grüne Folger fertilized with pollen from Purpurschote) "yellow," and asserts that the pods of the reciprocal hybrid P_{p+gr} (Purpurschote fertilized with pollen from Grüne Folger) contain seeds of a "purer green" or have "conspicuously more greenish" seeds than the self-pollinated mother plant.[39]

The 1897 Protocols

The experiments performed in the next vegetation period began in spring 1897. A table provides an overview of the new first-generation crosses (fig. 5).[40] The three varieties P_{BE} ("Bohnenerbse"), P_{gr}, and P_p were crossed once again, and the three new varieties P_π, P_{WH}, and P_{ggR} replaced the P_{pm} ("Pride of the Market"), P_w ("Alliance weisse Zwergerbse"), and P_{go} ("Späte Golderbse") of the previous year. The information in the table resembles that summarized in figure 3, but it has become somewhat more complex. It comprises the number of pollinations, the number of pods obtained, the number of seeds contained in the pods, the ratio of the number of these seeds to the number of pollinations, and the average number of seeds in a pod. Correns obtained many more seeds this year (1897), but this multitude of numbers represents rather an efficiency control for the individual acts of artificial pollination than a representation of the transmission of characters.

While the total number of seeds from the reciprocal crosses between P_{gr} and P_p was 14 (10 + 4) in 1896, it was 39 (23 + 16) in 1897. Again, the 23 P_{gr+p} seeds all looked yellow, while the P_{p+gr} seeds looked more orange, shading toward green.[41] Thus we have a total of fifty-three seeds in the first generation. In the table "Experiment I" in his first *Pisum* paper, Correns lists, it is true, only fifty-one yellow seeds, but the discrepancy is not relevant for these results. For reasons unknown, he included only fourteen of the P_{p+gr} hybrid seeds obtained in 1897 in the count, instead of sixteen.[42] On the other hand, he counted as "yellow" the four seeds from his 1896 P_{p+gr} crosses, which he had earlier described as "greenish," because the character to be compared later was the coloration of the cotyledons, not that of the seed as a whole (where he was looking at a combination of the coloration of the cotyledons and the seed coat).

The page on which we find the list of 1897 results ("Ergebnisse 1897," fig. 6) regarding the xenia question[43] confirms the assumption that it was the failed xenia of the previous year that led Correns to repeat the first-generation crosses and even to carry out various other crosses, rather than the wish to harvest larger numbers of seeds in order to obtain reliable statistics. Clearly there were still no observations relevant to the xenia question to be had. However, the list does suggest why Correns's experiments included so many different crosses: in one case and one alone, it

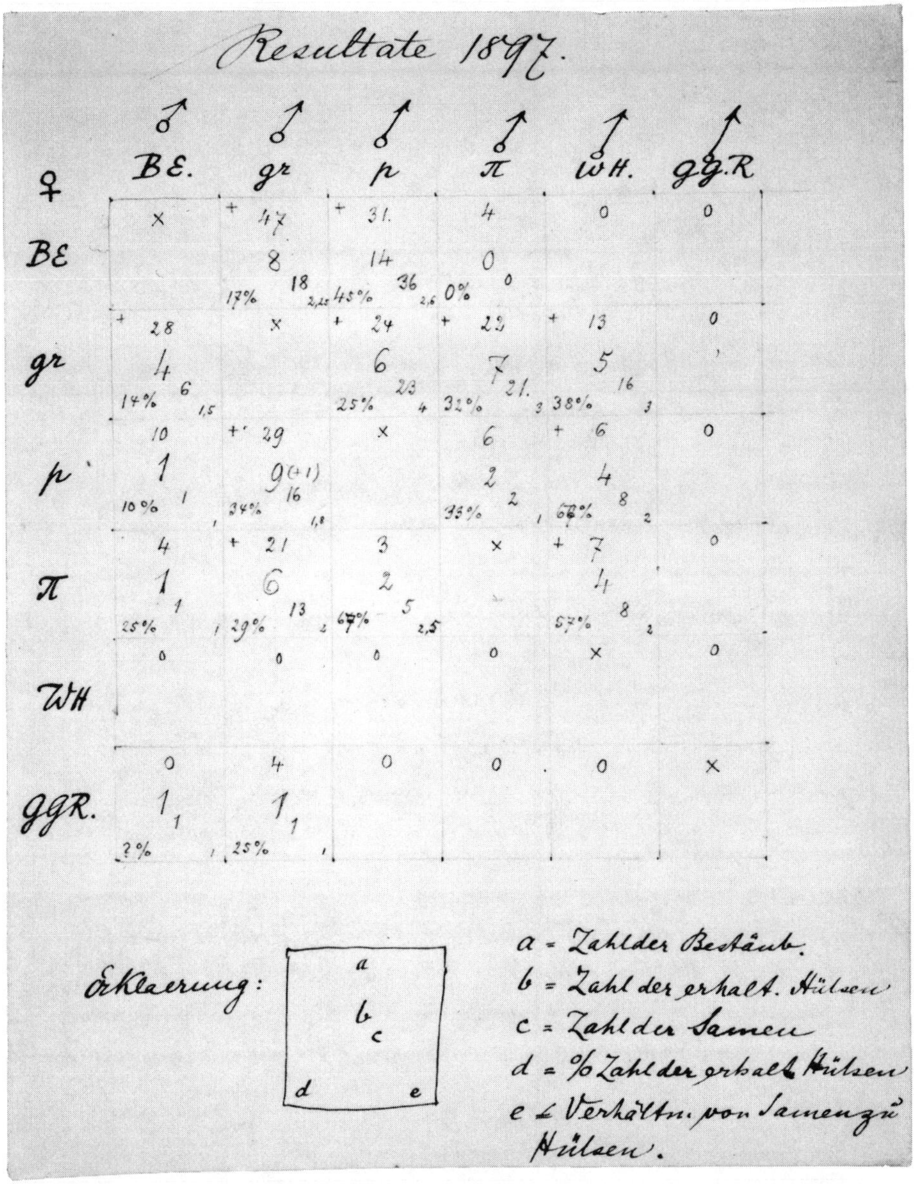

Resultate 1897.

Column headers (♂): BE. | gr | p | π | WH | ggR

Row labels (♀): BE | gr | p | π | WH | ggR.

♀ \ ♂	BE.	gr	p	π	WH	ggR
BE	×	+ 47 / 8 / 17% 18	+ 31. / 14 / 45% 36	4 / 0 / 0%	0	0
gr	+ 28 / 4 / 14% 6	×	+ 24 / 6 / 25% 23	+ 23 / 7 21. / 32%	+ 13 / 5 / 38% 16	0
p	10 / 1 / 10%	+ 29 / 9 (+1) / 34% 16	×	6 / 2 / 33%	+ 6 / 4 / 67% 8	0
π	4 / 1 / 25%	+ 21 / 6 / 29% 13	3 / 2 / 67% 5	×	+ 7 / 4 / 57% 8	0
WH	0 / 0	0 / 0	0 / 0	0 / 0	×	0
ggR.	0 / 1 / 3%	4 / 1 / 25%	0	0	0	×

Erklaerung:

a = Zahl der Bestäub.
b = Zahl der erhalt. Hülsen
c = Zahl der Samen
d = % Zahl der erhalt. Hülsen
e = Verhältn. von Samen zu Hülsen.

(Diagram: square with a at top, b/c in center, d at bottom-left, e at bottom-right)

5. Page of Correns's research protocol entitled "Resultate [Results] 1897."
MPG Archive, Section 3, Folder 17, No. 115, File "Pisum Resultate 1897."

Ergebnisse 1897.

Xenien.

$\begin{cases} BE + gr & \text{giebt normale } BE \text{ Samen.} \\ BE + p & \text{giebt normale } BE \text{ Samen.} \end{cases}$

$\begin{cases} gr + BE & \text{giebt gelbe Samen.} \\ gr + p & \text{giebt gelbe Samen.} \\ gr + \pi & \text{giebt gelbe Samen.} \\ gr + WH & g. \text{ normale } Gr \text{ Samen ??} \end{cases}$

$\begin{cases} p + BE & \text{giebt normale } p \text{ Samen.} \\ p + gr & \text{giebt normale } p \text{ Samen.} \\ p + \pi & \text{giebt normale } p \text{ Samen.} \\ p + WH & \text{giebt normale } p \text{ Samen.} \end{cases}$

$\begin{cases} \pi + BE & \text{giebt normale } \pi \text{ Samen.} \\ \pi + gr & \text{giebt normale } \pi \text{ Samen.} \\ \pi + p & \text{giebt normale } \pi \text{ Samen.} \\ \pi + WH & \text{giebt normale } \pi \text{ Samen.} \end{cases}$

$\begin{cases} ggR + gr & \text{giebt normale } ggR \text{ Samen.} \\ ggR + BE & \text{giebt normale } ggR \text{ Samen.} \end{cases}$

Bastarte.

$gr + p$ gab in der ersten Generation Samen, deren Grundfarbe zwischen grün und röthlich-orange schwankte und die alle mehr oder weniger (sparsam ja nur stellenweise bis stark) violett punktirt waren, sahen also weder den Xenien noch dem Vater oder der Mutter ähnlich, sondern beinahe ggR (bei der die ~~Schoten~~ Hülsen aber ganz anders aussehen

---

6. Page of Correns's research protocol entitled "Ergebnisse [Findings] 1897." MPG Archive, Section 3, Folder 17, No. 115, File "Pisum Resultate 1897."

was possible that the pollen-giving plant might have influenced the seeds of the mother plant. Next to this potential exception—gr pollinated with WH—Correns put two question marks.

At the bottom of the sheet, Correns noted, by way of comment on the coloration of seeds grown from self-pollinated, first-generation P_{gr+p} hybrids, that is, the second generation overall: "gr + p in the first generation yielded seeds whose basic color fluctuated between green and reddish-orange, while all of them were more or less spotted purple (lightly, as a rule, and up to heavily only in places). Thus [they] resembled neither the xenia, nor the father, nor the mother, but almost ggR (in which, however, the pods looked completely different)."[44]

This remark suggests that we should take a closer look at the protocols in which Correns recorded his findings about the first self-pollinated hybrid generation. On April 20, 1897, he sowed seven of the ten yellow seeds that he had obtained the previous year after pollinating P_{gr} with P_p. As a control, he sowed twelve seeds that derived from the same plants but had resulted from self-pollination.[45] All the hybrid plants and all but one of the control plants grew.[46] Correns carefully noted the form and coloration of each individual seed, with the result that each one could be traced back to the individual plant and even to the particular pod from which it came. He designated the seed colors as "orange yellowish green," "orange grayish," "nearly pure green," "nearly pure yellow," and a long series of other shades. In sum, the seeds' "basic color fluctuated between green and reddish-orange." There was no sharp division between green and yellow, but, rather, a range of different shades that "resembled neither the xenia, nor the father, nor the mother."

In the protocols for the first generation, Correns kept records of the self-pollinated control plants on the left side of the page and the cross-fertilized hybrids on the right; in the protocols for the second generation in contrast, he included detailed descriptions of the seeds on the right side of the page, leaving an empty margin on the left (fig. 7). In this margin, we read, for instance, about the results for the first of the seven plants, in small, cramped handwriting: "Σ was 22 and is 22. So *none* of these seeds was sown. 18 with yellow germ, 77.8%, 4 with green, 22.2%, medium-sized, moderately wrinkled, *lightly spotted*." About plant four, we find the following note: "Σ 18, only 15 remain. So 3 are missing. Of the rest (15), 3 have green germs = 20%, 12 yellow = 80%. Otherwise like I. Sowing, yellow 2 green 1, so 14 yellow, 77.8%, 4 green, 22.2%."[47] Correns thus went back to these protocols at a later date—at the earliest, some time after the sowing season of the following year (1898). The date can be deduced from the fact

Bastart

gr + p ♂ A 1., (gelb)

I. Pflanze I. Hülse
  {1. 1/2 gelblichgrün, wenig punktirt orange
  {2. wie 1.

∑ war 22 u. ist 22.
Hievon wurde also keines
ausgesaet.
18 mit gelbem Keim, 77,8%
4 mit grünem. 22,2%
, mittelgross, maessig faltig,
wenig punktirt.

3. 1/2 orangegrünlich, wenig punktirt

4. Sa. 5, alle schwach punktirt, 2 fast
rein grün, 3 orangegrünlich, grob
grubig.

5. Sa 4, 2 fastrein gelb, 2 orangegrün,
ganz fein punktirt u. sehr sparsam.

6. Sa 2, wie 1.
7. Sa 5, 4 wie 4, 1 kleiner, sehr stark
rothorange angelaufen

II. Pflanze II. Hülse
∑ war 19, ist es noch
also nichts ausgesaet.
15 mit gelbem Keim (26,7%)
4 mit grünem (73,3%)
aussehen wie oben

1. Sa 6, scherbengelb, sehr schwach punktirt, 1 fastrein (hell) grün, ebenso punktirt.

2. Sa. 5, davon 2 fastrein (hell) grün
   2. scherbengelb
   1 rothorange, alle ziemlich fein punktirt u. nicht reichlich.

3. Sa. 4, davon 3 grünlich-scherbengelb,
   1 rothorange, alle ziemlich wenig u. nicht stark punktirt.

4. Sa. 3, davon 2 mehr grünlich, 1 mehr
   gelblich scherbengelb, kenntl. wie gew. (1-3)

7. Page of Correns's research protocol entitled "Bastart [Hybrid] gr + p ♂ A 1, (gelb) [yellow]." MPG Archive, Section 3, Folder 17, No. 115, File "Pisum Resultate 1897."

that he must already have used the seeds that were now "missing," an exact count of which he now belatedly tried to produce so as to be able to calculate correct percentages. And we see that he now noted the coloration of the germ rather than of the seed. The ratio was calculated on the basis of small numbers and consequently subject to considerable fluctuation. Yet as far as the dominant yellow color of the cotyledons inherited from P_p is concerned, a shift toward a segregation ratio of 3:1 is clearly discernible in the first hybrid generation. What we have here is plainly a retrospective interpretation of the 1897 results, as is confirmed by the fact that this information is recorded on a separate part of each page. Let us now turn to the protocols of the following year (1898) to see whether they contain an indication as to what led Correns to abandon his research on xenia and to reinterpret his data in the light of the segregation law.

The 1898 Protocols

In 1898, Correns obtained a second generation of the hybrid $P_{gr♀ + p♂}$. He took nine seeds from plant III (A1) of the previous year's hybrid generation and sowed them in three pots. Six of these seeds were yellowish-orange, while the other three had a strong green tinge. Correns also sowed three seeds from plant IV (A1) (two yellow seeds and a green one), three from plant II (B1) (again, two yellow seeds and a green one), and three from plant I (B1) (all yellow) in another three pots.[48] When, later, he reconstructed the number of yellow and green germs that he had obtained from the harvest of the preceding year, 1897, the numbers exactly matched. There is, however, one interesting discrepancy. Correns had sown three green seeds from plant III of A1—in the 1898 protocols, we find the words "seeds conspicuously *green*."[49] Yet when he reconstructed the number of yellow and green germs in the margin of the 1897 protocol, he describes only two of the germs as "green." The reason is that in 1898 only two individuals of the second-generation hybrid plants produced germs all of which were green; one produced both green and yellow germs. Correns therefore concluded retrospectively that this greenish seed must have contained a yellow germ, and therefore counted it as yellow. This too implies that his supplementary marginalia must date from fall 1898 at the earliest.

Correns raised another set of second-generation hybrids from the reciprocal cross, that is, $P_{p♀ + gr♂}$. He observed that germs from the seeds of these plants were "sometimes green, sometimes yellow" and that most of the individuals represented "the normal hybrid." Two individual plants, however, were "pure p"; Correns therefore paid no further attention to them.[50]

Let me here emphasize once again that Correns was at all times thanks to the exactness of his protocols in a position to trace every single seed and the plant raised from it to a particular plant and seed of the preceding generation. He kept a virtually complete physical collection of *individual* seeds, which he also catalogued and annotated. This fact, too, bears out the hypothesis that the aim of the series of *Pisum* experiments was not to corroborate a statistical regularity that he might have noticed on his first, spring 1896 reading of Mendel's paper, and set out to test. He was looking rather for unknown seed characters that would only appear in the course of these hybridizations. The protocols were thus not merely passive data storehouses.[51] Although their design and structure were exactly tailored to the particular interest that Correns took in his object, they could also bring other, diverging aspects of that object into the light. And they contained an excess of information, allowing for a reorientation of the experimental gaze later on.

At the top of the page in the 1898 protocols is a short, summary description of the material sown and the form and coloration of the harvested seeds. Like the protocols of the preceding year, the 1898 protocols contain supplementary notes that could only have been added later. Located in the left margin of the page in 1897, these notes appear in 1898 at the bottom of the page. Here Correns noted the total number of seeds obtained from each plant, the number of green and yellow embryos, and the ratio of the embryos, expressed as a percentage. As before, he added to the actual number of seeds left in his boxes the number of those he had sown to obtain the next generation. For instance, we read on the protocol for pot three, under point b, about individual plant number two (fig. 8): "Pods with 3, 4, 4 seeds, green to light reddish-orange. Of the 30 extant, 10 have green germs, 20, yellow germs; of the seeds sown, 1 had a green germ, 2, yellow germs, making a total of 33 seeds, of which 11 = 33.3% had a *green* germ, 22 = 66.7%, a *yellow* germ."[52] Another remark on the protocol for pot 6, about individual plant number 2, reads: "Σ now: 7 green 20 yellow. Sown: 1 green 2 yellow. ΣΣ 30, of which 8 green = 26.7%, 22 yellow 73.3%."[53] In light of these calculations again including the seeds sown out, we can conclude that Correns made his supplementary annotations—and, therefore most probably those found in the 1897 protocols as well—only after he had sown the next generation of plants in spring 1899.

Let us now turn to another series of experiments that Correns carried out in 1898. In two more pots, he raised six more individuals of the first

III. Topf.

Bastart gr + p. ♂ a. gelb III.

zweite Generation.

I. Indiv. Aussaat: Samen auffallend grün.
Ernte: Samen zwischen roethlich orange u.
rein grün stehend, die roethlichen orangefarbenen
auch mit grünem Keimling, soweit untersucht
Punktirung vorhanden.

II. Indiv. Aussaat wie oben
Ernte: wie bei I, aber Keimlinge überwiegend
gelb, auch wenn die Samen ziemlich deutlich
grünlich sind.

III. Indiv. Aussaat wie oben
Ernte wie bei I, ausgesprochen roethlich orange
farbige Samen fehlen, alle untersuchten mit
grünem Keimling.

a. Hülsen mit 6, 4, 7, 5, 3, 5 Samen
grün und roth und gelben Hülse. S. mittelgross, etwas faltig,
nicht fastglatt, meist rein(?) häutlich grün oder etwas roth über
laufen.
Σ 64, mit dem Aussaatmaterial 67, alle mit grünem Keim.

b. Hülsen mit 3, 4, 4 Samen
grün bis hell roethlichorange.
Von den Vorhandenen 30 sind 10 grünkeimig, 20 gelbkeimig, vom
Saatgut waren   1 grünk.   2 gelbk.
also zusammen 33 S, davon mit grünem Keim 11, mit gelben 22 = 66,7%
                                          (= 33,3%)

C. wie a. alle Samen mit grünem Keim.

generation of hybrids, obviously using seeds obtained from the fertilization of P_{gr} with pollen from P_p in 1897. Again he included a short description of the resulting seeds in his protocol. About one plant, he noted that he had made a backcross with pollen from P_p: "1 ovary pollinated with p (wool red) (the other pollinations produced no results)." This pollination produced five seeds, all with yellow embryos.[54] The protocols for these experiments do not reveal which individual plant of 1897 provided the seeds in question. Correns does, however, state that the resulting plants yielded "yellow or green germs," but he omits any more precise counts.[55] Here too we find addenda on the bottom of the page about the number and percentage of green and yellow germs; manifestly, they were written only after some of the seeds from this crop were sown the following year (1899).

Correns does not say why he cultivated these additional hybrids. We cannot exclude that he wanted to produce more seeds for a statistical evaluation—but if so why did he not raise many more plants? At this point, he could easily have carried out a large-scale screening to test the 3:1 ratio among the descendants, concentrating on a few select hybrids. But the protocols speak a different language: Correns cultivated together with the appropriate control plants more different first-generation hybrids, namely, BE + gr (pots 8 and 9), BE + p (pot 10), gr + BE (pots 14 and 15), gr + π (pots 18 and 19), gr + WH (pots 20 and 21), p + WH (pot 26), π + gr (pot 28), and π + WH (pot 29).

The 1899 Protocols

In the protocols of the last year in which Correns experimented with peas, we see a fairly systematic extension of the experiments of 1898. Correns performed a first series of experiments with seeds from the second-generation hybrid gr + p. The result was a third generation of hybrids and the fourth generation of seeds overall. Correns sowed fifty-four seeds, of which forty-nine developed into seed-bearing plants. It appears that he took three seeds from each of the plants that he had obtained the previous year, a total of eighteen in all. Twenty-four green germs (ten of them from a plant which, by the second generation, already produced only green germs, and fourteen from a plant with only green germs in the first generation) and twenty-five yellow germs (seven of them from a plant with only yellow germs in the first generation) developed into plants, bringing forth a considerable number of seeds. In contrast to the 1897 and 1898 protocols, those for 1899 show no sign of having been revised; formally, they all look alike. They also differ from the previous protocols in that Correns records, immediately below the identifying description of the plant (for example, "1.a. B, gr + p, II. Gen. I. 1. seed *green*"), the number of seeds contained in the pods, distinguishing them according to whether they have green or yellow germs. Only then does he provide further information about the form, size and additional characteristics of the seeds, rather than proceeding the other way around, as he had earlier (fig. 9).[56]

A second group of similarly designed protocols (pots 46 to 48) describes a second generation of gr + p hybrids derived from the first-generation hybrids of 1898; a third group of protocols (pots 55 to 58) contains the results from a second generation of reciprocal p + gr hybrids. These plants yielded an impressive number of seeds, all of them classified according to whether their cotyledons were yellow or green. The question of the xenia makes itself heard one last time, in connection with two hybrids raised in 1899 and labeled "BE + gr ('xenia')." This was in all probability an ambiguous case, since the term "xenia" now appears in inverted commas and inspires no comment other than Correns's remark on the results: "Obviously gr + BE!"[57] In the paper to which the *Pisum* experiments ultimately gave rise, Correns laconically says: "my experiments on the formation of xenia here [yielded] only negative results."[58]

As usual, the individual pages of the protocol bear no dates, but there can be no doubt that they were written in the fall of 1899. As has already been said, it can be safely assumed that the addenda to the 1897 and 1898 protocols also date from this period: Correns was now fully aware of the segregation pattern with its pure lines on the one hand and the 3:1 segregation ratio for hybrids on the other hand. The set of *Pisum* protocols offers no straightforward indication as to whether Correns began to move toward a solution after thinking about the results of his 1898 harvest, or even earlier. In theory, he could have understood after observing the second generation of hybrids that all the germs resulting from his green peas remained green, whereas only some of those resulting from his yellow ones remained yellow, the others dividing up into yellows and greens. But the addenda to the 1898 protocols clearly suggest that the numbers and percentages were inserted only after the 1899 crop had been sown. Yet the tremendous energy that Correns expended on raising the hybrids of the second and third generations with gr + p and p + gr combinations as well as a number of others also suggests that he must have had good reasons for engaging in such an extensive, demanding experimental program in spring and summer 1899—at the same time as he was completing his book on mosses.

99.

1. a. B, gr + p, II. Gen. I. 1. S. grün

1,2,3 m. blüht roth, Hülsen

Hülsen, Samen, grün gelb. [keim]   7,1 : 1,6
                                     6,2 : 1,55

1/2	2	2	—
	3	3	—
3/4	3	3	—
	5	5	—	2 stark rothorange.
5/5	3/16	3/16	0/0

∑ 16, alle mit grünem Keim.

S. mittelgross, p Form
fast glatt bis mac[?]ig faltig?
bläulichgrün, fast unpunktirt bis deutlich punktirt,
zuweilen etwas rothorange angelaufen; 2 stark rothorange, angefressen! auf denangefr. Kotyl. gelblich.
Nabel bräunlich.

9. Page of Correns's research protocol entitled "99. 1. a. B, gr + p, II. Gen. I. 1. S. *grün* [green]." MPG Archive, Section 3, Folder 17, No. 115, File "Erbsen 99."

The Impact of the Zea mays Series

The other notes accompanying the set of *Pisum* protocols likewise fail to close this gap in our information. There is, however, a crucial note in Correns's protocols on the experiments he was simultaneously conducting with maize, contained in a folder labeled "Theoretical Matters etc." (fig. 10).[59] The note is dated January 2, 1897; an eight has been written on top of the seven. It is possible that early in the new year Correns first wrote the date of the old year by force of habit; the accompanying notes, insofar as they are dated at all, all bear dates from late 1897 or early 1898. In the note that I shall cite at length here, Correns made the following observation about one of his maize crosses: "If it is a fact that caesia ♀ (impure) + alba ♂ produces *more white* kernels than caesia ♀ (impure) + caesia ♂ (impure), this is more easily explained by the influence of caesia ♂ impure on caesia ♀ impure than by the influence of alba ♂ on caesia ♀ (impure)." The quote indicates that he was inclined to exclude xenia effects of *alba* pollen. But he had not yet made up his mind about the ratio of white to blue kernels in the two cases. "Caesia impure" is a hybrid between the two varieties *caesia* and *alba* that yields somewhat impure blue grains. Correns was comparing the result of self-fertilization of "caesia impure" and a backcross of the hybrid with the *alba* parent. He followed his observation up with a hypothetical alternative explanation:

> If half the ovaries in caesia (impure) are disposed to white and the other half to blue, alba pollen will change nothing here, whereas caesia (impure) pollen will increase the number of caesia grains. *Let us therefore assume* that half the pollen grains of caesia (impure) are disposed to white, and half to blue (and that blue *always* alters!); then, because "white" pollen grains will never only meet "white" stigmas and blue pollen grains will never only meet "blue" stigmas, and because "caesia" ovaries are directly [*sic*] not directly influenced by alba pollen and the crosses a ♀ + a ♂ yield white grains, where a ♀ + c ♂, c ♀ + a ♂ and c ♀ + c ♂ yield blue grains, the blue grains must amount to approximately ¾ rather than ½.
>
> One would first have to find out how many grains can turn out blue when *a* + caesia *pure* is crossed with *a* + caesia *impure*. But because it is already certain that blue does *not* always alter, the number of blue grains . . . could come to slightly less than ¾.[60]

Here we have, for the first time, a rather clearly formulated, albeit hypothetical supposition that there is a 3:1 split in the first hybrid generation and concomitantly an underlying disjunction of the factors responsible

2. I. 98

Denn es Thatsache ist, dass caesia (unrein) + alba ♀ ♂
mehr weisse Koerner giebt als alba ♀ caesia unrein + caesia
unrein, so erklaert sich das leichter als durch einen Einfluss der alba bei caesia (u) + alba, durch den Einfluss
der caesia (u) auf caesia (u).

Ist die Haelfte der Fruchtknoten bei caesia (unrein)
auf weiss, die andre auf blau gestimmt, so wird
alba - Pollen daran nichts aendern, caesia (u) Pollen dagegen die Zahl der Caesia Koerner vermehren. Dann
nehmen wir an, ½ der Pollenkoerner von Caesia (u) sei
für weiss, ½ derselben für blau (u. blau ändere stets!) gestimmt, so wird,
da sie gerade die weissen "Pollenk. auf die "weissen"
Narben, die blauen Pollenk. auf die "blauen" Narben
gelangen werden, u. caesia Fruchtk. durch alba Pollen
direkt pinkt direkt beeinflusst werden u die Kreuzun-
gen a♀ + a♂ weisse, dagegen a♀ + c♂, c♀ + a♂ u. c♀ - c♂ also
blaue Koerner als geben, die blauen Koerner etwa ¾
statt ½ betragen müssen.

Zunächst waere festzustellen, wieviel Koerner bei der
Kreuzung a + caesia rein und a + caesia unrein blau
werden Koennen. Da aber blau pinkt stets aendert, wie
schon sicher, Kann die Zahl der blauen Koerner bei der
Kreuzung caesia (u) + caesia (u) ♂ nicht ganz ¾ betragen.

a + caesia u. a + caesia unrein waere aber doch ein Versuch, bei dem sich zeigen würde, wieviel alba in caesia (u) steckt

10. Note on maize, 2. I. 98. MPG Archive, Section 3, Folder 17,
No. 85, File "Theoretisches etc. [theoretical matters etc.]"

for the blue or white color of the corn grains. It is obvious that Correns rules out the influence of xenia in the present case—for he observes that "'caesia' ovaries are . . . not directly influenced by alba pollen"—and, consequently, thinks that something altogether different must be taken into consideration in the case of these maize crosses. He comes back to this in a note of February 24, 1898: "As far as a dissociation of characters *in the pollen* is concerned, experiments would have to be carried out with the hybrid vulgata + dulcis. (Because the grains are *either* smooth *or* wrinkled, whereas intermediate formations occur with alba + caesia)."[61] These intermediate formations showed a less intense blue coloration of the seeds. It appears thus possible that, as Correns prepared the crosses of the 1898 season, he was already on the track of the future explanation for his findings.

As is suggested by these tentative formulations, the first suspicion may have occurred to him thanks to crosses between the maize varieties *caesia* and *alba* in which one of the crossing partners turned out to have been "impure." On closer inspection of the relevant maize protocols, we find a few of the elements that played a role in the above explanation. As early as 1894, Correns had observed that when the variety *alba* was pollinated with *caesia*, "a short, but full cob" resulted, with "*all* grains but one, +/- 'caesia', partially spotted."[62] That is, blue—almost—"always alters," as Correns would later put it in the note we have already cited. In 1896 again he found that the "influence of *caesia*" was "no doubt certain," although he continued to entertain doubts about the "purity of the *alba*"[63] (the mother plant). A self-pollination experiment which he then carried out with another caesia plant led him to the conclusion that "the caesia was *not pure*."[64] The reciprocal cross, in which he fertilized a *caesia* plant with pollen from *alba*, showed him that, with the exception of just one experiment, in which one of the crossing partners of the *caesia* had once more not been pure, "*no influence of the pollen on the ovary is recognizable.*"[65] Hence he knew that *alba* pollen did not give rise to xenia and that blue suppressed white more or less completely. In a series of further pollinizations of *alba* plants with *caesia* pollen, however, the expected alteration to blue grains did not occur in all cases. Instead, Correns obtained a mixture of blue and white grains. Yet his previous findings had so thoroughly convinced him by this time that he concluded that the *caesia* he had used "was not pure, but c + a!"[66] The reciprocal cross between a *caesia* mother plant and an *alba* pollen plant likewise yielded "grains approximately ½ *alba*, ½ *caesia*."[67] Inadvertent backcrosses were at issue here—one of the crossing partners happened to be a hybrid itself. Thus we come to the surprising result that unintended backcrosses with varieties of maize together with the impression of a rather well-established dominant character—in a context in which xenia could be ruled out—precipitated the turn in Correns's later experimentation.

Merging Lines

At this point, maize in which intermediary characters and xenia made it considerably harder to evaluate results yielded the scientific stage to the *Pisum* crosses, which consistently failed to exhibit xenia. In an undated note probably written nearly two years later in winter 1899–1900, Correns finally comes back to the crosses between the maize varieties *alba* and *caesia*—this time, however, with a precise conception of the regularities that had emerged in the *Pisum* crosses of 1898 and 1899. As if he had forgotten the starting point for his thinking, Correns now emphatically asks: "Does the behavior of the hybrids of the races of peas also apply to the hybrids between races of maize??"[68] In another note from the same period, he speaks directly of the possibility of "applying Mendel's theory to the hybrids between races of maize."[69]

The additional notes on the *Pisum* series, presumably also written in winter 1899–1900, are unfortunately all undated and nothing else allows us to establish their chronology. On one of these pages (fig. 11), Correns engages in statistical speculation about the generative process underlying his results:

> Combination of these cells with one disposition each in *pairs* (sexual act). Comparison with a sack of 2000 balls (1000 yellow, 1000 green). Probable: 500 unequal pairs, 500 equal ones, of which 250 yellow and yellow, 250 green and green. Thus 250 yellow + yellow, 500 green + yellow, 250 yellow + yellow [this should read "green + green," H.-J. R.]. . . . *Repeated* hybridization 250 green, 750 yellow. 500 green + yellow and 250 yellow + yellow only distinguishable through *experiment* (progeny). Veritable genealogical tree. Does this also hold for intermingling characters?[70]

Correns next proceeds to make comparable calculations for backcrosses with the parent plants. These reflections provided the basis for the description in his *Pisum* paper, which he was to submit to *Berichte der Deutschen Botanischen Gesellschaft* on April 24, 1900.[71] On another page we find a drawing (fig. 12) of a hypothetical "sequential order" of "dispositions" in hybrids differing in more than one character pair.[72] This drawing became the basis for the schematic explanation of a chromosome theory of inheritance that Correns published in 1902.[73]

(1000 gelb, 1000 grün)

Combination dieser Zellen mit je einer Anlage zu
Paaren (Sexualact) / Vergleich mit einem Sack mit 2000 Kugeln.
Wahrscheinlich,: 500 ungleiche Paare, 500 gleiche, davon
250 mal gelb ü. gelb, 250 grün u. grüne.
also 250 ge+ge, 500 gr + ge, 250 ge + ge.
Folge bei 1   2   „   3   „
weiteres Verhalten
~~halte zu die Folleng. d. E. Btj~~

Wiederholte Bastardirung / 250 gr, 750 ge.

500 gr + ge ü 250 ge + ge nur durchs (Nachkommen an)
Experim. zu unter-
scheiden.
/ Rechtiger Stammbäume   ge (gr)
gilt es auch für erbmischende   ² 1 ge   2 ge (gr)   1 gr
     Merkmale?
                                  ge   1 ge 2 ge (gr) gr   gr
                                                              etc.

                    d. Haushalten (Rückkreuzg).
Bestäubung mit fremdem Pollen: 1000 Sex. f. d. B.
    2) 500 gr, 500 ge   a. + 1000 gr (rec.)
                        giebt   500 gr + gr   500 gr + ge
                                grüne reine Rasse   gelb Bast
    1.) 500 gr, 500 ge , + 1000 ge (dom.)
                        giebt   500 gr + ge , 500 ge + ge
                                gelb Bast,   gelb, reine Rasse

    Bedeutg der Selbstbefruchtung. (weil gr + ge ü. ge + ge
         ~~von~~ nur an der Nachkommensch erkannt
              werden koennen   b. Aussaat aller Samen
    b. selbstbest   d)
c. gleiche Fruchtbarkeit   Aussterben des Bastartes / bei gleichtl Indx x1. Zahl
m. gleicher reget. Stärke des Bast ü. d. veg. Ar Eltern.

11. Statistical considerations and a genealogical tree.
MPG Archive, Section 3, Folder 17, No. 115.

Für die Art der (Reduction oder Halbirung ist sehr wichtig, daß die Merkmale unabhängig von einander sind, daß eines schon constant geworden sein kann, das andere nicht.

bei der BE z. B. die grüne Farbe des Keimes, aber die Nabelfärbig noch nicht.

Es ist keine constante Anordnung der Anlagen denkbar durch die (durch stets) dieselbe Theilung, wenn verschiedenen Anlagen zu der gewünschten Weise gelegt würden, die Anordnung muß verschieden sein.

Keim
Schale
Nabel
etc.

Man kann sich vorstellen, daß die Anlage-Paare hintereinander gereiht liegen und daß bei der Befruchtung die Anlagen von A u. die von B nicht alle mit einander zu liegen kommen.

Freilich ist kein Grund einzusehen, warum nicht alle die Form III die seltensten, statt der häufigsten.

Sind die zwei Rassen durch N Merkmale verschieden, so ist die Wahrscheinlichkeit, in der zweiten Generation die reine Rasse zu erhalten $\frac{1}{4}$, bei 1 Merkmal also $\frac{1}{4}$, bei 2 en $\frac{1}{16}$, bei 3 en $\frac{1}{64}$ etc.

aa'bb'cc'dd'ee'

12. Sequential ordering of hereditary dispositions.
MPG Archive, Section 3, Folder 17, No. 115.

Correns was thus well prepared to put together his *Pisum* manuscript in the span of two days when on April 21, 1900, he was caught unawares by an offprint of Hugo de Vries's "Sur la loi de disjonction des hybrides."[74] All the relevant parts of Correns's manuscript, including the calculations, had been formulated already. Until this date, he does not seem to have been in any particular hurry to publish his results. We can see from his notes that he was busy making a detailed revision of the voluminous protocols of his experiments with maize varieties between December 1899 and March 1900 with an eye both to the calculation of the segregation ratio wherever it seemed obvious and to xenia as well.[75] It was only when he received de Vries's paper that he suddenly felt the need to publish his results as quickly as possible. This may have been the reason that he did not pursue the additional experiments with peas that he had been planning for spring 1900.[76] For he had now to position himself with respect not only to his predecessor Gregor Mendel, but to his contemporary and competitor Hugo de Vries as well. He proceeded subtly. In the introduction to his *Pisum* paper, we read: "When I discovered the lawful behavior and the explanation for it . . . I reacted the way de Vries is obviously reacting now: I took all this to be something new. But then I was forced to conclude that, in the [18]60s, the abbot Gregor Mendel in Brünn . . . arrived at the same result. . . ."[77] This is a multifaceted formulation. Correns not only dissociates the realization that hybrids behave "lawful[ly]" from the "explanation" for it; he further implies that he had already discovered both before he "was forced to conclude" that Mendel had arrived at the same results several decades earlier—although he of course avoids saying anything precise as to how and when he stumbled upon Mendel. Moreover, he suggests that he discovered the solution before de Vries by affirming that he "reacted the way de Vries is obviously reacting now," thereby implicitly criticizing his competitor for "obviously" considering (and giving out) his findings as something new when they were in fact preceded by Mendel's—something which he implies de Vries may even have known yet passed over in silence. Correns heightened the contrast between his presentation of matters and de Vries's by publishing his own paper under the title "G. Mendel's Law about the Behavior of the Progeny of Hybrids;" he also rushed two more papers on hybridization into print before the year was out.[78] On the other hand, his voluminous monograph on maize, *Hybrids between Races of Maize with Special Attention to Xenia* appeared only in 1901,[79] as the sequel to a preliminary publication on xenia in maize dating from 1899.[80] In the field of xenia too de Vries had been a little quicker on the draw than Correns. He published similar results in a preliminary account in 1899 and in the form of a more extended report in 1900,[81] whereupon Correns sullenly decided to "postpone" his own publication on the subject for another year.[82]

Conclusion

My reconstruction of Correns's experiments has led to the conclusion, contrary to what is suggested by his notes on reading Mendel's paper on hybrids in April 1896, that the meaning of the Moravian abbot's observation and explanation of the segregation behavior of varieties of garden peas did not become immediately clear to him. As a result of an unintended backcross of varieties of maize, however, he must at least have suspected the prospective significance of both sometime around the turn of 1898, and pursued his hunch. At the same time, he began to lose interest in the xenia problem; ultimately, he dropped it altogether as far as *Pisum* was concerned, since all his experiments in this direction had negative results. His protocols do not allow us to exclude the possibility that the puzzle finally fell in place for him only in the fall of 1899. But we should no more lend credence to his declaration that the solution hit him "in a flash" one fine day in autumn of that year than we should credit his assertion that he first read Mendel's paper toward the end of 1899. On the most charitable interpretation, this claim can be taken to mean that he *re*read it late that year, this time with new eyes.

My motive for reconstructing the trajectory of Correns's investigation has not been to "unmask" yet another rediscoverer of Mendel's laws. My main interest in this case study and others like it grows rather out of a desire to trace the way experimental systems develop their own dynamics and to point out the unexpected directions in which they can lead scientists.[83] Correns appears as a particularly interesting example in this respect precisely because he *could* have known "it" in advance. He even had important information to hand when he was first setting up his experimental system. From the standpoint that was his at the time, however, this information made no sense to him. For his original intention was not to discover the rules of hybridization, but to shed light on the process that leads to the formation of xenia.

It can of course be deemed a lucky side-effect of this original intention that Correns had to focus on characters which he expected to become visible in the germs. This turned out to be crucial for the further course of his experiments. Pointing to the paradox, we might say that the xenia first prevented him from recognizing the transmission ratios, but ultimately *enabled* him to do so. The seeds that he collected and set aside to sow the next generation served him as a physical and so to speak natural digital proto-col: they had either green or yellow cotyledons. Correns could reopen this "protocol," his boxes of peas, whenever he liked, and so reconsider his findings years after first arriving at them—including findings to which he had at first paid no mind. Of course, if what was involved had not been seeds, then these traces would have vanished long since, together with the plants that produced them. Only after Correns had spent some four years familiarizing himself with the hybridization system of *Zea mays* and *Pisum* did this characteristic of his system become relevant. And it was only then that he realized that one possible interpretation of his results pointed in a new direction. The recursive reorientations and assurances made possible by certain material attributes of experimental systems constitute a basic, generalizable feature of research. Researchers rely on them in their labori-ous endeavor to assign meaning to the data they obtain in their experimen-tal efforts.

5.   Eudorina

Max Hartmann's Experiments on

Biological Regulation in Protozoa, 1914–21

In 1917 the protozoologist Max Hartmann began writing a series of scientific communications that he collected under the general title "Investigations on the Morphology and Physiology of Change of Form (Development, Propagation, Fertilization, and Inheritance) of the Phytomonadines (Volvocales)." In the first of these papers, publication of which was delayed until 1919 by the turbulent conditions surrounding the end of the First World War,[1] Hartmann evoked the early beginnings of his investigations. Hartmann's higher education began at the Bavarian Forest Academy in Aschaffenburg. In 1897, he transferred to the University of Munich, writing a doctoral dissertation on the maturation of animal eggs with the protozoologist Richard Hertwig that earned him a doctoral degree in 1901. He then became the assistant of Johann Wilhelm Spengel at the Zoological Institute of the University of Gießen. In 1902, he met the protozoologist Fritz Richard Schaudinn, who brought him to the Imperial Institute for Infectious Diseases in Berlin. Setting out from the findings of Hertwig and the botanist Georg Klebs of the University of Halle, he started to outline his own research program. However, after assuming the directorship of the Department for Protozoological Research in 1905, he set his research on the propagation of unicellular organisms aside for almost a decade in order to specialize in the study of pathogenic protozoa, in line with the mission of the Imperial Institute. This work led him to Rio de Janeiro in 1909 at the invitation of the Brazilian government. He spent six months at the Instituto Oswaldo Cruz in Manguinhos, where he established a Department of Protozoological Research. After returning to Berlin he was appointed extraordinary professor for zoology at the University of Berlin, and in 1914 accepted the offer of a directorship at the Kaiser Wilhelm Institute for Biology.[2] In this chapter I would like to show how the concept of biological regulation in Hartmann's later vision of a general biology emerged from his early experiments.

The Beginnings: Propagation and Fertilization

In 1902 Hertwig presented a paper to the Society for Morphology and Physiology in Munich in which he laid out a theory of the interrelationship between nucleus and cytoplasm as a physiological explanation of the causes of propagation—essentially, of cell division. Hertwig argued that there was a fixed relationship or relational norm between the mass of a nucleus of a cell and the corresponding cytoplasm. Expansion of the nuclear mass at the expense of the cytoplasmic mass during growth acted as a signal precipitating cell division, in the course of which the normal relationship between the nucleus and cytoplasm was re-established.[3] A year later, he also explained the process of sexual differentiation and fertilization in terms of his scheme of the relationship between nucleus and cytoplasm. In this context he described the sexual cells as the most interesting examples of a "re-regulation [*Umregulierung*] of the usual nucleus–plasm relation."[4] The sperm could exhibit enormous diminution of the protoplasm, whereas differentiation of the egg generally proceeded in precisely the opposite direction. When fertilization occurred the tension relaxed and the usual relationship between the masses of nucleus and plasm was re-established. Hertwig leapt to the conclusion that the nucleus–plasm relationship had to be considered the physiological basis for the phenomenon of sex in both the animal and plant kingdoms.

Controversies: Rejuvenation, Death, and Immortality

Silently informing Hertwig's hypothesis was a long-standing controversy. The themes intertwined in this debate featured prominently in the protozoological discourse of the day: aging, death, and the function of sex. Let me briefly sketch the main lines of this controversy. It is of interest not only because of the role the burgeoning field of protozoological studies played, but also because it gives us an idea of the way the theoretical and general problems of biology were approached in the last quarter of the nineteenth century.

As early as 1876 the Heidelberg protozoologist Otto Bütschli had put forward the thesis that the phenomenon of conjugation in single-celled infusoria should be interpreted as a fertilization process, and that the significance of the act of conjugation lay in the periodic "rejuvenation" of the animals that carried it out.[5] In the late 1880s Émile Maupas, who had investigated unicellular organisms as a conservator at the National Library in Algeria, took Bütschli's studies of conjugation in single-celled organisms a step further. The results of his impressive experiments with ciliates, which reproduced in an alternating cycle, sexually and asexually, indicated that these organisms deteriorated when they were prevented from reproducing sexually and forced to reproduce by division alone over a certain number of generations. In a paper entitled "Recherches expérimentales sur la multiplication des infusoires ciliés," Maupas described his experiments with the cultivation of different species. About his first sustained, methodologically controlled attempts to cultivate the ciliate *Stylonichia pustulata*, he reported: "On 28 February 1886, I undertook a new preparation, isolating a *Stylonichia* from a culture that had attained its seventy-seventh division. . . . It finally died on 10 July after an uninterrupted series of 316 divisions, notwithstanding all my efforts to keep it alive. All the other *Stylonichiae* died of atrophy as well; they were no longer capable of nourishing themselves or reproducing."[6] Only an act of sexual reproduction would allow them to "rejuvenate," as Maupas put it with Bütschli and the Belgian cytologist Edouard Van Beneden, and against the "hypothesis" defended by the zoologist August Weismann from Freiburg, for whom on Maupas's description of his viewpoint the fertilization process was simply an act in which hereditary characters were transmitted, "the major agent of individual variations."[7] Thus according to Maupas processes resembling aging and degeneration occurred not only in the life cycle of an individual, but also over a series of generations; these latter processes could be reversed by occasional sexual reproduction. Maupas claimed to have provided experimental confirmation of this phenomenon. Let us note here only in passing that the concepts of "degeneration," "depression," and "rejuvenation," used by all the authors just mentioned, reflected a broader, biopolitical public discussion of the dangers of degeneration in the closing decades of the nineteenth century.

To understand the significance of Maupas's reply to Weismann we have to look at the broader discussion on the biological phenomena of "life" and "death" that fueled the late nineteenth-century debate concerning rejuvenation in the decade between Bütschli's observations and Maupas's experiments. Early in the 1880s August Weismann in a book about the duration of life called *Über die Dauer des Lebens* (On the Duration of Life) defended the thesis that the death of the individual should be considered an adaptive trait acquired by higher organisms in the process of evolutionary selection. In unicellular organisms, in contrast, death did not yet exist as a physiological necessity: such organisms were therefore, Weismann thought, potentially immortal.[8] Alexander Goette, a zoologist from Rostock, objected strenuously to this idea: aging, he said, was a process inherent in life itself. Even in single-celled organisms germ formation—whether or not it was combined with sexual acts in higher organisms or conjugation in infusoria—represented a necessary process of "rejuvenation" that checked the wear and tear of the life process and periodically restored the "original condition" (*Urzustand*) of the species.[9] Weismann, however, stubbornly clung to the view that natural death was a form of adaptation acquired by higher, multi-cellular organisms alone and that the original mode of propagation observable in unicellular organisms was potentially unlimited cell division. In particular, he insisted, there was no reason or justification for a kind of "mystical interpretation of an intrinsically dubious 'process of rejuvenation.'"[10] To be sure, Weismann acknowledged the necessity of sexual propagation in lower organisms as well, but he assigned it a completely different meaning, explaining it in terms of his theory of "amphimixis." The biological significance of sexual propagation on this theory lay in the effects, advantageous from the evolutionary standpoint, of a regular mixture of individual germ plasms.

Hertwig, who had published his first studies on the fertilization of *Paramecia* in 1889,[11] was familiar with these explanations. Rather than pursue the question of rejuvenation he sought to replace it with the notion of a measurable, periodic "reorganization" of the living substance[12] supposed to forestall physiological cell death. As he saw it, his observations and experiments on the quantitative relation between cytoplasm and nucleus of a cell during the growth of protists provided proof of this "reorganization," or "regulatory influence."[13] His student Hartmann would later take up Hertwig's concept.[14]

The situation was further complicated by a series of additional observations, first and foremost the experiments with algae and lower fungi that Georg Klebs had been conducting for decades. Early in his career while still a professor in Basel Klebs had tried to learn about, create, and maintain conditions favorable to the growth and reproduction of his experimental organisms. His conclusions flatly contradicted the sexual rejuvenation theorists, for he succeeded in propagating single-celled plants and fungi for generation after generation with no signs of deterioration: all that was required was to provide adequate living conditions and maintain the purity of the cultures. For instance, Klebs said about his *Chlamydomonas* cultures that,

> For the sexual process to occur, lack of nurturing salts is as necessary as light, the prerequisite being that, prior to the experiment, the cells have been able to grow vigorously in the presence of nurturing salts. If they are

provided with a steady supply of such salts in abundance, on the other hand, all my experience to date goes to show that copulation never takes place. The first pure culture was established on 13 December 1895. . . . Since then, the descendants of this culture have been in permanent, lively vegetation. . . . I have not the slightest doubt that asexual generations can be maintained in this way for years.[15]

Elsewhere Klebs concluded: "To date, neither vegetative growth (of whatever duration) nor permanent reproduction in one form has led necessarily to the occurrence of the other."[16]

Klebs's sustained success supported in a certain sense Weismann's clear-cut distinction between multicellular organisms for which the death of individuals with their differentiated soma was a necessity and the perpetuity of the species was ensured exclusively by the germ line, and unicellular organisms, which divided without natural individual death and could accordingly be considered immortal.

Hartmann's Experiments

Hartmann focused on the second aspect of Klebs's observations. If it was true that single-celled organisms could dispense with sexual reproduction altogether, the idea that sexuality had a rejuvenating, reorganizing, or regulatory function for the life of the organism could be dismissed. But then why did sexuality and the phenomenon of fertilization occur at all? Weismann's amphimixis theory was in Hartmann's opinion no solution. For Hartmann the mixture of germ plasms was not the physiological *cause* of fertilization, but its *result*. Whereas Weismann reasoned at the level of a theory of heredity, Hartmann sought physiological explanations.

Looking back he wrote that, to attack these questions, he had drawn up a plan to combine, experimentally, the research project of the botanist Klebs—who aimed at defining the external conditions for the exclusion of sexual reproduction—with the attempt of the zoologist Hertwig to find a physiological explanation for propagation by cell division alone. Whether interesting theoretical perspectives would open up as a result was something he hoped to discover in the course of his experiments; at the outset, he confessed, he did "not yet have any clear ideas at all about that."[17] He needed an experimental organism that met two conditions: "One was easy cultivability and external conditions that were simple to oversee and control, both of which obtain in forms exhibiting vegetal metabolism; the other was the occurrence of both sexual reproduction and propagation by means of the simple division of single-celled individuals, because it was only here that the basic problem of reproductive physiology . . . could be tackled."[18] After choosing an order of green algae (*Volvocinea*) he obtained a species of *Stephanosphaera* and *Chlamydomonas* from the freshwater marshes of his native region, the Rheinland Palatinate. But all his attempts to cultivate *Stephanosphaera* or induce sexual differentiation in *Chlamydomonas* were in vain. It was not until summer 1914 that his project began to advance.

In the years during which Hartmann was working on protozoological parasites various things had been learned about the general morphology of the nucleus of protists, especially about the role of nuclear dimorphism in certain single-celled organisms and the function of the centrosoma, the cellular organelle involved in nuclear division. Victor Jollos, a fellow student of Hertwig, was among the first of Hartmann's colleagues to follow him to the Kaiser Wilhelm Institute for Biology in 1915, after working for three years as an assistant at the Berlin Institute for Infectious Diseases.[19] Jollos's experiments with infusoria led him to postulate the existence of two independent material *factors*, a division factor and a growth factor, on whose "correlative coupling," in Hartmann's phrase, the relation between cellular growth and cell division was based.[20] This postulate allowed Jollos to take a big step toward a causal physiology of propagation and to move well beyond Hertwig's hypothesis about a cellular nucleus–plasm relationship and his distinction between an "excitatory" function of propagation and a "regulatory" function of fertilization.[21] Everything else, however, remained open and as Hartmann was forced to acknowledge, "even the simple question as to whether sexual propagation could be eliminated without causing any detrimental consequences or giving way to some other type of regulatory process remained unsolved."[22]

Hartmann decided to address this question first. To do so he needed a suitable organism with the simplest possible nuclear conditions; there could be no question of using infusoria, because they had a generative micro-nucleus and vegetative macronucleus. He also needed "more transparent external conditions," that is, controllable cultivation media.[23] A batch of a *Closterium* species that Hartmann had cultivated in the Biological Station on Lake Lunz in lower Austria in summer 1914 was damaged in the first weeks of World War I and later perished. Ultimately he settled on *Eudorina elegans*, a species of green algae (fig. 13). Because this single-celled organism was rather sensitive to changes in the concentration of its nutrient solution as well as to contamination by protococci and bacteria, it required much patience to find the right cultivation conditions, and a

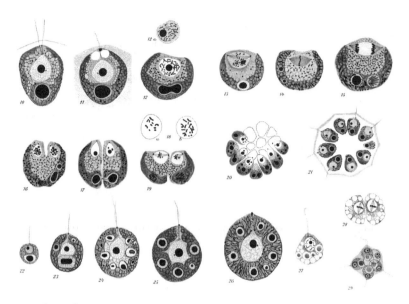

13. *Eudorina elegans*. Figs. 10–16: cuts. Figs. 17–29: immersion. Magnified 1900 times (Fig. 18: 2500 times). Figs. 10–21: Normal division of colonies collected in the wild in the Hundekehlensee near Berlin. Figs. 22–25: Culture forms from colonies of different ages. Fig. 26: Giant form (6 days old). Figs. 27–29: Precipitous division (after 7 hours of uninterrupted exposure to constant light). Hartmann 1921.

few "somber experiences" occurred along the way.[24] Once these difficulties had been overcome, however, Hartmann seemed able to cultivate his model organism *Eudorina* in an agamic way (that is, without intermittent cycles of sexual differentiation) for quite possibly an unlimited number of generations. When he reported on these experiments for the first time in 1917 he could point to a long, uninterrupted history of agamic propagation cycles (fig. 14) going back to summer 1915 and already running to 550 generations "without depression," as he proudly reported.[25] By the time he wrote the sequel to his first report on these experiments in 1920, he had arrived at better than 1200 agamic generations. It therefore seemed to him that the "final verdict on the rejuvenation hypothesis about fertilization" was in.[26] As it turned out light intensity had a considerable influence on the length of the division cycle. In the following series of experiments Hartmann therefore tried replacing natural sunlight with artificial light. In his third communication about this series of experiments he listed 1300 unbroken individual generations of *Eudorina* (he raised the number to 1500 in a footnote to the galleys), which he had been cultivating in artificial light since November 1919 (fig. 15).[27]

Hartmann was not the first to obtain such a long string of asexually reproduced generations; Lorande Woodruff had already raised thousands of generations of *Paramaecium*.[28] But it turned out that the process was, in the case of *Paramaecium*, periodically accompanied by a reorganization of the nucleus that was however not followed by conjugation, justifying the inference that something like an intermediate parthenogenetic phase was involved.[29] Hartmann therefore believed that this organism was not an appropriate model for solving the problem he had set himself.

He drew two conclusions from the long series of agamic generations of *Eudorina*. First, he deemed his findings decisive proof that fertilization was unnecessary for survival. It followed that the two processes of propagation and fertilization had at the analytic level to be rigorously separated. Each was informed by principles peculiar to it alone and the physiology of sexuality had to be studied by itself. Furthermore, the biological significance of fertilization and sexuality could no longer be located in a kind of periodic "rejuvenation," as Hertwig had assumed. Some other function had to be ascribed to it. Second, it now became possible to ask whether periodic propagation was an absolute necessity for the organism. If a single-celled organism could be propagated indefinitely without any intervening fertilization, might it not also be possible to keep biological systems, as Hartmann put it, "permanently thriving without signs of aging or degeneration and without reduction of the system by division (propagation) or other forms of regulation?"[30] It was true that all attempts of the sort had so far failed. For the time being, Hartmann concluded that "the question of rejuvenation . . . has been displaced from fertilization to propagation, and the latter is approached as a process not only of multiplication, but

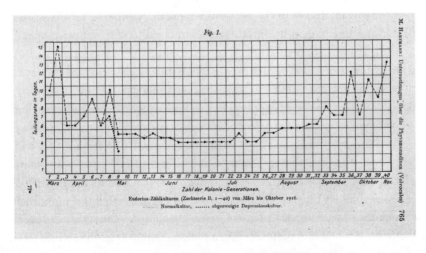

14. Plot of a cultivation series (Series B) of *Eudorina elegans*. Hartmann 1917, Fig. 1.

Zuchtserie B.

Nr. der Generation	Datum der Teilung und Isolierung	Dauer der Generation (Tage)	Bemerkungen
1	10. 3. 1916	10	
2	20. 3. „	15	Für 3. Generation auf zurückgebliebene Kolonie gegriffen, da Schwesterkolonie verunreinigt
3	4. 4. „	6	
4	9. 4. „	6	
5	15. 4. „	7	
6	22. 4. „	9	
7	1. 5. „	5	
8	6. 5. „	10	
9a	16. 5. „	5	
10	21. 5. „	5	
11	26. 5. „	5	
12	31. 5. „	4	
13	5. 6. „	5	
14	10. 6. „	4	
15	14. 6. „	5	
16	19. 6. „	4	
17	23. 6. „	4	
18	27. 6. „	4	
19	1. 7. „	4	
20	5. 7. „	4	
21	9. 7. „	4	
22	13. 7. „	4	
23	17. 7. „	5	
24	22. 7. „	4	
25	26. 7. „	4	
26	30. 7. „	5	
27	4. 8. „	5	
28	9. 8. „	5½	
29	15. 8. „	5½	
30	20.—21. 8. 1916	5½	
31	26. 8. 1916	6	
32	1. 9. „	6	
33	7. 9. „	8	
34	15. 9. „	7	
35	22. 9. „	7	
36	29. 9. „	12	
37	11. 10. „	7	
38	18. 10. „	11	
39	29. 10. „	9	
40	7. 11. „	12	
41	16. 11. „	21	Parallelkulturen in künstlichem Licht (100 Watt Nitralampe)
42	7. 12. „	18	
43	—	18	18. 11. 11 Tage 29. 11. 8 Tage 7. 12. 9 „ 16. 12. ? „
44	12. 1. 1917	17	
45	29. 1. „	20?	Degeneriert, abgebrochen

Nr. der Generation	Datum der Teilung und Isolierung	Dauer der Generation (Tage)	Bemerkungen
46	—	25?	
47	—	25?	
48	31. 3. 1917	12	
49	12. 4. „	13	
50	25. 4. „	10	
51	7. 5. „	5	
52	12. 5. „	6	
53	18. 5. „	5	
54	23. 5. „	6	
55	29. 5. „	7	
56	5. 6. „	9	
57	14. 6. „	5	
58	19. 6. „	5	
59	24. 6. „	5	
60	29. 6. „	6	
61	5. 7. „	6	
62	11. 7. „	7	
63	18. 7. „	7	
64	25. 7. „	8	
65	2. 8. „	6	
66	8. 8. „	7	
67	15. 8. „	6	
68	21. 8. „	6	
69	27. 8. „	etwa 8	
70	Zahlen verloren gegangen durchschnittlich	„	
71		„	
72		„	
73	27. 9. 1917	8	
74	5. 10. „	9	
75	14. 10. „	9	
76	2. 11. „	13 (19)	In Klammern Parallelkultur, geriet bei Generation 79 in Depression
77	15. (21.) 11. 1917	20 (13)	
78	5. (4.) 12. „	(8—9)	
79	11. 1. 1918	26	
80	6. 2. „	16	
81	22. 2. „	11	
82	5. 3. „	7	
83	12. 3. „	6	
84	18. 3. „	7	
85	25. 3. „	7	
86	1. 4. „	5	
87	6. 4. „	5	
88	12. 4. „	6	
89	18. 4. „	8	
90	26. 4. „	8	
91	2. 5. „	8	
92	7. 5. „	6	
93	13. 5. „	5	
94	18. 5. „	6	
95	24. 5. „	5	
96	29. 5. „	6	
97	3. 6. „	6	
98	8. 6. „	4	
99	12. 6. „	6	
100	18. 6. „	5	

15. Cultivation series of *Eudorina elegans*. Hartmann 1921, 254, 257.

also of rejuvenation"—in the sense of a regulatory process.[31] Let us note that in the Hertwigian tradition Hartmann continued to conceive of cell division as a form of regulation. Accordingly, he assumed, as he put it elsewhere, "that aging also occurs in protozoan individuals and that propagation means not only multiplication, but also rejuvenation of living matter."[32] With his *Eudorina* experiments, he had succeeded in showing that *generationally* conditioned aging of his protists could be forestalled; they seemed capable of propagating indefinitely in asexual fashion. But it had also become clear that there was an *individual* aging process that was periodically checked by propagation. Interestingly, it was thus the exclusion of one regulatory phenomenon—sexual rejuvenation—that threw another regulatory phenomenon—propagatory rejuvenation—into sharper relief. With these subtle distinctions, Hartmann was not just playing an idle conceptual game. His attempts to achieve greater precision helped him think about how to frame experimental procedures so as to represent these concepts and facilitated new experimental arrangements.

Hartmann now revisited the question of the physiological mechanism of propagatory regulation. He sought to discover experimentally whether "the rejuvenating effect of propagation [could] be replaced by other forms of regulation."[33] To do so, he opted for systematic, repeated diminution of a biological system by resecting the cells before the onset of the division process. It turned out that the infusorian *Stentor coeruleus* could be "maintained permanently in growth" if parts of its cell body were periodically removed. Untreated cultures used as controls divided; the infusoria that had been subject to resection regenerated, but did not divide.[34] These experiments drew Hartmann's attention back to Hertwig's reflections on the relationship between nucleus and plasm as well as to Jollos's work on the relationship between growth factors and division factors.

Regulation and General Biology

To conclude, let us look briefly at the way this process of experimental and conceptual clarification was reflected in Hartmann's notion of a general biology. In 1921, Hartmann delivered a series of lectures on the subject as an honorary professor at the University of Berlin, publishing them in 1927 in the form of a comprehensive monograph termed *Allgemeine Biologie* (General Biology). In this monograph, Hartmann distinguishes three basic biological processes that make up an organism's "fabric of life" (*Lebensgetriebe*).[35] The first comprises stationary processes or "dynamic equilibria" resulting from the permanent flux of matter and energy into and out of biological systems. The second comprises "irreversible changes," Hartmann's term for the phenomena of growth, development, death, fertilization, and heredity, all subsumed under the category of "metamorphosis" (*Formwechsel*). The third process consists of physiological oscillations: regulatory phenomena that, after an imbalance induced by excitation, lead back to the point of departure. Hartmann subdivides this third process into two main categories. The first is related to the phenomenon of dynamic equilibria. Such equilibria can be perturbed to a certain extent yet remain capable of returning to their original steady state. All stimulation phenomena and all higher expressions of life and adaptive behavior belong in this category, according to Hartmann. The second category in contrast is bound up with the "irreversible changes." Irreversible changes—and, with them all differentiation of an organism—inevitably lead to the extinction of the life form subject to them in the absence of mechanisms that periodically re-regulate the system, setting it back to zero, as it were. Since differentiation phenomena could by definition be seen as irreversible, a kind of super-regulation was required to forestall the consequences of their irreversibility. For Hartmann one of the basic forms of such regulation was propagation, which also resulted biologically in the periodic multiplication of organic bodies. Another basic form of super-regulation was sexuality, considered from the physiological standpoint of fertilization. Hartmann devoted a good part of his later work on sexuality to explaining the physiology of fertilization.[36] He began experiments on this question around the time our account comes to a halt. These investigations were deeply influenced by his mentor Fritz Schaudinn's ideas on sexual polarity and the factors compensating for it, which had been presented *in nuce* in an essay on the fertilization of protozoa that Schaudinn had published a year before his death in 1906.[37]

Hartmann was convinced that the basic forms of these different regulations and the factors that played a role in them, which he described ever more resolutely as specific substances, could be elucidated in experiments on protozoa. He believed that it was possible to find among these lower, almost infinitely varied organisms an appropriate model organism for every one of these regulatory processes, one that would lend itself to illustrating the most basic form of the process in question. Hence, in addition to his work in the laboratory Hartmann carried out fieldwork at various zoological stations, such as the Stazione Zoologica in Naples, in an ongoing search for possible new models. Protists were particularly well-suited to serving as objects for the combination of developmental biology and histomorphology that he favored, whose potential he saw in a "generalizing induction." Yet they were supposed to lead in the long term to experimental systems that would allow for "exact induction," systems he now attempted to construct for the first time.[38]

With these concepts, Hartmann distinguished systematic observation from experimentation properly so called. He firmly believed that given the state of the life sciences at the beginning of the twentieth century the logical form of generalizing induction in combination with the accompanying practices of regular observation would lead to valid formal generalizations; thus it could help define the basic problems of a general biology; to solve them would require further experimental investigation. To what extent this way of looking at things represented a rationalization of Hartmann's own scientific education—that is, to what extent it should be ascribed to the limits of his own field of action or accepted as an accurate assessment of the state of general (and theoretical) biology around 1920—

is open to debate. Hartmann's laboratory experiments in this period seem to be grounded in systematized observation and thus obey the logic of "generalizing induction" rather than that of "experimental induction" in the strict sense. When we compare them with the experiments of his contemporaries and immediate predecessors, they do not exhibit particularly novel features. Yet they show that Hartmann was aware of the importance of creating controlled culture conditions and choosing appropriate organisms for the study of each specific question. These organisms only became experimental models because of protracted observation and on the basis of a consolidated natural history.

6.   Ephestia

Alfred Kühn's Experimental Design for a
Developmental Physiological Genetics, 1924–45

Historians of science have rightly praised the significant role in
relating genetics to biochemistry played by Boris Ephrussi's and George
Beadle's work on *Drosophila* and, later, Beadle's and Edward Tatum's work
on *Neurospora*.[1] Much less attention has been paid to a contemporaneous
research enterprise.[2] In this chapter I chart its decisive experimental way-
stations, retracing the path that led Alfred Kühn and his coworkers at the
University of Göttingen and later the Kaiser Wilhelm Institute (KWI) for
Biology in Berlin-Dahlem to their conception of a developmental physiological genetics that concentrated on the notion of "gene-action chains"
and their relationship to "substrate chains." My aim is not to establish who
first identified the connection that under the name of the "one gene—one
enzyme hypothesis" was eventually ranked as a milestone in the history
of molecular biology. None of the groups involved took it as a guiding
principle in its early work. But neither the history of biochemical genetics in particular nor the history of developmental genetics as a whole can
be written properly without examining the work of this group, including,
besides Kühn, Karl Henke, Ernst Caspari, Ernst Plagge, Hans Piepho, and
Erich Becker. Here, I focus on the path they took in setting up an experimental system that productively combined genetics, physiological chemistry, and questions of development. To that end I trace the crucial experimental reconfigurations that led to the emergence of this system and
examine how the system's hybrid nature or the variable weight of its distinct epistemic strata was reflected in the group's conceptual terminology.

The Scientist: Alfred Kühn

Alfred Kühn was born in 1885. He studied zoology and physiology with
August Weismann and Johannes von Kries at the University of Freiburg,
where he also took courses in philosophy with the neo-Kantian Heinrich
Rickert.[3] After graduating, he spent a summer at the Zoological Station in
Naples. In 1908 he became Weismann's assistant. Two years later, he re-

ceived his *Habilitation* and in 1914 was named extraordinary professor. He was drafted shortly thereafter and served from 1915 as a medical orderly in the German army headquarters in Berlin-Falkenhausen. His responsibilities were to control parasites and to organize other military hygiene measures. At the end of the war, he began working with the noted zoologist Karl Heider at the University of Berlin, where he remained for two years. In 1920, he succeeded Ernst Ehlers in his post as chair of the Zoological Institute at the University of Göttingen.

In his Freiburg years, Kühn worked on research projects of very different kinds. Among them were the development of reproductive cells in parthenogenetic Cladocera,[4] ontogenetic and phylogenetic relationships among Hydrozoa,[5] cytological investigations of protozoa such as amoebae and trypanosomes,[6] and animal morphology.[7] He was unable to carry out sustained experiments during the war; in this period, he wrote his two early monographs, *Anleitung zu Tierphysiologischen Grundversuchen*, the first manual of animal physiology for scientists in Germany, and *Die Orientierung der Tiere im Raum*.[8] After conducting morphological and physiological studies of cell division in Berlin at war's end,[9] he concentrated in Göttingen on the comparative physiology of spatial orientation in animals. In the interdisciplinary climate that reigned at the University of Göttingen in this period, he collaborated with the physicist Robert Pohl on the perception of color in insects, and also had the opportunity to lecture to "mixed" audiences. To his older colleague Hans Spemann, he reported in 1921: "This winter, I discussed the 'problem of morphogenesis' in evening lectures for an audience of people from all faculties. . . . I quite enjoyed giving the lectures; my audience was made up of some one hundred fifty listeners from different disciplines, including philologists, jurists, theologians, as well as a number of colleagues."[10] In this period, Kühn also wrote his *Grundriß der allgemeinen Zoologie*,[11] which ranked for decades as the standard introduction to zoology and went through fifteen editions in his lifetime. After the botanist Fritz von Wettstein joined the Göttingen faculty in 1925, he and Kühn joined forces in an effort to define the basic features of a "general biology," as Hartmann was doing at the time in Berlin. Kühn regarded this as a necessary complement to the ongoing disciplinary specialization that he practiced in his own research.[12] By this date, his leading position in German zoology was uncontested: he no doubt exercised greater influence on the academic affairs of his discipline than any other zoologist in the Weimar Republic except for Spemann.

When he was almost forty, Kühn decided to change the direction of his work. In a curriculum vitae that he wrote in connection with his election as a corresponding member of the Academy of Sciences in Vienna in 1936, he reviewed this period of his life:

> The problems that consistently held the strongest claim on my attention . . . lay . . . in a completely different field, developmental physiological genetics. My interest in them had already been awakened in Freiburg. Between 1907 and 1910, Weismann let me take part in the experiments that he had been performing for decades on the modification of the design pattern of butterflies as a result of temperature stimuli.[13] He could no longer find the resolve to work out the results. The questions that were then being raised in the field he himself had opened up had become alien to him. He had set out from a phylogenetic standpoint; but his experimental results strongly tended to mandate questions of developmental physiology.[14]

Thus Kühn's reasons for reorienting his research agenda and focusing his attention on developmental physiological genetics go back at least in part to his formative years with Weismann. Kühn was also influenced by Richard Goldschmidt's work on the genetics of wing coloration in the gypsy-moth *Lymantria*.[15] Kühn acknowledged the value of Goldschmidt's effort to arrive at an overarching theory of gene action in pattern development. But he also had reservations about it. In the introduction to his genetic and developmental physiological studies, he stated that Goldschmidt's "effort also reveals how obvious it is that the facts we need to found a rigorous developmental physiological theory are still lacking."[16]

The Organism: *Ephestia kühniella* Zeller

Drosophila—the fly that had become the organism of choice for Thomas Hunt Morgan's school of transmission genetics—was manifestly not in Kühn's view the ideal object with which to link genetic analyses to physiological processes and developmental patterns.[17] But what other organism fit the bill? This was among the most important problems to be resolved before he could begin his research work. To help make the decision, Kühn's doctoral student Karl Henke began a dissertation on the experimental manipulation of the coloration and design of the firebug *Pyrrhocoris apterus*,[18] while another doctoral student, Friedrich Seidel, investigated its sexual organs during embryonic development.[19] It turned out, however, that firebugs were not cultivable in the laboratory. At about the same time, yet another of Kühn's doctoral students, Egon Schlottke, analyzed the variability of the pigmentation of the parasitic wasp *Habrobracon juglandis* and its

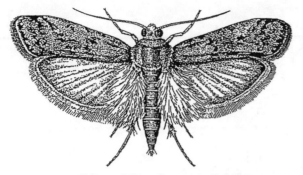

Mehlmotte, Weibchen. Bergr. 4 mal. (Nach Winter.)

16. The flour-moth
Ephestia kühniella
Zeller. Dingler 1925,
65, Fig. 36.

experimental manipulation by means of temperature stimuli,[20] work that was subsequently continued and extended by Hans Kaestner.[21]

Kühn never explained why he finally settled on the flour-moth *Ephestia* (fig. 16); there are reasons for supposing that the decision was not the result of a carefully premeditated choice. The curriculum vitae mentioned above contains only a laconic commentary on the subject: "In 1924, I found an experimental animal that was in many respects ideally suitable for my purposes: *Ephestia kühniella* Zeller."[22] It is quite possible that Kühn and his student Schlottke happened to hit on a few striking mutants of this moth, large numbers of which must after all have been present in their laboratory, since *Habrobracon juglandis* feeds on *Ephestia* caterpillars. *Ephestia kühniella* Zeller was not named after Alfred Kühn as one might suppose, but after Julius Kühn of the Agricultural Institute at the University of Halle, who had sent specimens of it to the entomologist Phillip Christian Zeller in 1877. Zeller's description of the insect subsequently appeared in 1879.[23]

Ephestia had a decisive advantage: it had already been "tamed" and utilized as a laboratory animal.[24] Kühn credited the applied entomologist and pest control specialist Albrecht Hase with having made *Ephestia* accessible as a "laboratory object."[25] The flour-moth had been one of the main targets of pest control during the First World War: the giant mills that appeared as a result of the centralization of food production provided a new habitat for it.[26] *Ephestia* could, moreover, be bred year round and kept (reasonably) free of parasites under simple laboratory conditions.[27] Its generation period was three and a half months. "What we attempted and finally accomplished," Hase had emphasized, "was an 'industrialized' cultivation of certain species." With reference to the interdependencies between different species—Hase's example was the relation between parasitic wasps and flour-moths—he added, "I do not think I am exaggerating when I say

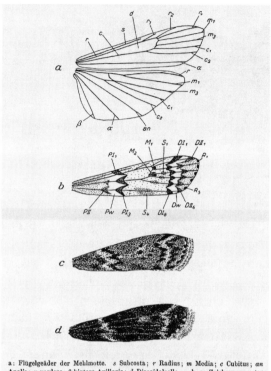

17. Wing of *Ephestia
kühniella* Zeller. a) wings
and their veins; b) and
c) design patterns,
schematically represented
in b); d) a design pattern
of black animals. Kühn
and Henke 1929, Fig. 1.

a: Flügelgeäder der Mehlmotte. *s* Subcosta; *r* Radius; *m* Media; *c* Cubitus; *an*
Analis; *α* vordere, *β* hintere Axillaris; *d* Discoidalzelle. — b, c: Zeichnungsmuster
des Vorderflügels, in b schematisch zur Demonstration der Zeichnungssysteme.
M₁, M₂ Mittelflecken; *S₁—S₄* Schatten; *DI₁—DI₆* erste, *DII₁—DII₆* zweite Distal-
binde; *PI₁—PI₃* erste, *PII* zweite Proximalbinde; *Dw* weiße Distalbinde; *Pw* weiße
Proximalbinde; *R₁—R₅* .Randflecken. — d: Zeichnungsmuster schwarzer Tiere. —
Sämtlich 6 ×.

'industrialized' cultivation, because one sees [from this example] the kind
of constant interaction typical of the factory."[28] Kühn may have met Hase
during the war,[29] since as a medical orderly in Falkenhausen Kühn had
been assigned the task of setting up a parasite laboratory.[30]

The Initial Epistemic Object: Wing Patterns

Together with his assistant Henke, Kühn decided in 1924 to focus on the
coloration and design pattern of the wings of *Ephestia* (fig. 17).[31] The two
men began their work with a population that they had obtained from Hase
of the Biologische Reichsanstalt für Land- und Forstwirtschaft in Berlin-
Dahlem and with another from an American source.[32] *Ephestia* thus became the experimental object for a six-year-long series of wing experiments. Between 1924 and 1928 alone, Henke and Kühn bred and screened
more than 100,000 moths from mass cultivations developed especially for

the purpose. The number makes palpable the difference between Weismann's "phylogenetic perspectives," based on observation and comparatively small-scale experimentation, and Kühn's "questions of developmental physiology," which called for "industrial" cultivation in large numbers and examination of many individuals of different mutant strains over a long series of generations. Such extensive studies became possible after Kühn was offered chairs at the Universities of Munich and Tübingen (in 1924) as well as the University of Vienna (in 1926). He turned them down in return for the means to reconstruct the Göttingen Zoological Institute to his needs, a project completed only in the early 1930s.

In the introduction to the collection of papers that issued from this long-term collaboration, Kühn and Henke declared their intention to "help link questions and facts pertinent to both breeding analysis and developmental physiology"; "the current state of development of the problem of heredity," they stressed, was "a pressing reason for forging such links."[33] Such a need was felt much more strongly in the German context of Corrensian and Goldschmidtian physiological genetics than in the United States, where transmission genetics was at the height of its development. But how did Kühn and Henke plan to carry out their intention? How did they proceed to experimentally realize their claim that physiology, development, and genetics could be linked? How were they to come by the "facts" lacking in Goldschmidt's theory? There was no question but that a combination of different techniques would be required to set up the kind of integrated experimental system envisaged. At this early stage of their research, Henke and Kühn thought that a dual approach was required. First, they systematically selected individuals with altered wing patterns and produced genetically pure mutants through breeding. Second—an equally classical procedure—they varied external factors such as temperature and diet and observed the influence of these stimuli on the different parts of the wing pattern in sensitive periods of development. Along this way, they also discovered new possibilities for improving cultivation conditions.

With its high variability wing pattern was a classical model for the study of questions of morphogenesis. From the first attempts to uncover hereditary processes through systematic breeding, the color pigments of a variety of body parts had been used as "special labels" to provide pointers to the underlying physiological processes. Wings were advantageous in part because they were (virtually) two-dimensional, reducing the complexity of the pattern by a dimension, "they had relatively simple components, . . . their development could be easily followed, and . . . it was possible to experimentally influence this development at different stages and in different ways."[34] This epistemic object—consonant with the former research agenda that Kühn had pursued with Weismann in Freiburg—played a decisive role in his early years in Göttingen. In the latter half of the 1920s, Henke and Kühn singled out eighteen different characters and twelve genetic factors by collecting and selecting mutations whose hereditary traits obeyed Mendel's transmission rules. What mattered to Kühn was not to make a genuine contribution to transmission genetics by, say, trying to establish *Ephestia* as an alternative to *Drosophila*: there could be no question of promoting *Ephestia*—with its thirty or so chromosome pairs versus just four pairs in *Drosophila*—to the rank of a research object for transmission genetics. Nor was his main aspiration to discover non-Mendelian, more complicated, or plasmatically localized patterns of inheritance, although he was generally interested in patterns of inheritance not controlled by the nucleus. As late as 1938 when his research on *Ephestia* was at its peak, he declared in a letter to Henke,

> The crucial task today seems to me to discover experimentally *whether* there are genetic elements in the plasm, *how* they behave in their continuity and constancy, *what* determines the changes in them, and how the determination proceeds. . . . But plasmon, predetermination, and so on must, I presume, simply take and retain, for the time being, the form "if X, then Y." I think that v. Wettstein was initially inclined to invest too much in the concept; that gives it too fuzzy a form, and its content, under certain circumstances, is walled off from other factual domains.[35]

As an alternative, Kühn decided to start by using simple Mendelian traits related to the wings that were manifestly governed by nuclear inheritance. He then sought to detect the laws of development of wing patterns by influencing their expression through systematic variation of experimental conditions.

We need not look more closely at the details of this immense, truly "factory-like" experimental infrastructure here with its regularly maintained and monitored insect houses and their impressive thermostatic equipment. But in the present context it is worth noting that although the laboratory generated a steady stream of dissertations and publications throughout the 1920s, not a single dissertation based on investigations of *Ephestia* was produced between 1924 and 1931, although three were turned out on *Habrobracon*. Publication on *Ephestia* began only with Kühn's and Henke's seven *Untersuchungen* of 1929; but in the 1930s, work that was directly concerned with *Ephestia* amounted to about forty percent of dissertations (fourteen out of thirty-two) emerging from Kühn's Göttingen laboratory and publications (twelve out of twenty-seven) in which he had a direct hand between 1930 and 1936.[36] Thus the large-scale *Ephestia* system had by then acquired a priority status in the lab that it did not have from the beginning, when it was one option among several that were being explored. Throughout the 1920s the *Ephestia* studies were due largely to Henke; Kühn's other students and coworkers continued to work and publish on a wide variety of other physiological topics, especially the physiology of vision, and questions of developmental biology.

A New Mutation: Eye Color

On May 30, 1929, Kühn's technician Veronika Bartels found a red-eyed moth in one of her cultures for the first time.[37] Kühn's former student Hermann Hartwig recalls that "Kühn used to promise considerable premiums for *Ephestia* mutants important to him—in my day, for example, mutants 'with white eyes.' The result was that the coworkers watched out not only for specific characters (for example, in wing patterns), but for mutations in general. The technicians, who were highly esteemed in the Institute, participated in this competition if they were responsible for flour-moth cultures."[38] Although this was by no means the first eye-color mutation observed in an insect—vermilion (v) and cinnabar (cn) were known in *Drosophila* and had been the object of genetic studies for many years[39]—this mutant was particularly striking and was therefore rapidly characterized with greater precision. Here a relationship emerged between an allele and a simple character:

$$(1929) \quad A/a \text{ gene} \quad \rightarrow \quad \text{dark/red eye color}$$

In subsequent breeding experiments, Kühn and Henke found that the relevant allele followed a Mendelian inheritance pattern and was recessive. Kühn observed with interest that his recessives were somewhat less viable than the homozygotes or heterozygotes of the wild phenotype and that they had a slightly prolonged developmental cycle. With the help of such statistically significant, albeit fairly unspecific pleiotropic effects—one single gene affecting several different characters—he hoped to account experimentally for certain features of the developmental physiology of the mutants, making it possible to link analysis based on transmission genetics with developmental physiology.

However, just as the wing experiments of the preceding years had failed to match Kühn's "expectations in certain respects,"[40] the observed correlation between eye pigment reduction, viability, and the length of the breeding cycle did not prove to be an adequate experimental tool for exposing the hidden connections linking gene action, development, and physiology. Despite the untold statistical tables and figures that filled Kühn's and Henke's publications up to 1936, there emerged from this correlation no single, clear-cut differential characteristic promising enough to warrant pursuit at the cost of neglecting other experiments. This is where matters stood when Henke received a Rockefeller Fellowship that took him to the United States for a year. Kühn urged him to feel free in choosing his place of residence and his next research project,[41] but it was probably no accident that Henke finally went to Yale University to work with Ross Harrison. Harrison was a pioneer in the field of animal tissue culture. It seemed likely that he would provide Henke with new ideas for developing the Göttingen experimental system further.

A New Technique: Transplantation

Meanwhile, observations and experiments had also showed that the pleiotropic effect of the pigment mutation affected not only the general viability and length of *Ephestia*'s breeding cycle, but also the coloration of the testicles (as well as the body color) of the *Ephestia* caterpillar. Other eye mutations appeared that were "noteworthy for their developmental physiological effects."[42] Kühn wrote to Henke in the United States to share his excitement about these additional observations. In May 1930, he reported that Werner Blaustein had undertaken histological investigations of the process of pupation in *Ephestia*;[43] he himself was trying to learn more about the moth's sexual organs.[44] Other doctoral students were conducting classical developmental and morphological studies of *Ephestia*: Helmuth Wagner was working on the maturation of ova and sperm, especially chromosome patterns; Else von Gierke was studying molting and the pace of larval development; and Wilhelm Köhler was investigating wing development.[45] In August 1930, Kühn informed Henke of radiation experiments designed to excite phosphorus and sulfur atoms. He was "very curious about the effect this would have . . . since P must play a particular role in the nucleus."[46] He equipped his laboratory with a whole new set of thermostats that functioned between 18° and 30° Celsius, as well as a cold thermostat and an ice machine. Another of his doctoral students, Karin Magnussen,[47] was performing a series of operations on different butterflies "for training purposes," as Kühn also reported to his correspondent in the United States. He added: "I would be all too glad to be able to pursue the idea of tissue culture and tissue transplantation. . . . Even if we have had no tangible results yet, it appears to me that we can achieve a great deal even with smaller butterflies." In another letter in which he gave Henke a report about a visitor from Copenhagen, he summed matters up this way: "It is evident that we are so far ahead that no-one can ever catch up with us; for he [Lemmke] was overwhelmed by what we know already, everything we are working on now, and the results that are coming in from week to week."[48] Kühn had obtained funds from the Notgemeinschaft der Deutschen Wissenschaft; around the same time, the reconstruction of the Institute in Göttingen was completed. Until 1930 he had had to make do with an average of ten staff members, doctoral students included; in early 1933, the number shot up to forty.[49]

Ernst Caspari, who had joined Kühn's laboratory in 1929, soon held "an undoubtedly exceptional position among Kühn's coworkers."[50] Caspari explored the potential for transplantation surgery in further analyzing the pleiotropic color effect in the red-eyed mutant. The comparatively large size of the flour-moth facilitated Caspari's experiments. This was one of the reasons for the success that he achieved several years before Ephrussi's and Beadle's joint work on the transplantation of imaginal eye disks in *Drosophila*.[51] Caspari carried out his first successful transplant in May 1932.[52] In 1933 he observed in his seminal work, "On the Action of a Pleiotropic Gene in the Flour-Moth *Ephestia kühniella* Zeller," "Nobody has yet attempted to attack the question of the mechanism of gene action with the experimental methodology used most commonly elsewhere in developmental physiology, namely, experimental surgery."[53] Caspari, who was seconded in 1931 by a young researcher from Portugal, Alberto da Cunha,[54] decided to explore the effects of implanting pale testicles of red-eyed mutants (aa) into the black-eyed wildtype (AA), and colored testicles of the black-eyed wildtype (AA) into red-eyed mutants (aa). Although many of the individuals operated on died, the surgical intervention had a surprising effect on those that survived. Caspari summarized his main results as follows: "When aa-testicles are transplanted into AA-animals, one observes . . . an *influence of the host on the implanted organ.* . . . When AA-testicles are implanted in aa-animals, the opposite effect obtains, that is, *the implanted organ has an influence on the host.*"[55] In other words, what Caspari observed was that wildtype testicles had the capacity to darken the eyes of mutant moths (fig. 18), and that mutant testicles became colored when implanted into wildtype caterpillars. He concluded that "the A-testicle secretes a substance into the blood that has the capacity to induce intensive formation of pigment in testicles or eyes in whose cells gene A is lacking."[56] A first intermediary link, a gene-dependent diffusible substance, had now been introduced into the reaction chain:

$$(1933) \quad \text{A-gene} \quad \rightarrow \quad \text{'A'-substance} \quad \rightarrow \quad \text{pigment}$$

"With the appearance of a red-eyed mutant," Kühn later affirmed, "a new research field was opened up."[57] That was not quite accurate, for the mutant was only a necessary, not a sufficient condition for the "opening up" of the new field: its pleiotropic character was another determining feature. Moreover, before it could become a component of a productive experimental machinery, the transplantation technique had to be developed. Yet the new experimental arrangement was enticing and seemed to hold out countless possibilities for further development. It was based on an organism suitable for transplants, it generated unanticipated research questions embodied in new epistemic objects, and it accommodated the introduction and development of new techniques. In short, it displayed all the features of a productive experimental system.[58]

Support from the Rockefeller Foundation

Kühn and Caspari were aware of these qualities of their biological-genetic experimental system and its potential scope. Thanks to the transplantation technique, their system allowed direct access to physiology. They accordingly set to work extending it, seeking out ways to biochemically characterize the soluble substance that had come to light in the transplantation experiments. Kühn now also made contact with one of Adolf Windaus's students, the biochemist Adolf Butenandt, who had obtained his doctoral degree in 1927 in Göttingen, where he had worked on the chemical composition of sexual hormones. Butenandt, who took a professorial post at the Technical University in Danzig in October 1933, had from the early 1930s been in contact with Wilbur E. Tisdale, a staff member of the Rockefeller Foundation's European Bureau, in connection with a possible U.S. fellowship. After a two-day stay in Danzig in June 1934, Tisdale handed in a report to the foundation:

> B[utenandt] has just started collaborating with Prof. A. Kühn, a geneticist in Göttingen. They are working on a moth (*Ephestia kühniella*). This moth has a red-eyed and a black-eyed mutate [*sic*]. They have discovered the following: if the testicles of one strain are transplanted into the other in the

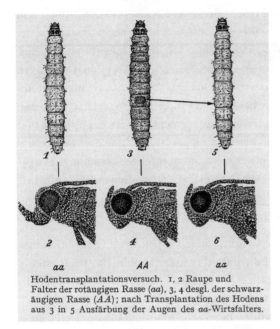

18. Testicle transplant system.
1 and 2: caterpillar and imago
of the red-eyed mutant (*aa*);
3 and 4 caterpillar and imago
of the wildtype (*AA*); 5 and 6:
after transplantation of the
testicle from 3 into 5, the eyes
of the mature host animal
become dark. Kühn 1936a,
1–10, Fig. 9.

Hodentransplantationsversuch. 1, 2 Raupe und
Falter der rotäugigen Rasse (*aa*), 3, 4 desgl. der schwarz-
äugigen Rasse (*AA*); nach Transplantation des Hodens
aus 3 in 5 Ausfärbung der Augen des *aa*-Wirtsfalters.

larval stage, the moth's eye color changes. B. and K. have the impression
that a hormone is involved here, and B. is in the process of setting up a
breeding facility for moths in order to collect enough material to be able to
go to work on isolating the hormone.[59]

Just days later on June 20th Butenandt was promised $3,000 for September
of that year "for apparatus & animals for research in the chemistry of hormones," a sum that was also to cover "a project carried out in cooperation
with Kühn."[60] According to the grant description, the goal of this collaboration was "to isolate and crystallize" the hormone thought to be responsible for eye coloration in *Ephestia kühniella*.[61]

On July 3rd Kühn reported to Dr. Wildhagen of the Notgemeinschaft
der deutschen Wissenschaft about the current state of his work. "We have
been able to break down complex developmental processes that could
hitherto only be taken as a whole into individual component processes
that it is possible to influence." He informed Wildhagen that the Deputy
Director of the Rockefeller Foundation's European Bureau, Dr. Miller,
whom he had often advised in the past about the selection of Rockefeller
fellows, had come to see him a few weeks earlier. Kühn, politically a member of the German Democratic Party that had just been dissolved, asked
cautiously whether such contact was desirable under the given political
circumstances, soliciting Wildhagen's approval:

I was very surprised to hear [Dr. Miller] offer Rockefeller Foundation aid in accelerating and expanding these activities. . . . I have never sought support for my work from the Rockefeller Foundation or any other foreign agency, and I naturally prefer to obtain the means needed to carry on my research from the sources of support available in my fatherland. . . . With the aforementioned observations, we have scored a success of the sort that American research schools have been trying in vain to achieve. I am happy to see that the Rockefeller Foundation prefers to support the continuation and expansion of this work in Germany rather than promote the creation of a large-scale biological-chemical research team abroad. Creation of such a research team is well within the realm of possibility; with the right equipment and management, it could catch up to and surpass us.

Nevertheless, to cover his bets, Kühn added, "if you are of the opinion that it is at present undesirable [sic; should read "desirable"] to reject even a voluntary and in no way binding Rockefeller Foundation contribution to German research work, I am of course prepared to turn it down."[62]

Tisdale wrote soon thereafter—before the month of July 1934 was out—to Warren Weaver, the Director of the Natural Sciences Division of the Rockefeller Foundation. "[Kühn] feels that genes control hormone action," he reported, adding that he needed assistants to help him carry out his research.[63] A few weeks later, he decidedly stated that "his [Kühn's] School of Genetics is without doubt the best in Germany, and he has nothing to do with racial problems." Weaver's reaction was favorable: "The cooperative researches of Prof. Kühn and Prof. Butenandt," he wrote, "seem most interesting and promising."[64] Meanwhile, Tisdale had also sought and obtained advice from a close acquaintance, Richard Goldschmidt of the KWI for Biology in Berlin. Weaver was informed of this, too: "I asked G. what he knew of A. Kühn in Göttingen. He said that without doubt K. has the best school of genetics in Germany and that while he is not noted for new and imaginative ideas, he does seem to have the facility to pick up important problems suggested by others, and has a unique organizing and administrative ability to carry them through."[65] The reserved personal assessment of Kühn that Goldschmidt expressed here was plainly matched by Kühn's equally reserved opinion of Goldschmidt.[66]

There were, however, other reasons for the Rockefeller Foundation's reluctance to make a long-term commitment to Kühn. Because of the "uncertain political situation in Germany," the foundation decided to support Kühn for only one year, while holding the option of further collaboration open.[67] For a project on amphibian hormones (in cooperation with the organic chemist Gottwalt Fischer in Freiburg) and a project on gene and hormone action in *Ephestia* (in cooperation with Butenandt in Danzig), Kühn received an initial grant of $3,000 in September 1934. This support was to be released in installments every three months on reception of progress reports (which today make it possible to reconstruct Kühn's work step by step).

Tisdale declared in no uncertain terms that "there is nowhere on the Continent or in England that we can find chemists, embryologists and geneticists willing to cooperate among themselves, as are these German scientists."[68] The Deutsche Forschungsgemeinschaft (DFG) obviously did not object to the financial support given to Kühn by the Rockefeller Foundation. The DFG in its turn funded the experiments on heredity carried out under his lead "as well as cooperative efforts to breed experimental animals with particular constitutional characters, including cancerous mice, and to utilize them in various areas of medical research, such as X-ray induced hereditary damage." Kühn took this description to encompass "studies on the triggering of mutations and their effects in insects as 'model experimental animals,'" as his report to the DFG indicates.[69]

Kühn wrote his first report to the Rockefeller Foundation in March 1935 from Naples during one of his usual spring trips to the Stazione Zoologica. By 1935 it was no longer easy for him to travel. As he reported in a letter to Reinhard Dohrn in Naples at the very end of January, his application for a table in Naples had been approved by its sponsor, the Prussian Ministry, only after a long delay.[70] Dohrn for his part had asked Kühn in September 1934 to intervene with the Gesellschaft Deutscher Naturforscher und Ärzte in support of a resolution against the termination of scientific travel. Kühn demurred on the grounds that the "people in charge" would be "particularly unhappy about being told what to do by such a Society." Like many of his colleagues he had renounced open resistance to the regime in favor of individual interventions.[71] As far as his stay in Naples was concerned he asked Dohrn to arrange to have "an already medium-sized octopus and a few rocks" placed "in each of the indoor pools [of his laboratory of the previous year],"[72] since he wanted in the course of his spring visit to pursue his study of the adaptation of these animals to the color of their rock nests. This can be seen as a way of counterbalancing with traditional biology the efforts being made in chemistry in Göttingen, about which he issued the following report to the Rockefeller Foundation in Paris:

> To begin elucidating the chemical nature of the gene hormone, it was first necessary to conduct an exact investigation of pigmentation reactions; this

task has been achieved thanks to the work on eye pigmentation performed by several of our collaborators. Work on testicle pigmentation and skin pigmentation is underway. In April, an assistant of Dr. Butenandt's [Dr. Ulrich Westphal] will be coming from Danzig to Göttingen in order to cooperate in my laboratory with Dr. Caspari in working out . . . suitable methods for obtaining hormone from the body of the caterpillar and for testing the hormone effect.

Kühn also informed the foundation that "a series of further experiments [on] the problem of hormone action in insect metamorphosis" was in progress. "Along with the attempt to obtain A-gene hormone from the body of the caterpillar, an attempt to isolate the metamorphotic hormones will simultaneously be carried out." It was above all Kühn's co-worker Hans Piepho who focused on this second project. The third project that Kühn addressed in this initial progress report bore on the way hormones controlled ontogenesis in vertebrates, a project involving experimentation with salamanders, carried out jointly with Gottwalt Fischer in Freiburg; the member of Kühn's team taking part in it was Hermann Hartwig.[73] Before, Hartwig had written his doctoral dissertation in Kühn's laboratory on the influence of temperature variations on the development of salamander larvae.[74]

A series of related investigations concerned morphological and histological aspects of the development of the eye in *Ephestia* and its sensory physiology. Thus, in 1934 Herbert Brandt worked on the moth's light orientation, Antonette Busselmann on the structure and development of ocelli in the caterpillar, and Wilhelm Umbach on the structure and development of the complex eye.[75] Other doctoral students of Kühn's continued the studies on which he and Henke had worked intensively in the late 1920s when investigation of the flour-moth commenced. Wolfgang Feldotto explored the sensitive periods in the development of the wing pattern and Eckhardt Hügel studied the genetics of the white distal band; Johannes Behrends worked on the development of the lacunal, tracheal, and circulation system in the wing of the flour-moth chrysalis.[76] Kühn himself had focused since 1930 on determining the symmetry system of *Ephestia*'s anterior wing, working with his technician Melitta von Engelhardt.[77] In a series of experiments they explored a phenomenon that Goldschmidt had called "phenocopying," meaning that the specific traits of genetic mutants could also be triggered by external stimuli in certain phases of wild-type development.[78] As we learn from a later letter of Caspari's, in 1932 Kühn had also begun stocking cultures of a new model organism, *Ptycho-*

poda seriata, a species of geometric moth. Although his coworkers initially greeted this effort with smiles, *Ptychopoda seriata* was to take on increasing importance as an object of comparison with the *Ephestia* system.[79]

Conceptualization I: Hormone Action

The suspected relations between genes and hormones, that is, the hypothesis that genes acted by way of hormones, comprised so to speak the leitmotif of these various projects; and increasingly the flour-moth, as *the* experimental animal of physiological genetics, occupied a position at their center. Since the 1920s, hormones had been one of the preferred objects of physiological and biochemical investigations. In addition to Butenandt, Leopold Ruzicka, a chemist who worked first in Utrecht and then in Zürich, had achieved major breakthroughs in the field of sexual hormone research in the late 1920s and early 1930s. Thus the developmental physiological geneticists in Göttingen were taking a fashionable approach, moving in step with their time when they began to interpret gene action in their insects in terms of hormone activity. In their studies of *Drosophila*, Ephrussi in Paris and Beadle at the California Institute of Technology and Stanford from 1937 on also quickly adopted this view.[80]

In 1935, Kühn, along with Caspari and Ernst Plagge, who the year before had also received temporary funding from the Rockefeller Foundation, reported two new findings about "hormonal gene action" in *Ephestia* in the journal *Nachrichten von der Gesellschaft der Wissenschaften zu Göttingen*.[81] This was the first appearance of the hormone hypothesis in print. One of the reported findings had to do with what the authors called a "matrocline inheritance" of the dominant phenotype. When wildtype testicles were implanted in mutant, red-eyed female moths whose eggs were then allowed to develop, their offspring displayed pigmentation, although they remained genetically speaking defective mutants, which is to say that they were not capable of synthesizing the A-substance by themselves. Besides publishing this paper, Kühn also issued a report to the Rockefeller Foundation in which he stated: "Thus the substance that is diffused into the blood stream by the A-testicle implant penetrates the egg cells." The fact that "this tiny quantity of A-substance suffices, when it is taken up by the egg, to trigger pigment production for several weeks in the first caterpillar stages" corroborated the Göttingen scientists' assumption that the A-substance "was a hormone effective in very tiny amounts, one that triggered certain production processes."[82] They therefore conceived of gene action from now on as a "reaction chain" between the gene and its phenotypic expression in which a hormone acted as an intermediary:

$$(1934) \quad \text{A-gene} \quad \rightarrow \quad \text{'A'-hormone} \quad \rightarrow \quad \text{pigment}$$

In addition to the effect of matrocline inheritance, Kühn in his July 1935 report to the Rockefeller Foundation listed no fewer than nine more "new findings" relevant to the question of the "hormonally mediated effect of certain genes in the development of insects," among them the coloration of the brain, the coloration of the eyes of the caterpillar, and the emergence of a new allele (a^k) that caused eyes to take on a brown coloration—in other words, brought in its wake a somewhat attenuated pigmentation defect. Following this report, the foundation unhesitatingly decided to renew its financial support for another year. Interestingly enough, its written justification of the renewal stated that "the eye pigmentation of the flour-moth [was] *controlled* by a certain gene,"[83] whereas Kühn had only spoken of the "effect" of certain genes in his reports. This is a small detail that nonetheless marks an interesting difference in the approaches to the investigation of gene action on either side of the Atlantic.

Kühn, Caspari, and Plagge lengthily described yet another finding in their 1935 paper. This finding was based on a fine quantitative observation: implanting organs of the black homozygous and heterozygous wild phenotype in a red-eyed mutant had quantitatively the same effect on coloration, whereas implanting two organs, whether homozygous or heterozygous, had an additive effect. From this fine distinction, the three scientists drew a far-reaching conclusion: there had to be a sort of "primary reaction in the plasm" that was decisive in determining whether the A-substance was produced and then diffused: "the process of dominance A>a," they wrote, "is displayed in the primary reaction in the cell."[84] The term "primary reaction" was merely a stand-in for a still unknown process. However, it snapped the unmediated connection between the gene and the hypothetical gene-triggered hormone production.[85] These experiments made possible a first glimpse into the space between the two poles "gene" and "character," a space in which determination and development were decided (fig. 19). The nucleus represented the genome. The gene-triggered "primary reactions" were now deemed to take place in the cytoplasm. These cytoplasmic entities or events led in turn to diffusible substances, hormones that left the cytoplasm and in trace amounts acted as signals in different tissues and organs, where they induced the expression of certain characteristics. Whence the following formula:

(1935)   A-gene   →   primary   →   'A'-hormone   →   pigment
                            reaction
     (nucleus)   (cytoplasm)   (circulation)   (organs)

The general framework within which these findings took on meaning was still conceived as a series of "intermediar[ies] in the reaction sequences that lead from a particular gene to the mature characters of organization," as Kühn, Caspari, and Plagge put it in their paper.[86] PR$_A$, the primary reaction, was considered the first of these intermediate products; it was thought to proceed from the A-gene and to occur in the cytoplasm of the "hormone producing cell." The process was believed to continue with the diffusible A-hormone, the circulation of which was ultimately thought to result in five different pigmentation patterns. Three of them, the coloration of the hypodermis, eyes, and testicles, were displayed by the caterpillar, the other two, the color of the complex eye and brain, by the imago.

Together with Plagge, Caspari had improved the transplantation technique in "preliminary experiments with big butterflies" to the point that quantitative experiments of this kind would be likely to succeed as Kühn reported to the Rockefeller Foundation in December 1935.[87] What he passed over in silence, however, was the fact that his assistant no longer worked on their joint research project. Unable to assume a paid post at the University of Göttingen because he was of Jewish descent, Caspari, who had had "a successful career before him," emigrated to Istanbul in 1935.[88] In

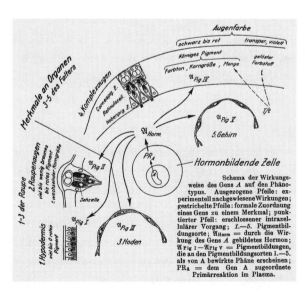

19. Scheme showing the mode of action of the A-gene on the phenotype. Kühn, Caspari, and Plagge 1935, Fig. 22.

1937 while still in Istanbul he contacted Ephrussi in Paris to ask whether he could work with him on *Drosophila* transplantation experiments. Ephrussi reacted with reserve, writing in a letter to his friend Beadle, "Will see in Paris what I can do for him. It sounds interesting, I think, but I would first like to find out what kind of a fellow he is."[89] In 1939, thanks to the intervention of Leslie Dunn at Columbia University, Caspari managed to come to the United States, traveling by way of Paris.[90] In the United States he first found work at Lafayette College in Easton, Pennsylvania. After unsuccessfully attempting to gain access to the research team working on *Penicillium* with Beadle,[91] he joined Curt Stern at the University of Rochester in 1944. He finally settled at Wesleyan University in Middletown, Connecticut, in 1946.[92] As a result of his emigration forced by the German Nazi State, he lost ten crucial years before he could take up regular and sustained experimental work again.

In Göttingen, meanwhile, work on the A-gene hormone continued; it was now mainly in Plagge's hands. Kühn's report to the Rockefeller Foundation in December 1935 was the first to note a significant distinction between the transplantation of genital organs and brains. After the implantation of genital organs, the coloration of the host animals' eyes and testicles proceeded in correlated, parallel fashion; the brains, however, exhibited an effect only on the eyes, not on the testicles. "From this, there follows the second compelling conclusion of importance," Kühn wrote; "namely, that the same A-gene stimulates the production of qualitatively different hormones in the testicles and ovaries on the one hand, and the brain on the other."[93] Kühn soon detected an "analogy to sexual hormones" in vertebrates here; "under the influence of the constitution of the gene of the male or female sex . . . there are produced, by different organs of the same organism, sexual hormones exhibiting different effects in the tests."[94] Plagge and Kühn found further support for their hormone hypothesis in the fact that again in analogy to the reaction-specific but not species-specific behavior of vertebrate hormones implants were active across species boundaries. The corresponding experiments, however, proved considerably more difficult to perform, since "the mortality rate of the experimental animals is much higher than with transplants between strains of the same species."[95]

As part of his doctoral thesis, Plagge had carried out a series of extensive comparative histological studies on the different strains of *Ephestia* that had accumulated in the laboratory over the years. However, these studies provided no significant, new information about the hormone.[96] The same holds for Hans Piepho's work on the effects of hormones on the pupation of caterpillars.[97] The many transplantation experiments and constriction experiments were also funded by the Rockefeller Foundation; ultimately they served as the point of departure for another long-term project of Kühn's that I shall not go into here.[98] In line with the general observation that it is "characteristic for the hormones of vertebrates to be reaction-specific, but not species-specific," Plagge performed transplantation experiments with the testicles of different butterfly species and showed that the reactive A-substance was, as expected, active across species boundaries for a wide variety of insects.[99] This was weak yet altogether welcome support for the hormone hypothesis, which was pursued further along various lines of investigation.

The Approach to Biochemistry

The considerable refinement and partial quantification of the coloration tests, which went hand in hand with the improvements in transplantation methods, comprised one facet of the experimental program by means of which Kühn hoped to elucidate the action of the A-gene. The investigation had matured to the point that the obvious next step was to try to learn more about the chemical composition of the hormone-like substance. As Kühn had already reported to the Rockefeller Foundation in July 1935, Ulrich Westphal from Danzig had visited the laboratory in Göttingen in April and May of that year in order to hold discussions with Kühn's research team on how to "make the transition from the transplantation experiment with living tissues to a chemical-physiological attempt to obtain extracts from the blood and organs as well as the incorporation of organ extracts." A parallel attempt would also have to be made to "work out measurable test reactions of the kind that can be carried out as quickly as possible."[100] The goal according to Kühn's report was to replace living tissue with "previously deadened tissue"—for example, by freezing—as well as by the "infusion" of fresh and thickened blood. For all the "great losses" sustained in the first stages of these experimental attempts, Kühn was initially unable to report any new findings to the foundation; all he could claim was "good technical progress."[101] The establishment of *in vitro* experimentation aimed at here proved difficult to achieve; however, it was the precondition for quantitative assays on dosing the substance and the "transition to chemical investigation."[102]

Meanwhile, Ephrussi and Beadle had demonstrated by means of imaginal disk transplantation experiments with *Drosophila melanogaster* that the substances corresponding to the v⁺ (vermilion) and cn⁺ (cinnabar) genes in eye pigment formation comprised an ordered sequence: thus the substance corresponding to cn⁺ was the transformation product of the substance that corresponded to gene v⁺.[103] Plagge immediately began making preparations to transplant imaginal disks in *Ephestia* as well, and he was soon successful.[104] But his first injection experiments with hemolymph had "no effect," as he reported in a short note in 1936.[105] He was obviously not aware at this point that the injection experiments that Ephrussi and Beadle were conducting at the same time had been successful.[106]

The Rockefeller Foundation held out the prospect of an additional year's support for Kühn beginning in October 1936. Tisdale in the resume contained in his report to the foundation said that "the genetic and experimental research was at a well advanced stage" and that "the chemical investigation" was now underway.[107] For its part the DFG had dispatched to Göttingen different kinds of experimental equipment, including thermostats, thermographs, and lenses for the microscope that "had been loaned to him" for his "investigations of heredity and modifications in insects,"[108] as indicated by Kühn's report on its uses.

Kühn's "Berlin affair," as he put it in an April 24th letter to Tisdale, was also becoming clearer in 1936.[109] Despite the support extended by the Rockefeller Foundation and DFG, the situation in Göttingen had grown more tense. "I could no longer stand it at the changed University," he wrote to Caspari after the war (as far as I can see, Kühn never expressed himself more explicitly about this move).[110] Among other things, the number of his coworkers, which had risen sharply until early 1933, now fell drastically, shrinking from forty to eleven between 1933 and late 1936. The "Berlin affair" began with Richard Goldschmidt's forced retirement at the end of 1935; he emigrated to the United States in 1936. In his memoirs, Goldschmidt writes: "Only when I was already in the United States did I receive the official document, which read, 'You are pensioned beginning January 1, 1936.' No 'Dear Sir' or 'Yours truly' or 'I regret to inform you.' Just a small, untidy slip of paper. I had been on the job for thirty-five years."[111] Kühn's close friend and former colleague Fritz von Wettstein, who had already moved from Göttingen to Berlin in 1934 in order to succeed Carl Correns as director of the KWI for Biology, played a decisive role in effecting Kühn's transfer to Berlin-Dahlem. With great diplomacy and against the explicit wish of the educational bureaucracy he managed to place Kühn in the position vacated with Goldschmidt's forced retirement and emigration. Kühn kept the Rockefeller Foundation informed of developments. "The K. W. Institute and I are working very diligently on arranging my transfer to the Institute for Biology in Dahlem," he wrote to Tisdale, "and I very much hope that it will ultimately prove possible to obtain permission to go from the educational authorities." Whereas he informed the DFG about his "transition" to Dahlem only in February 1937,[112] he wrote to Tisdale about it already at the end of 1936: "On the first of April, I will be transferred to the K.W. Institute for Biology. . . . I am very glad that in the future I can devote all my time and strength to pursuing my research."[113] Kühn's friend and student Henke, who had since 1930 been working in Goldschmidt's section of the KWI for Biology, subsequently took over Kühn's post in Göttingen.

After the transfer to Berlin-Dahlem, Kühn intensified his work on the biochemical components of the A-substance. Among the collaborators who followed him to Berlin were Ernst Plagge, Hans Piepho, Victor Schwartz, Erich Becker, and his technician of many years Melitta von Engelhardt. Becker had been trained in Darmstadt by Clemens Schöpf, a professor of organic chemistry; in spring 1937, he succeeded in extracting the A-gene product from homogenates of *Ephestia* ovaries, using ethanol and acetone; he also successfully carried out a positive injection test with hemolymph, something Plagge had proved unable to do in 1936. In June 1937, Kühn reported to the Rockefeller Foundation Bureau in Paris: "Our attempts to inject the A-gene hormone using active extracts has [*sic*] succeeded with ovarial extracts."[114] On the basis of these extraction experiments, Becker concluded that the substance could be neither a protein nor a lipid-soluble compound and held out the prospect of its "further enrichment and purification."[115] Later that year, Becker and Plagge by injecting extracts from *Drosophila* into *Ephestia* and vice versa succeeded in proving that their effects were interchangeable. In other words, the extracts were equivalent.[116]

The Rockefeller Foundation's financial aid ceased in 1937. At the end of the year Kühn sent a report to the foundation's Paris bureau to announce that his attempts to obtain the A-gene hormone in active extracts had "been successfully continued," and expressed his thanks "for the help without which [he] could not have made progress in a field of research the experimental difficulties of which were so great that the tools and the staff available in ordinary university institutes would not have sufficed to overcome them."[117] Wasting no time, he applied on December 15, 1937, for further support from the DFG for his "investigations of the mode of action of hereditary traits, especially the active components of genes." On May 18, 1938, he was awarded the sum of 3,000 Reichsmark (RM) for that purpose.[118] This was about one-quarter of his annual Rockefeller Foundation grants, but it allowed him to carry on.

In Paris Ephrussi and his coworker Yvonne Khouvine had already extracted the substance active in eye pigmentation from *Drosophila* in 1936.[119] They pursued their purification efforts as speedily[120] as Beadle at Stanford.[121] The need to collaborate with chemists was becoming increasingly important for Ephrussi, Beadle, and Kühn. Early in 1937 Ephrussi had still been convinced that it was "not desirable to push the chemists into cooperating with us";[122] that summer, however, he expressed his satisfaction that his friend Beadle had managed to "get [himself] a chemist [Edward Tatum]."[123] In August Ephrussi wrote to Beadle, "I have re-read all the Kühn and C[aspari] papers. There are a lot of interesting findings in them, but I think that most of their interpretations are full of mistakes."[124] Only a few days later he expressed his great interest in the differential behavior observed by the team in Göttingen of implanted testicles and brains in *Ephestia*, a finding Kühn had repeatedly underscored in his reports to the Rockefeller Foundation.[125] Over the next few weeks Ephrussi tried to obtain *Ephestia* strains from Kühn's stock for his own experiments through Nikolai Timofeev-Ressovskii, who was scheduled to travel from Berlin to Paris in October in part to see Tisdale with Ephrussi and find out how much Rockefeller Foundation support they could count on for a planned series of conferences "on problems of genetics and comparable problems in other fields."[126] But Ephrussi's hopes were dashed. In October, he wrote to Beadle: "As I wrote to you Kühn promist [*sic*] to send *Ephestia* stocks with Timofeyev, but Timofeyev came without stocks and I got a letter from Kühn who sais [*sic*] that his cultures are poor because infected etc and that for a month or two he will not be able to send me any. In other words he is a s. of a b."[127] In early 1938, Ephrussi invited Kühn to take part in the April Klampenborg conference on chromosome structure and conjugation in Copenhagen. Kühn accepted the invitation and then changed his mind, inviting Ephrussi to come visit him after the meeting in Berlin instead.[128] Ephrussi "after mulling it over" finally declined.[129]

For some time now little progress had been made in Berlin. In July 1938 Becker could report nothing new in his survey article in *Die Naturwissenschaften*.[130] Toward the end of 1938 he joined forces in Dahlem with Wolfhard Weidel, a doctoral student of Butenandt's.[131] Butenandt had come from the Technical University in Danzig to the KWI for Biochemistry in Berlin-Dahlem in November 1936, replacing Carl Neuberg as director of the institute after Neuberg's forced retirement.[132] As a result of Becker's and Weidel's co-operation the project got moving again. In 1939 as in the previous year the DFG granted Kühn 3,000 Reichsmark.[133] Moreover, on 21 September 1939, the *Reichsforschungsrat* "recognized the importance for the war effort of the work being carried out with the allocated funds." Kühn was therefore able not only to continue his work on the same financial scale as before, but also to revoke a number of recent dismissals and create new posts with part of the money he had been granted.[134] In December 1939, a few months after Nazi Germany's invasion of Poland, which marked the beginning of the Second World War, he reported to Henke in Göttingen that "recently, Becker and Weidel, a doctoral student of Butenandt's, have made good progress with the 'a$^+$'-substance and the pupation hormone." He continued, "I would be happy if we in Germany succeeded in elucidating the chemical nature of these substances."[135] According to his letters of this time, Kühn clung to the hope that achieving success under the conditions of the Nazi regime would show the world that there were still scientists in Germany who were pursuing good basic research.[136] He wrote to Henke, "I cannot and will not restrict myself to literature-based research again for years, especially now, when the experimental work of many institutes and researchers is coming to a standstill. In this situation, it seems to me important, above all, to use my position to promote such work to the greatest possible extent."[137] In a letter to Hans Spemann, Kühn explained his worries about the independence of the Kaiser Wilhelm Society as a whole, which was in danger, in his view, of "falling completely into the hands of the ministry," and he insisted that "in a period like the present, marked by a dramatic world-historical process, the value of a universally oriented German science that is conscious of its autonomous mission should surely be upheld."[138] Shortly before his death Spemann replied, "What you tell me of your concerns about the Kaiser Wilhelm Institute, and also about the development of biology and of our culture in general, is something you should say to as many people as possible, both privately and in public. . . . We have to hatch a conspiracy of the spirit against the demon [*Ungeist*] of our time."[139]

In Paris work was grinding to a halt. In March 1939, Ephrussi wrote Beadle that "the purification of pigment seems . . . to have reached a dead point: we have it quite pure, but not absolutely pure and cannot get it purer: it does not want to cristallise [*sic*]."[140] A few months later he remarked that "chemistry did not do the slightest progress."[141] In October 1939 Ephrussi was drafted. It was not until February 1940 that he responded to a letter that Beadle sent him from Stanford to report on a recent success.

"Naturally, the most important thing is the crystallization of the hormone. I congratulate you and am happy to see that all this is coming from your lab."[142]

Meanwhile an observation of Tatum's at Stanford had given the group in Berlin the decisive hint. Tatum had demonstrated that in the presence of tryptophan certain bacteria synthesized a substance that produced the same effects as the v[+] substance in *Drosophila*.[143] In the end Becker and Weidel proved a little quicker than their French and American colleagues. In their investigations they used a bacterium, *Coryne mediolanum*, that Luigi Mamoli had successfully cultivated at the KWI for Biochemistry in his experiments with steroids: *Coryne* reacted to tryptophan in the same way as Tatum's strains.[144] Becker and Weidel eventually identified their *Ephestia* "hormone" as kynurenin, a derivative of the amino acid tryptophan. The structure was published in January 1940.[145] Shortly thereafter Rolf Danneel, another of Kühn's coworkers, showed that kynurenin also colored mutant *Drosophila* eyes in vitro.[146] By the time Tatum and Arie Haagen-Smit published their results in 1941,[147] scientific contact between Germany and America had ended.

Conceptualization II: "Das Wirkgetriebe der Erbanlagen" (The fabric of hereditary dispositions)

The hormone hypothesis that had led Kühn and his working group into the borderland between genetics and biochemistry had to be buried there before the gates of biochemistry—aided and abetted, ironically, by the laboratory of Adolf Butenandt, one of the greatest hormone specialists of his day. No hormone was found, but instead a modified amino acid. From then on Kühn no longer spoke about "hormonal" effects; he used the term "humoral" effects instead. The fact that a biochemical approach had been grafted as it were onto the *Ephestia* experimental system brought a major shift of direction. A diagram taken from the extensive review paper that Kühn published in 1941 in *Nachrichten der Akademie der Wissenschaften in Göttingen* documents the change.[148] We see here (fig. 20) a new distinction between what Kühn now called a "gene-action chain" on the one hand and a "substrate chain" on the other.[149] What had earlier been conceived as a "reaction sequence" with "intermediaries"[150] was now represented as a two-dimensional network. Genes such as a[+] (*Ephestia*) or v[+] (*Drosophila*) were now thought to be responsible for "ferments" or ferment systems; these ferments intervened as enzymes in a series of metabolic re-

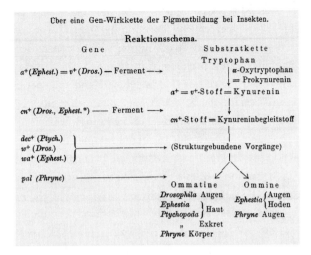

Über eine Gen-Wirkkette der Pigmentbildung bei Insekten.

Reaktionsschema.

Gene   Substratkette
                                        Tryptophan

a^+(Ephest.) $= v^+$(Dros.) — Ferment ——→ | α-Oxytryptophan
                                        | = Prokynurenin
                              $a^+ = v^+$-Stoff = Kynurenin

cn^+(Dros., Ephest.*) —— Ferment ——→
                              cn^+-Stoff = Kynureninbegleitstoff

dec^+ (Ptych.)
w^+ (Dros.)   ——————————————→ (Strukturgebundene Vorgänge)
wa^+ (Ephest.)

pal (Phryne) ———————————→
                              Ommatine   Ommine
                              Drosophila Augen   { Augen
                              Ephestia   }   Ephestia { Hoden
                              Ptychopoda } Haut   Phryne Augen
                                   „   Exkret
                              Phryne Körper

20. Reaction scheme showing the relationship between gene-action chains and substrate chains. Kühn 1941, 255.

actions leading from the amino acid tryptophan through kynurenin to the chromophores of the insect eye. Becker in 1939 had called them "ommines" and "ommatines."[151] The A-substance was no longer a "gene hormone" or "biocatalyst," but a "substrate [in the] process of pigment formation."[152]

On the other hand, the "primary reaction," which had so far gone un-explained—since 1935 it had been regarded as a mere supplement stand-ing in for some unknown relation between nucleus and cytoplasm—now turned out to be another reaction chain; as mentioned at the outset, this "primary reaction" would later come to be known as the "one gene—one enzyme" relation. As the diagram indicates, this reaction chain meshed with another: the substrate reaction chain, consisting of a great many pre-liminary stages occurring prior to the expression of the phenotypic char-acters. The interaction of the two chains also reflected the experimental integration of biochemistry and genetics: *biochemical genetics* had attained a molecular resolution of one of its dimensions, the substrate chain:

(1941)   substrate 1
        A-gene   →   ferment system 'A'   →   ↓
                                                substrate 2
        B-gene   →   ferment system 'B'   →   ↓
                                                   (+)
        (+)   (+)   →   ↓
                                                pigment

Kühn concluded his 1941 paper with the following words:

We stand on the verge of a vast research domain: exploring it will call for a great deal more biological-biochemical collaboration. Yet general features of the way heredity works have already swum into view. Our conception of the expression of hereditary traits is changing from one that is, in a certain sense, static and preformistic to one that is dynamic and epigenetic. The formal attribution of certain characters to individual genes situated at specific loci on the chromosomes makes only limited sense. Every step in the realization of characters is, so to speak, a knot in a network of reaction chains in which many different gene actions are brought together. One character appears to stand in a simple correlation to one gene only as long as the other genes of the same action chain and other action chains tied up in the same knot remain the same. Only a methodically conducted genetic, developmental physiological analysis of a large number of single mutations can gradually reveal the fabric of hereditary dispositions [*das Wirkgetriebe der Erbanlagen*].[153]

Thus Kühn now recognized that the assumption of a direct correlation between one gene and one character was invalid. There were numerous, even bifurcating physiological reactions, as well as genes that—to use the terminology of 1935—led by way of a "primary reaction" to ferments (enzymes). The "one gene—one enzyme" relation, although it is not explicitly mentioned in the text, is present in this scheme as an integral component of a network. Kühn clearly wanted to avoid highlighting one side of the picture to the detriment of the other; his overriding concern was to promote an understanding of the complexity of the situation as a whole. He did not claim that a particular gene led to one and only one particular ferment.[154] The important thing for Kühn was to recognize how genes influenced pigment formation in the insect eye through the mediation of ferments. "Eye pigments," he maintained even many years later, "lead us relatively close to the threshold of morphological differentiations."[155] The development of the organism remained his chief concern; biochemistry was a means of homing in on it.

The epistemological message of the 1941 paper is made explicit in a letter Kühn had written Henke three years earlier: "Johannsen and Bauer [*sic*][156] were quite right to stress again and again, in the first stages of gene research, that we know nothing about 'the gene,' that we know only about 'fundamental differences.' The reality known as a 'gene' obtains whenever this or that crossing produces this or that result. The concept of the gene has acquired the rich content it has today because a series of different 'experimental conceptual definitions' of that kind intersect in it."[157] These

few sentences encapsulate the basic features of an experimental epistemology at a far remove from both a reductionist and a holistic philosophy.[158] One might say that Kühn implicitly rejected holistic alternatives—
in favor of a proliferation of reductive procedures. He acknowledged the
necessity of operational definitions that remained closely bound up with
the experimental process. By themselves, however, such "experimental
conceptual definitions" have merely syntactic value: they allow us to draw
"basic distinctions," that is, to keep the experimental game of producing
differences going. All experimentation is grounded in the production of
reliable "basic differences." But how do they acquire "rich content"? They
do not do so because we add one more theoretical level to an existing hierarchy of generalizations. We may take Kühn to mean that the proliferation
of experimental conceptual definitions enables "overlaps" and reciprocal
interactions with already existing ones. These in turn generate scientific
content. For Kühn in sum a good experimental system produces differences that acquire meaning through overlap and juxtaposition. Thus the
generation of differences at the empirical level is ultimately reflected at the
level of the concepts that enter into the characterization of the scientific
object under investigation.

The War Years

Kühn was able to pursue his work during the Second World War, albeit
on a limited scale. International contacts virtually ceased to exist; foreign guests now came only rarely to Dahlem—Torbjörn Caspersson from
Stockholm, who visited Hans Bauer as late as March 1941, was one of the
few to pass by during this period.[159]

In a report about his research to the *Reichsforschungsrat* (RFR), which
was closely linked to the DFG at the organizational level, Kühn described
the tasks before him in 1941. He wanted to discover "whether the active
substances emitted by implants, or the substance that has been found to be
active, kynurenin, act as catalysts or as raw material for the eye pigments
and body pigments that are formed."[160] In this connection, he emphasized
the unique opportunity "to make further progress toward resolving questions of gene action, the cutting-edge field of research on heredity today.
This opportunity must be explored further with the greatest possible investment in personnel, applying all the methods that we have worked
out, so that we can maintain the leading position that we now hold in
this field of research. Our responsibility to do so is all the greater in that
my Institute is the only place in Germany in which work of this kind is

going forward."[161] In response, Kühn received another 4,000 Reichsmarks for "investigations into the mode of action of hereditary traits, especially gene action."[162] Yet he was unable to keep his coworker Becker from being drafted into the army on June 15, 1941,[163] after Weidel, who had been called up for military service in December 1940.[164] Becker's experiments had tended to show that kynurenin acted not as a biocatalyst, but as raw material in the pigment production chain. He died in the war before the publication announcing this finding saw the light.[165] His death was a "heavy blow," as Kühn said in a January 1942 report to the RFR. His laboratory was not to recover from it to the end of the war, despite a renewal of its annual financial grants (8,000 Reichsmarks for 1942,[166] 7,000 Reichsmarks for 1943–44,[167] and 6,680 Reichsmarks for 1944–45[168]).

Approaching his sixties, Kühn had only a few coworkers left; most of the others had been conscripted. He continued his work with Victor Schwartz. Their main objective now was to establish the gene sequence involved in pigment formation and the integration into the "gene-action chain" of the new mutations that played a role in the same process. In 1941 the sequence was thought to be a, cn, dec; the wa mutation was subsequently included as well.[169] Kühn and Schwartz pursued these investigations in 1942 and 1943. They aimed at finding out whether dec and wa intervened in the substrate chain directly through their ferments or enzymes, or indirectly by way of changes in the pigment's protein carrier; they also wanted to define the action of various modifier genes that were not directly linked to the gene-action chain.[170] In August 1943 Kühn moved his section of the institute from Berlin-Dahlem, which was exposed to bombardment, to an old mill in Hechingen near Tübingen in Württemberg. "This has not caused any interruption in our work," he informed the DFG in April 1944.[171] As an additional precaution, he managed to find a haven for "the most important *Ephestia* strains in Henke's institute in Göttingen and with [Georg] Gottschewski in Vienna."[172]

In his report for 1943, Kühn informed the DFG about newly discovered "antagonistic gene actions"; "further exploration" in this direction, he said, would take place the following year, since antagonisms of this kind opened up "a new and promising field in the effort to penetrate gene physiology." He also planned to carry out phylogenetic comparisons: "Comparison of the fabric of hereditary dispositions in different species should provide insight into the transpositions that must take place in the processes of species change." The biological investigations required were to be pursued in the "fully operational laboratories" in Hechingen, whereas the "specifically chemical aspects of the investigation" would continue to fall within the purview of the KWI for Biochemistry in Dahlem.[173]

The international isolation of German science, which had commenced with travel restrictions imposed by the Nazis and culminated in the wartime embargo, was taking its toll. What Kühn referred to as late as 1942 as the "lead we hold over the United States and Japan, where, setting out from our results, a great deal of money is being spent in pursuit of the same research goals" was rapidly melting away.[174] At Stanford, Beadle and Tatum had introduced a new model organism, the filamentous fungus *Neurospora crassa*; with their experiments on its metabolically defective mutants, they had established a comprehensive biochemical genetics by war's end.[175] From then on, the *Ephestia* system was of merely historical interest.

Late in 1944 Kühn confessed to the botanist Otto Renner in Jena: "Work of the most intensive kind is the only thing that allows me to bear times like these, which we can do nothing at all to influence. Work distracts me completely for hours and even days at a time."[176] He withdrew from the leadership of the virus research group that he had helped found in 1937.[177] He was "sad and almost despondent," as he said in his letter to Renner, "over the deaths of so many young people." Shortly after the war, Kühn remarked in a letter to his former collaborator Rolf Danneel, who had moved to Göttingen in the last months of the war: "It is only by working that one overcomes one's sorrow over everything that madness has wrought."[178]

Rather than examine Kühn's work between 1941 and 1945 step by step here, I shall quote a fairly lengthy summary of it written by Kühn himself. It is contained in a late 1946 letter addressed to the London zoologist Vincent B. Wigglesworth, a friend of Kühn's who had last visited his Dahlem laboratory from January to March 1939. In 1946 international contacts could be resumed. Wigglesworth was, with Caspari and Goldschmidt, one of the first foreign colleagues with whom Kühn renewed relations after the war.

> Let me tell you just a little about what we worked on. Our investigations of gene action in the pigment formation process were continued. Becker was able to biochemically characterize the eye pigments and skin and excretion pigments with still greater precision. In the eye, the derivatives of kynurenin are built up into compounds that have a higher molecular weight than do those in the skin or excretions. In the eye [of *Ephestia* and *Ptychopoda*], the pigment precursor is bound to protein. I found a mutant 'weißäugig' [white-eyed variant] of *Ephestia* that produces the precursor pigment and diffuses it as an implant, but is itself incapable of forming pigment. The

reason is that the mutation "wa" inhibits the formation of protein granules, as was confirmed by histogenetic investigations. In *Ptychopoda*, we observed a few interesting pigmentation mutations that influence the formation of melanine and also, to a certain extent, Becker's "ommochrome pigments." Because the reciprocal transplantation of organs, including eye discs, is possible between *Ephestia* and *Ptychopoda*, the effects they have can be nicely studied. By injecting kynurenin, I could also induce the predetermination of the pigmentation in the caterpillar, which it had earlier been possible to do by implanting a⁺-tissue into an a-mother. By means of extraction from meat fly cocoons and, recently, *Bombyx mori* cocoons, it has proven possible to greatly enrich the pupation hormone. In collaboration with Prof. Butenandt's KWI for Biochemistry, which is also located in Tübingen, we are now trying to isolate this hormone and test its effects in different insects. Obviously, a mixture of several substances is responsible for the different stages of metamorphosis; the same holds for pupation. With *Ephestia* and *Ptychopoda*, I have progressed in the analysis of pattern determination by different mutations that can be phenocopied by means of modifications. While still in Dahlem, I examined, with the help of an electron microscope from Siemens, the fine structure of the scales of *Ephestia* and *Ptychopoda*, and the changes produced in them by various mutations. Unfortunately, we have no electron microscope in Tübingen.[179]

As this resume shows, Kühn had expanded his experiments in different directions. They included the characterization of new intermediate stages in the pigment substrate chain, particularly the "wa" mutation, which brought the biochemistry of pigmentation into relation with the morphology of eye pigments—work on which Victor Schwartz collaborated;[180] the characterization of new mutants;[181] further analysis of wing-pattern formation;[182] the ultra-structural representation of wing scales;[183] investigation of hormone action in metamorphosis[184] as well as the introduction of a new model organism, *Ptychopoda seriata*, which was now systematically compared with *Ephestia*.[185] It had cost Kühn and his technician von Engelhardt almost a decade of work to transform *Ptychopoda* into a laboratory animal. Model organisms are not to be found ready-made in nature; their conditioning takes effort. On this subject, Caspari remarked in a letter: "I am glad to see that *Ptychopoda*, too, has developed into such a good laboratory animal. I can still remember the doubts that we doctoral students had when you began working with it in 1932, because the animal needs so much care and such carefully regulated humidity."[186] Kühn replied: "*Ptychopoda* has really become a very good experimental animal.

Cultivation methods have now been so well worked out that it takes little effort to raise it. . . . The fact that it is possible to carry out xenotransplantations between the two rather widely separated genera [*Ephestia* and *Ptychopoda*] enriches the exploitation of mutations with a humoral effect. . . . The short generation time, six weeks, is also a plus."[187]

All this work, however, merely consolidated past achievements without going beyond them or taking them in new directions. Unable to cultivate international contacts, cut off from discussion and competition, and deprived of his best coworkers, Kühn had for years been engaged in a desperate effort to continue the kind of pioneering research that had characterized an earlier stage of his career. Yet he failed to arrive at new, far-reaching "experimental conceptual definitions." His wing studies moved closer to the center of his concerns again; this will surprise us less if we recall his long-standing interest in the laws governing the development of the butterfly wing and the experimental experience he had accumulated as a result of it. Indeed it would appear that he was never really interested in genes as entities, that is, in their structure and expression per se. His interest bore rather on the *effect* genes have on developmental processes; for this reason, he was also loath to adopt the straightforward strategies of nascent molecular biology with its predilection for "simpler" model organisms such as molds, bacteria, and phages as well as simple model reactions. It should be added that Kühn lacked regular access to the international literature throughout the 1940s. He also failed to switch to *Drosophila*, although his coworker Rolf Danneel did carry out a few comparative in vitro experiments on *Drosophila* eye pigmentation in the early 1940s.[188] Even biochemistry remained in a certain sense a foreign body in Kühn's developmental physiological genetics. He never believed that a consistent in vitro biochemistry based on fractions of homogenized organisms or organs would prove to be of great use. For him the intact organism was and remained the "laboratory for specific chemical events."[189] When Caspari wrote him in 1946 about his observation that an *Ephestia* homogenate would produce pigment in the presence of tryptophan, he replied, "I have thought a lot more about your observation that an *Ephestia* a+- or a-homogenate forms pigment from tryptophan. Can we be sure that the pigment is an ommochrome, specifically, skotommin? I find it hard to believe that the whole synthesis—which depends in vivo upon the many different links of a gene action chain, and which not only polymerizes kynurenin, but also utilizes completely different substances and ultimately takes place on protein granules, to which it is bound—can occur in a tissue homogenate."[190] Homogenates were precisely the kind of experimental substrates being opened up for research in the American context in which Caspari now found himself. Albert Claude and his coworkers at the Rockefeller Institute in New York had been doing pioneering work in the field of tissue homogenization since the late 1930s. In vitro systems gained the upper hand after the massive introduction into biochemistry of radioactive phosphorus, sulfur, carbon, and hydrogen as tracer elements immediately after the Second World War.[191]

Conclusion

In order to present a manageable picture of the main lines of the *Ephestia* experimental system as it developed from 1924 to 1945, I have deliberately concentrated on Kühn's eye-pigmentation experiments and the role they played in his research on developmental physiological genetics. This choice has undoubtedly led to a certain one-sidedness in my historical reconstruction. Thus for example a proper description of Kühn's work on the effects of hormones on the metamorphosis of insects is missing from my account.[192] Moreover, my presentation of the development of Kühn's system may well appear all too linear. Like most histories of experimental systems, the history of Kühn's experiments is replete with blind alleys, bifurcations, and unexpected turns. Kühn himself once remarked that the *Ephestia* system had led him "over the course of the years, into problem areas completely different" from those his "original question" would have led one to think.[193] He had set out to understand the genetic and physiological secrets of pattern formation in butterfly wings. Instead of resolving this question—which he never completely lost from view, even if he pursued it only tangentially for years at a time—he was unexpectedly led in his search for useful mutations of *Ephestia* onto the terrain of pigment formation. The experimental techniques introduced by his collaborator Caspari to trace the diffusible substances involved in this process brought him to yet another field of investigation: the effects of hormones on insect metamorphosis. But if Kühn, after a number of years, had to drop the hypothesis that hormones were involved in pigment formation, the further investigation of metamorphosis—which was carried out by Butenandt as well as his doctoral student and later coworker Peter Karlson—led to remarkable results in the field of hormone biochemistry. The eye-pigment system turned out to be simple and complex at the same time: simple enough to be further explored in both genetic and physiological experiments, yet complex enough to display at least some of the nonlinear characteristics that Kühn suspected went into the making of *das Wirkgetriebe der Erbanlagen*.

With its thirty chromosomes and relatively long generation period of three and a half months, *Ephestia* remained a limited system, useful only in resolving the questions confronting a genetically based developmental physiology of the kind that Kühn investigated in Göttingen and later Berlin-Dahlem. The flour-moth did not lend itself to the sort of genetic mapping as *Drosophila*, with its small set of chromosomes and a generation period of just two weeks. In contrast to *Ephestia*, *Drosophila* could serve as the basis for an extensive system of genetic knowledge production, as Robert Kohler has shown in *Lords of the Fly*.[194] The fruit-fly became a "generic" organism for classical genetics.[195] But this does not mean that it represented the organism of choice for the molecular genetics that emerged in the late 1940s. For a period of thirty years the major experimental systems of the burgeoning field of molecular biology were based on bacteria, phage, and viruses. But *Drosophila* has made a comeback: in recent years, it has been used to systematically study, on a molecular basis, the regulatory processes of embryonic development. Not so *Ephestia*: after making its entrance onto the experimental stage as a model organism in the 1920s and then acting as an ideal experimental animal in the pioneer period of physiological genetics and developmental genetics, it became obsolete in the 1950s. As a model organism it did not survive this one very narrowly circumscribed historical constellation.

Yet Kühn remained faithful to his flour-moth, to the style of investigation of complex gene action chains that he had developed, and even to the terminology in which he formulated it.[196] One of his last papers on genetically determined mosaic formations in *Ephestia*, a problem with which he had wrestled conceptually and experimentally since the early 1940s, bears witness to this fidelity.[197] At the same time we find Kühn around 1960 at the age of 75 taking notes on cybernetics as he prepared the 15th edition of his *Allgemeine Zoologie*.[198] Cybernetic feedback circuits harmonized better with his conception of gene action chains than did the molecular details of pure gene expression. That both options would only a year later enter into extremely close combination in François Jacob and Jacques Monod's operon model of gene regulation was a development that Kühn had certainly not anticipated.[199]

Tobacco Mosaic Virus

Virus Research at the Kaiser Wilhelm Institutes

for Biochemistry and Biology, 1937–45

In the 1950s the Max Planck Institute for Virus Research and the Max Planck Institute for Biology in Tübingen were driving forces behind the development of molecular biology in Germany.[1] The main lines of the founding myth of the Institute for Virus Research were laid down by Adolf Butenandt. Butenandt often told the story about how the news of Wendell Stanley's success in crystallizing tobacco mosaic virus (TMV) reached him at the memorable annual meeting of the *Gesellschaft Deutscher Naturforscher und Ärzte* (Society of German Natural Scientists and Physicians) in September 1936 and immediately convinced him that virus research had become a vital necessity. "The announcement of Stanley's findings," Butenandt wrote, "made a deep impression on the conference participants. His results were discussed again and again at the conference. I, too, was among those who heard the announcement there, and, to this day, I have not forgotten the feelings that came over me. I never for a moment doubted that Stanley's findings had opened up an utterly new field of biochemistry."[2]

Points of Departure

On October 31, 1936, a year after declining the offer of a professorial post at Harvard, Butenandt moved with his staff from the Technical University in Danzig to the Kaiser Wilhelm Institute (KWI) for Biochemistry in Berlin-Dahlem.[3] In his first report to the institute's board of trustees in April 1937 he not only explained his scientific research plans, but also explicitly declared his "basic willingness . . . to see to it that, in addition to work of a purely biochemical nature, experiments are also carried out in the framework of the four-year plan."[4]

Butenandt was Carl Neuberg's successor as director of the institute; Neuberg, of Jewish descent, had been dismissed from his post in 1934, but continued to conduct the institute's affairs as acting director until Butenandt took over. A memorandum by Friedrich Glum, secretary general of the Kaiser Wilhelm Society for the Promotion of the Sciences (*Kaiser-*

Wilhelm-Gesellschaft, KWG), about a May 1936 conversation involving himself, Butenandt, Rudolf Mentzel (then Reichsminister Bernhard Rust's commissary on the Verwaltungsausschuss of the KWG), and Ernst Telschow of the General Administration, reads, "Prof. Butenandt has requested that Prof. Neuberg maintain no further relations with the Institute. This is the wish of the Ministry as well. Prof. Neuberg will accordingly have to be informed that he cannot be accorded special working privileges in the Institute or on its premises."[5] All Butenandt allowed his predecessor to keep was equipment and chemicals from the institute's stock; Neuberg initially used this material to set up a private laboratory for himself in a foodstuff factory in Steglitz.[6] In 1939, he emigrated by way of Holland, Palestine, Iraq, Iran, India, the Dutch East Indies, New Guinea, and Hawaii to the mainland United States, where he ultimately earned his livelihood as an industrial consultant.[7]

As we saw in chapter six the zoologist Alfred Kühn became one of the directors of the neighboring KWI for Biology in spring 1937. He replaced Richard Goldschmidt, who, of Jewish descent as well, had been forced to retire and had also emigrated to the United States. Kühn, be it recalled, had managed to obtain his transfer to Berlin thanks in part to a massive lobbying effort by his former Göttingen colleague, the botanist Fritz von Wettstein, who for his part had succeeded Carl Correns as director of the institute in Dahlem in 1934, a year after Correns's death. Georg Melchers, who had been a doctoral student of von Wettstein's when the latter was in Göttingen, then a coworker of his in Munich, and finally one of his first assistants in Dahlem remembers, "I started out as a coworker of Fritz von Wettstein's in 1934. I arrived in the phase in which the most interesting question was how one could defend oneself against the Nazis. Those of us in [the Institute for] Biology succeeded marvelously well at this, for Wettstein happened to be an Austrian."[8] In conducting his political negotiations in science policy circles, Wettstein had recourse to Aesopian arguments, as is attested by an exemplary passage from a letter in which he pleaded for Kühn's transfer to Dahlem:

> Essential aspects of the life of our state are now being constructed on the basis of the science of heredity, which is being carried down to the deepest levels of the population. But this will continue to be the case only if we remain in the forefront of the science of heredity. We Germans [wrote the Austrian von Wettstein in 1936] built this science. Today, the Americans are incontestably in the lead, and we have even enhanced their superiority by handing over top-notch people such as Goldschmidt or Stern. The task

of winning our old position back is an immensely difficult one. Hence the only man who can make a decisive contribution here must be put in the right position.[9]

The tone of Wettstein's obituary of Carl Correns was less ambiguous. In a passage with no relevance to the topic whatsoever he spoke about "the fresh courage with which, today, an effort is being made to bring out, to the full, the best of the people's genetic heritage . . . wage a titanic struggle to stem the ongoing, shocking loss of valuable patrimony, a danger constantly confronting civilized peoples . . . [and] mobilize all available means to avoid an accumulation of unfavorable genetic material in a people— indeed, push it back as far as possible."[10]

Wettstein and Kühn had been friends since they had jointly introduced courses on General Biology into the curriculum at the University of Göttingen. Kühn's acquaintance with Butenandt dated back to Butenandt's time in Adolf Windaus's laboratory in Göttingen. Since 1934 the Rockefeller Foundation had been funding the two scientists' research on the physiological genetics of eye pigmentation in the flour-moth *Ephestia kühniella*.[11]

Virus research in the KWG has been described on several occasions.[12] However, my aim in this chapter is not primarily to discuss, using virus research as my example, whether it was possible to conduct relatively "neutral" basic research under National Socialism in Germany and in particular in the KWG.[13] Rather, I would like to reconstruct how the leading actors in the KW Institutes of Biology and Biochemistry, in establishing the new research field, arrived at and implemented specific science policy decisions determining the scientific strategies, the power relations, and the networks internal and external to the institutes involved in this program.[14] One of my theses is that in the case of virus research a versatile research instrument, the ultracentrifuge, significantly contributed to binding disparate lines of research and research activities in an alliance that extended into the military-industrial domain. In the latter connection I shall also look at the career of Gernot Bergold, a member of the virus group who worked on the premises of the I.G. Farben plant in Oppau between 1940 and 1943.[15]

The Informal "Arbeitsgemeinschaft Virusforschung"

In summer 1937 the three newly appointed institute directors Kühn, von Wettstein, and Butenandt agreed to found a "working group for virus research"—the *Arbeitsgemeinschaft Virusforschung*. Their long-term goal was to create a new KWI for Virus Research. As a preliminary step each director would give one of his coworkers—a biochemist, a botanist, and a zoologist—a chance to become acquainted with the new field. Heinrich Hörlein, head of I.G. Farben's factories in Elberfeld, hailed this initiative, declaring that he was prepared to help launch the effort with substantial funds from private industry. The 1938 budget of the KWI for Biology shows that the institute received 76,300 Reichsmark (RM) from I.G. Farben that year.[16] Under prevailing conditions, this was a sizable sum, roughly equivalent to the annual budget of the director of an institute.[17] Butenandt also managed to bring in funds from Schering AG[18] in the amount of 65,000 RM for 1938.[19] For special projects unrelated to virus research, he had additional funds from private industry at his disposal; this money, however, was not posted in the institute's budget and was kept in separate accounts that Butenandt refused to allow the KWG's auditors to audit.[20] He insisted on his right to manage these accounts as only he and the donors saw fit. An American visitor, Thorfin R. Hogness of the University of Chicago, reported in 1937 to the Rockefeller Foundation that Butenandt, known for his work on steroid hormones, was also working on other matters that he did not wish to divulge.[21]

Two windowless, air-conditioned chambers were installed in the basement of the KWI for Biochemistry for virus research purposes. Their temperature was held constant and could be brought down to 0°C. They were to accommodate an analytical ultracentrifuge (60,000 rpm), a preparative ultracentrifuge (40,000 rpm), and a Tiselius electrophoresis apparatus. Butenandt reported to the institute's trustees in spring 1939 that "when construction of the new laboratories is completed in the near future, the Institute will doubtless rank among the best-equipped biochemical research facilities in Germany and the world."[22] The KWI for Biology also acquired new facilities: construction of a greenhouse for virus-infected plants began in spring 1938. On April 11 Kühn asked for an allocation of more than ten tons of pig-iron, and on April 12 he wrote to the Reichsminister for Food and Agriculture, "Extensive private funding has been put at the disposal of the Institutes [of Biology and Biochemistry] to enable them to carry out these experiments. To make a successful start on the work on plant-virus diseases, an experimental greenhouse will have to be built, among other things." He did not miss the opportunity to add that "the work . . . is of interest to German agricultural science and, no less, to the German chemical industry." It was all the more desirable in that "abroad, especially in America, very useful results have been attained as a result of cooperation between biologists and biochemists."[23]

In von Wettstein's department Georg Melchers was to be put in charge of virus research. He found this assignment "very daring," since he had hitherto been working on the induction of flower formation in higher plants,[24] and on his own self-assessment "had not the faintest notion and absolutely no specialized knowledge of the field." He "could have gone to America with a Rockefeller fellowship, but said at the time: 'one cannot do that—work in a field that one doesn't really know at all.'"[25] Melchers decided to stay in Dahlem and teach himself about working with tobacco mosaic virus and similar viruses; to that end he wrote Stanley to ask him to send isolated TMV and a few experimental plants.[26] At the same time he continued to pursue his experiments on flower induction hormones with his coworkers Anton Lang and Hedwig Claes.[27]

In the first years following the National Socialist assumption of power, the Rockefeller Foundation continued to support projects and award stipends in Germany. Wilbur E. Tisdale wrote the director in charge of the foundation's section for natural science, Warren Weaver, in August 1934, "G[oldschmidt] posed the question by asking whether we were still willing to consider German fellows after the Nazi hiatus. I assured him that dependable candidates, especially in the field of genetics, would certainly be given every consideration that they had been given heretofore."[28] Butenandt as we have seen was granted Rockefeller Foundation funds for research in 1934 and in 1935 a traveling fellowship that enabled him to familiarize himself with many different American universities.[29] Between October 1935 and December 1936 Hans Bauer, who belonged to the section of the KWI for Biology headed by the protozoologist Max Hartmann, spent time in Thomas Hunt Morgan's lab at the California Institute of Technology, at Woods Hole, and in Cold Spring Harbor with Milislav Demerec.[30] In 1936–37, Georg Gottschewski of Goldschmidt's department traveled to the United States on a Rockefeller fellowship.[31] Hans Gaffron, who had lost his post as an assistant when Carl Neuberg was forced to retire and then been given a position by von Wettstein in his section,[32] visited the Hopkins Marine Station beginning in April 1938 and decided to remain in the United States.[33] As late as January 1939 Butenandt's assistant Ulrich Westphal went to Columbia University on a foundation fellowship, returning to Germany only after the outbreak of the Second World War.[34] In New York he had been able to procure seeds for Melchers's TMV work,[35] but the extended stay that Butenandt tried to arrange for him in August 1939 with Stanley at Princeton finally was cancelled.[36]

In October 1939 Rolf Danneel came from the Zoological Institute of the University of Königsberg to work as a new member of Kühn's department;[37] at the same time, the *Drosophila* specialist Georg Gottschewski went from Dahlem to Königsberg. The exchange took place in the framework of attempts to forge closer relations between the KW Institutes and German universities. In Königsberg, Danneel had worked on hair pigments in rabbits since the early 1930s;[38] his interest in the genetics and physiology of pigmentation processes coincided with one of Kühn's main concerns at the time. However, he joined the virus group only in 1941 and even in this new research context stuck with his old experimental animal. In consultation with Butenandt he decided to investigate a virus that causes papilloma. Danneel thus found himself working in the border region between virus research and cancer research.

Gerhard Schramm, one of a number of Butenandt's coworkers who came with him from Danzig to Berlin, had previously worked on the enzymatic modification of steroid hormones and the importance of cholesterol for the resorption and transport of fatty acids. Butenandt put him in charge of the biochemical aspects of their TMV research. Schramm therefore went first for a few months to Uppsala to learn advanced techniques of analytical ultracentrifugation and electrophoresis from Theodor Svedberg and Arne Tiselius. After returning to Dahlem he worked with the engineers of the *Physikalische Werkstätten* (Phywe) in Göttingen to develop an air-driven ultracentrifuge "in order to make such analyses possible in Germany as well."[39] Despite his work with Svedberg, then, Schramm opted for the American model of the air-driven high-speed centrifuge of Jesse Beams and Edward Pickels in place of Svedberg's costly instrument with its oil turbine-drive.[40] The electrophoresis apparatus was built by F. Hellige & Co. in Freiburg on Tiselius's specifications.[41]

The core group also included Schramm's coworker Hans Müller and Melchers's coworker Anton Lang. Among the broader network of scientists involved in the virus research project were Heinz Dannenberg of the KWI for Biochemistry, the biologist Hans Friedrich-Freksa, who had been working with Butenandt since the summer of 1937, Hans Joachim Turnit, and Manfred von Ardenne, whose "supermicroscope" (*Übermikroskop*) in Lichterfelde-Ost was also used to perform joint electron microscopy. Although the members of the research team worked in different laboratories and facilities, a multifaceted form of collaboration quickly sprang up. By 1940 it had resulted in a first wave of publications.[42]

Just as the interdisciplinary make-up of the virus group was the product of deliberate choice, so its research program too was oriented from the beginning toward overarching biological questions. Schramm and Müller opened their 1940 paper on the chemistry of TMV with the words: "The aim of chemical virus research is to throw as much light as possible on the composition and structure of viruses in order to obtain insight into the decisive process of their reproduction."[43] Virus reproduction in turn was seen as a model of the multiplication of genetic material. Melchers's primary interest was in isolating and characterizing mutations as analytical tools, using methods analogous to those experimental geneticists applied to higher organisms. The whole research team came together around a common model "organism," TMV and its variants.[44] In the span of two years it developed an array of complementary biochemical, biophysical, and biological procedures and techniques, at the center of which stood analytic and preparative ultracentrifugation.

According to an undated memorandum that was most probably written before the summer of 1940, von Wettstein had the hypothetical sum of 300,000 RM at his disposal, which he intended to invest in an Institute for the Collection and Study of Wild and Primitive Forms of Cultivated Plants, already approved by the KWG Senate, and in expanding the Virus Research Facility. He also wanted to increase the budget of the KW Institutes for Biology and Biochemistry "for the particular promotion of research on active components from a genetic and developmental-physiological standpoint in combination with research on biochemical problems." In this context, von Wettstein stated, "Virus research is a field that, especially in the United States, has been magnificently developed in recent years as pure and applied science. Giving special attention to promoting it seems to us, from a purely scientific standpoint, to be one of our most urgent tasks."[45] In the negotiations over expanding virus research the lead held by the United States was stressed again and again. The word "virus" soon came to stand as a general designation for state-of-the-art biological-chemical research. It signified a mysterious molecular entity lying somewhere on the dividing-line between the animate and the inanimate, an entity that only purposeful, interdisciplinary collaboration could grasp. "To laymen," Melchers recalled looking back, the term "virus" sounded "mysterious and dangerous" in this period.[46]

The Foundation of the "Arbeitsstätte Virusforschung"

In July 1940 von Wettstein, Kühn, and Butenandt sent the General Administration, that is, Ernst Telschow, a carefully formulated "Proposal for the Creation of a Virus Research Subdivision of the Kaiser Wilhelm Institutes for Biochemistry and Biology." "In the investigation of viruses," the three directors wrote, "the U.S.A., especially the research team in Princeton, has acquired a considerable lead. For three years now, England has been successfully striving to catch up. It therefore seems to be a most urgent, important task to create, in the German Reich as well, a research facility in which this field of research can be cultivated with an investment of energy and means that will allow us once again to attain to a position at the very forefront of research."[47] Von Wettstein, Kühn, and Butenandt highlighted their own efforts of the past few years to create the experimental conditions for accomplishing such a mission, among them the creation of operational teams or "germ cells." The agenda was therefore clearly set:

> A research facility of this sort must meet the following requirements. Virus research is a perfect example of modern research in a research group. Only through the right kind of cooperation among biochemists, botanists, and zoologists can this work be effectively promoted. Such a research group must be in a position to carry out research with the most modern equipment. Ultracentrifuges, electron microscopy, and the most modern experimental facilities are indispensable prerequisites. Virus research takes place in the border area between biology and biochemistry. Hence it can only be developed as a result of, and through extremely close cooperation between, biological and biochemical institutes. —The most crucial research must still be carried out in the domain of basic research, but, from the outset, fine-tuned cooperation and direct translation into practice are of paramount importance, which means that links must be forged with the major breeding institutes.[48]

Let us sum up these arguments. First, interdisciplinary teamwork and the newest research technology were indispensable to the success of modern virus research (this was Rockefeller Foundation philosophy at its purest). Second, only the initiators of virus research, that is, Butenandt, von Wettstein, and Kühn, could ensure that such teamwork would be carried out properly; they should therefore be provided with the requisite means. Third, while the need for practical applications was clearly stated, the petitioners wanted a free hand in this respect.

General Secretary Telschow immediately took the necessary political steps. On July 15, 1940, he sent the proposal to State Secretary Herbert Backe of the Reichsministerium for Food and Agriculture; Backe was also a Senator of the KWG and was soon to become its first Vice President (1941–45). Telschow pointed out in his cover letter that although "there [were] as yet no plans to construct a particularly big research institute with administrative offices, etc.," the KWG intended to solicit funds from private industry for a future building. However, the running expenses of approximately 200,000 RM a year for four sections should be covered by the Ministry for Food and Agriculture. To punch up his initiative, Telschow played on the rivalry among the different ministries for control over research: "With the construction and maintenance of the Institute for Breeding Research in Müncheberg, the Institute for Animal Breeding Research in Dummerstorf, and the Institute for Virus Research, the Ministry for Food and Agriculture would have, directly or indirectly, the leading role in the three most important areas of nutritional policy [*Volksernährung*], and would consequently be in a position to dominate and promote a vital sector of German research."[49]

Four months later, on November 15, 1940, Telschow addressed a letter to the General Director of the Deutsche Industriebank, Wilhelm Bötzkes. He reminded Bötzkes of earlier negotiations between the Industriebank, the late President of the KWG, Carl Bosch, and the President of I.G. Farben's Board of Directors, Professor Carl Krauch, who had been the First Secretary (*Schriftführer*) of the KWG since 1937 and was also the head of I.G. Farben's military liaison office in Berlin. Telschow also reported the positive negotiations with the Ministry for Food and Agriculture over the provision of funds for running costs and asked Bötzkes to donate 1.5 to 2 million RM toward the future construction of a new building for the institute. "Confidentially," he added that he planned to propose to the KWG Senate to name the new institute "The Carl Bosch Institute."[50] After receiving a further inquiry from Telschow, Bötzkes replied on January 15, 1941, that the Deutsche Industriebank was prepared to make a one-time donation of half a million RM toward construction of an Institute for Virus Research.[51]

With these accords between the General Administration of the KWG, the Industriebank, and the government of the Reich, the next steps to be taken were set in advance. Wettstein quickly proceeded to make a preliminary estimate of costs for 1941, arriving at a figure of 140,000 RM for the three sections headed by Schramm, Melchers, and Danneel. He noted that they could "make a good start . . . with the amount initially requested," but made sure to add that it was "of course to be hoped that reinforcements would arrive later, since the range of tasks to be accomplished [was] a broad one."[52] Similarly Butenandt spoke of the "staggering breadth" of the new research field, notably in the commemorative speech he delivered on January 23, 1941, before the Prussian Academy of Science, "Die biologische Chemie im Dienste der Volksgesundheit" (Biological chemistry in the service of the people's health).[53] Wettstein also envisioned two possible construction sites for the planned expansion, both of them on the Dahlem campus.

On April 1, 1941, the existing working groups were united under the name "Arbeitsstätte Virusforschung der Kaiser-Wilhelm-Institute für Biochemie und Biologie." "Guidelines" for the implementation and operation of the new unit were issued in June 1941.[54] On July 31 at its seventieth session the KWG Senate voted unanimously to create the new research facility, approving a projected budget of 250,000 RM and projected construction costs of 800,000 RM for a new building. Addressing the Senate von Wettstein stressed the agricultural and medical importance of the new facility, declaring that Germany had now caught up with the United States in this field of research.[55]

Before the end of 1941 conversations and negotiations on the establishment of a fourth, entomological section of the research facility had been opened, but for various reasons they proved difficult. The planned head of the new section was Gernot Bergold, who had studied biology in Vienna and traveled throughout Europe on behalf of the European Parasite Laboratory of the U.S. Department of Agriculture and the British Imperial Institute of Entomology between 1931 and 1938: his mission had been to identify parasites of insect pests that had been carried from Europe to the United States and might be subject to biological pest control. In 1939 he was one of the recipients of a Deutsche Forschungsgemeinschaft (DFG) fellowship that enabled him and other scientists, among them the animal virologist Werner Schäfer of Gießen, to study coffee plant parasites on Heinrich Bueb's plantation in Tanganyika. When the Second World War commenced the group of scientists was interned by the British and finally sent back to Germany in January 1940. In Berlin-Dahlem Bergold met with Heinrich Bueb's father Julius, a former member of I.G. Farben's Board of Directors. Julius Bueb arranged for Bergold to be introduced to Telschow and Krauch. Krauch, Bergold recalled later, "asked several very reasonable questions about fighting the potato beetle and about [my] experiences from the northern tip of Finland to the southern Tirol, and from Holland to the Carpathian Mountains in Hungary."[56] Krauch and Telschow in turn smoothed Bergold's way to Kühn and von Wettstein and thus to the KWI for Biology.[57] Shortly thereafter in April 1940 Bergold received an offer to set up a KWI satellite in the I.G. Farben biology laboratory in Oppau near Ludwigshafen. In Bergold's estimation "the only reason I was given the job . . . was that the higher circles were unanimously of the opinion that Germany could not win the war without potatoes."[58] In his capacity as head of this satellite Bergold was paid from funds supplied by private industry: from 1940 to 1942 I.G. Farben donated an annual 40,000 RM for this purpose to the KWI for Biology.[59] State Secretary Backe was visibly unhappy over this arrangement.[60] The negotiations involving von Wettstein, Telschow, and the Ministry for Food and Agriculture regarding the new entomological section dragged on. The Ministry finally allocated funds for the proposed fourth working group, although the decision to finance it took effect only in April 1943.[61] Formal approval of a fourth, entomological section of the Facility for Virus Research was, however, withheld.

On May 9, 1944, Telschow noted, "The R.E.M. [the Reichsministerium for Food and Agriculture] holds the view that there is no justification for creating a separate section for entomology in the framework of virus research. Insect research, it maintains, falls under zoology. The Ministry also wants to collect information on Dr. Bergold's scientific qualifications. I informed the R.E.M. that I could understand the argument for endowing virus research with a tripartite structure—botany, zoology, and biochemistry."[62] Under these circumstances, Butenandt decided in August 1944 that the attempt to create a fourth section should be abandoned and that Bergold should be brought into the zoological section on an equal footing with Danneel.[63]

Already in a March 21, 1941, memorandum von Wettstein had diplomatically summed up the situation: Bergold's working group was "at present so productively integrated into [the laboratory in] Oppau" that Wettstein and his colleagues "did not at present wish to pull the group out of it."[64] In Oppau Bergold enjoyed the support of Carl Krauch and from I.G. Farben's Board Martin Müller-Cunradi and Otto Ambros.[65] But what kind of "productive integration" into the lab in Oppau had Bergold arranged for himself? His long-term interest was in the polyhedral insect disease, which was caused by a virus; as mentioned above, he had studied this virus and its possible use in biological pest control on behalf of the U.S. Department of Agriculture. With the help of Josef Hengstenberg of I.G. Farben's physics laboratories in Ludwigshafen[66] he adapted another air-driven Phywe ultracentrifuge in addition to Schramm's for his experiments.[67] In collaboration with Schramm in Dahlem and the physicist Rudolf Brill from I.G. Farben's research lab in Oppau he attempted to characterize insect viruses with additional methods based on chemistry and physics,[68] and he constructed superpolyamide tubes capable of withstanding centrifugal forces of up to 150,000 × g.[69] Bergold also developed a micro-syringe capable of injecting as little as 0.1 microliters of a test solution into his insects.[70] The overall relationship between Dahlem and Oppau was a good one, as he remembered it later: "I was in constant contact with Butenandt. He knew exactly what was going on."[71]

In Oppau Bergold's most important wartime task was initially to determine whether the polyhedral virus could be used as a biological weapon against the potato beetle, which at the time had started to invade the country from across the Rhine. When the assays he made in this connection failed he vainly sought to mobilize his earlier contacts to procure potato beetle parasites from the United States. In his final report to Albert Vögler, president of the KWG, he laconically noted that "correspondence with the relevant agency of the U. S. Dept. of Agriculture in Washington D.C. showed . . . that the Americans were not willing to send us the insects we wanted."[72] Attempts to use fungi rather than insect parasites as biological pesticides likewise produced no tangible results and experiments with food poisons and contact poisons were ruled out by the Reichsanstalt for Biology. What Bergold could do was to apply his knowledge of polyhedral viruses to the conservation of the huge silkworm cultures that the Luftwaffe maintained for the production of parachutes. In this connection he and Rudolf Brill conducted silk stretching experiments. His centrifuge was used in characterizing biological macromolecules as well as synthetic products of high molecular weight. "For I.G. Farben in Ludwigshafen, that is, Dr. Hengstenberg," he reported, "experiments were performed on high molecular weight polymers, such as polystyrols, kollidones [polyvinyl pyrrolidon], Opanol, and Buna. For I.G. Farben and Höchst, that is, Dr. [Franz] Patat, diffusion measurements of a series of chloroprenes were carried out, and, for Dr. Bockmühl, the molecular weight of Höchst-insulin was established."[73]

Bergold was only one of thousands of scientists who worked on war-related problems in I.G. Farben's laboratories; but because he quickly mastered and considerably improved the technology of his ultracentrifuge, he received assignments from neighboring labs. Thus very different projects and purposes revolved around his centrifuge. As a result of his "productive integration" into the team in Oppau he served as a link between basic and war-related research. Touching on the subject in a 1998 interview, Bergold did not mince words:

> What kept me out of the war was the silkworms in Celle. . . . The Luftwaffe's main silkworm culture was in Celle. It was a gigantic, a truly gigantic operation. Yes, that kept me out of the war. And then the blood substitutes . . . polyvinyl pyrrolidon . . . which I mention only in passing.[74] There were

absolutely no publications. . . . But all that is nothing but . . . in itself, when you look at it from the outside. . . . What a joke, what's it called? — basic research, which is supposed to be independent, at least in theory, isn't that how the story goes?[75]

On another occasion, Bergold said: "Distinguishing between 'basic' and 'applied' research is bureaucratic-semantic nonsense."[76]

Meanwhile in Berlin-Dahlem things were proceeding only in part according to the plan drawn out by Butenandt, von Wettstein, and Kühn in 1938. Danneel believed that he had isolated a papilloma virus similar to the Shope papilloma virus from the skin of rabbits[77] and began investigating the skin cancer caused by it. However, lack of standardized material of the sort he had originally received from Shope made it difficult to continue. After the war Danneel explained the situation: "Since, only two years later [that is, in 1943], as a result of the circumstances that had arisen in the meantime, I was running out of vaccine, which was available only in the USA, we had to continue our work using another object."[78] Moreover, he was unable to reproduce his initial finding and had to publish a retraction.[79] Thereafter, his interest shifted back to the pigmentation of the fruit fly *Drosophila melanogaster*.[80] As we noted earlier, his investigations of the subject, which complemented Kühn's *Ephestia* experiments, had already commenced when he joined Kühn's department in 1939. Later Danneel turned his attention to the experimental induction of cancer in mice. His group had never really become part of the "virus community" in Dahlem. On August 12, 1944, he wrote frankly to Kühn:

> It is becoming increasingly clear to me that I must abandon virus research and that this is also my own wish. Our work in this field has in fact come to a virtual standstill; I have started only the papilloma experiments up again, although I have little hope that something will come of our rabbit breeds. . . . I have no eye for the problems of virus research, at least as far as the zoological aspect goes, so that I would probably not even be able to accomplish much with "positive" results. Over the years, I have of course made countless isolated observations; but they don't tell me anything because I can't subsume them under a problem.[81]

During the evacuation and resettlement, in stages between 1943 and 1944, of the Kaiser Wilhelm Institutes for Biology and Biochemistry to Tübingen and the nearby area, Danneel went his own way. Eventually he and his group found a place for themselves in the Institute for Hygiene at the University of Göttingen.

Results on TMV

Melchers extended his research for spontaneous TMV mutations, especially the parallel mutations *flavum* and *luridum*, which he had derived from *Marmor tabaci var. vulgare* and *subsp. Dahlemense*.[82] He hoped to gain insight into the molecular nature of the mutations by systematically characterizing these variants on a serological and chemical basis, work that he carried out together with Schramm, Dannenberg, and Friedrich-Freksa.[83] He was also busy working out a purification procedure for the potato Y-virus.[84] In collaboration with Schramm and Karl Günter Zimmer from the Genetic Department of the KWI for Brain Research in Berlin-Buch and Hans Joachim Born from the Radiological Section of the Berlin Auer Society his assistant Lang investigated the incorporation of radioactive inorganic phosphate into the nucleoprotein of TMV. They came to the conclusion that the proliferation of the phosphate-carrying components of TMV had to be the result of a synthesis from simple building blocks.[85] Yet the virus group failed to follow up on these early experiments based on radioactive tracing. Late in 1943 Melchers transferred his group to the Botanical Institute of the University of Tübingen. His mentor von Wettstein helped him avoid conscription,[86] but under the restrictive conditions prevailing in Tübingen's Botanical Institute it was impossible to conduct extended experiments of the kind that had been possible in the Dahlem greenhouse. The result was that Melchers's experimental work came almost to a complete halt in Tübingen.

Butenandt too contributed an article to the description of TMV. In 1942 in a Festschrift in honor of the sixtieth birthday of the factory director at I.G. Farben, Heinrich Hörlein, he reported on a series of elegant UV absorption measurements carried out in collaboration with the Physico-Chemical Institute of the Technical University of Munich on TMV particles that had been oriented using an electric current. These measurements showed that the bases of TMV RNA had to be stacked up in the virus one on top of the other and horizontally with respect to the axis of the TMV rod "like 'coins in a roll of coins.'"[87]

Schramm meanwhile continued his physical measurements and chemical treatments of TMV. Together with Friedrich-Freksa he conducted serological analyses of the virus, visualized the precipitate under the electron microscope, and tried to determine its length.[88] The values reported in the literature varied: whereas Schramm found that his particles fixed with osmic acid measured about 200 millimicrons, Helmut Ruska of Berlin's

Charité Hospital together with Gustav Kausche and Edgar Pfankuch of the Dahlem-based Biologische Reichsanstalt für Land- und Forstwirtschaft had arrived at values of 150 or 300 millimicrons according to their first measurements. Stanley and Anderson for their part had come up with a length of 280 millimicrons.[89] In another communication dating from 1941 Schramm reported that he had been able to remove the nucleic acid from the TMV protein enzymatically.[90]

Melchers's biological control experiments indicated that "after the splitting off of the nucleic acid, the biological activity [of the virus] all but completely disappears."[91] Butenandt erroneously concluded from these experiments on digestion that "the nucleic acid [was] located exclusively on the surface of a connected protein scaffold."[92] This tallied with Friedrich-Freksa's hypothesis that the nucleic acid of viruses served as a negatively charged template making possible a precisely specified accretion of positively charged amino acids, so that the reduplication of the virus could be conceived of as a copying mechanism "comparable to the familiar processes of technical copying procedures."[93] However, the presumed splitting off of the nucleic acid did not accord with Schramm's further observation that the electrophoretic mobility of the nucleic-acid-free protein remained unaltered as compared to the protein associated with nucleic acid.[94] These experiments nevertheless greatly interested Stanley, who wrote to Schramm in October 1941 that "we have for some time been trying to achieve the result that you describe, so far unsuccessfully," and asked for samples of Schramm's enzyme preparation and nucleic-acid-free protein.[95] Because the United States entered the war against Germany soon thereafter his request went unanswered. Schramm himself soon realized that the nucleic acid had indeed been enzymatically split, yet had not been removed from the protein. In a review article on the constitution of TMV that appeared in 1944 he corrected his erroneous observation[96] but did not bring his mistake to Stanley's attention until 1946 when postal relations between the United States and Germany were resumed.[97]

Schramm took his experiments to be indicative of "the necessity of nucleic acid for the multiplication process" of TMV.[98] However, unlike Edgar Pfankuch and Gustav Kausche of the Biologische Reichsanstalt or Hans Stubbe of von Wettstein's department,[99] he continued to believe that the mutable and thus also the genetically active material of TMV was the protein. In an overview published in 1942 he affirmed, "Pfankuch's experiments further indicate that the nucleic acid is of particular importance in determining the biological characteristics of certain viruses. . . . Stanley

. . . in contrast . . . believes that the mutations are predominantly due to changes in the protein component. [Our] electrophoretic investigation of different TM virus mutants tends to corroborate the second hypothesis."[100] Schramm together with his coworker Müller had observed that the acetylation of amino groups of TMV was a two-step process. The easily accessible proteinic amino groups were modified first with no concomitant inactivation of the virus. Although Schramm concluded from this that "the virus's capacity to multiply is not dependent on the greater part of the amino groups," he did not consider his finding to be an argument against the replication function of the protein. He also failed to follow up the possibility that the second step, the acetylation that inactivated the virus, might have affected the RNA amino groups.[101] It was impossible to overlook the divergences in these interpretations, yet belief in proteins appeared to reign in the Virus Research Facility in Berlin.

Melchers in a report on German science between 1939 and 1946 written for the *FIAT Review*[102] observed with obvious reservations that Pfankuch, Kausche, and Stubbe "believed" that they had to interpret their results "as an indication that the mutational changes took place in the nucleic acid component of the molecule." In contrast "Schramm is of the opinion that it is not possible to seek the differences in the molecule's nucleic acid component; with Stanley, he locates them in the protein component."[103] With the benefit of hindsight it is astonishing to see how profoundly Schramm and the other members of the virus group were swayed by Stanley's authority; despite massive experimental indications to the contrary they flatly rejected the possibility that nucleic acid might be the genetically active principle. Schramm was nevertheless convinced that RNA had "decisive importance for the functioning of the virus"[104]—but of another, structural kind—and in cooperation with Dannenberg he devised a method based on UV absorption for measuring the nucleic acid content of the TMV particle.[105]

Another surprising result obtained in 1942 kept Schramm busy for the next few years. He observed that TMV particles could by means of alkaline treatment not only be split up into subunits either containing or lacking nucleic acid, but could also be reconstituted from the subunits if the pH was shifted to the acidic side. As far as could be seen, this spontaneous process of self-assembly fully restored the morphology of the virus; yet it remained biologically inactive, whether nucleic acid was present or not.[106]

To sum up: the members of the virus group in Dahlem had—in part, after major autodidactic efforts—welded themselves into a closely cooperating team which with the exception of Danneel practiced an effective division of labor. Melchers provided spontaneous TMV mutants; Schramm compared them in the ultracentrifuge and using the Tiselius apparatus characterized them physically and conducted biochemical investigations of their functioning; Bergold worked on improving the centrifuge; Dannenberg contributed UV measurements; and Friedrich-Freksa conducted serological examinations. According to Bergold, Friedrich-Freksa was "absolutely incapable of carrying out experiments, but was, otherwise, by far the most brilliant of all of us."[107] Von Ardenne's contribution to the joint effort was electron microscopy; the Auer Society and Karl Günter Zimmer from Berlin-Buch were responsible for radioactive tracing. Between 1940 and 1942 the number of publications to the group's credit rose steadily; moreover, Schramm endeavored to bring virus research to the attention of a broad public, lecturing to physicians and—like Danneel—writing newspaper articles aimed at the general public.[108] Then from 1942 to 1944 the number of publications produced by the group sank to a minimum. Toward the end of the war it became increasingly difficult to maintain the intensity and continuity that had characterized scientific discussion until then. From the early 1940s on the group had been working in deepening isolation. For one thing the decisive international feedback crucial to such research became ever harder to come by. "It was very hard," Peter Karlson said later. "During the war, we continued to receive the *Biochemical Journal*, which was delivered by way of Sweden. Butenandt received it personally, since he was a member of the Society. We also received Swiss literature and, once the military campaign in France had come to an end, French literature as well. But from America and England we either received nothing at all, or, if we did, then only very late."[109] Although the institutes in Dahlem were able to keep some Anglo-American journals on display,[110] the group now published exclusively in German journals. Thus, mutual, transnational criticism had largely ceased and Butenandt, Kühn, and von Wettstein controlled a substantial part of the domestic journal market.[111] Berlin was doubtless the only place in Germany in which all the components of the advanced technology required to do virus research—the ultracentrifuge, electrophoresis, UV-spectroscopy, radioactive tracers, and the electron microscope—were assembled in one institutional setting. Together, however, these factors created a monopoly that was not conducive to comparative discussion. Why there was virtually no interaction between the virus group around Schramm and Melchers and Pfankuch's and Kausche's team at the Biologische Reichsanstalt's Virus Research Agency is another question, one that calls for further research.[112]

Tübingen 1943–45

In summer 1943 the Speer Ministry ordered that the KW Institutes for Biology and Biochemistry, the Virus Research Facility included, be evacuated from Berlin. They were gradually transferred to Tübingen and the surrounding area in southern Germany.[113]

The move had no effect on the budget, which had risen steadily since the facility was founded in 1941. All the money for running expenses was provided by the Ministry for Food and Agriculture. The projected budget for 1942 was 107,200 RM, although it was eventually reduced to 76,450 RM, since the Oppau group had not yet been included. In 1943 the Ministry accorded the virus research group 124,500 RM and in 1944 as much as 141,100 RM.[114] Thus the facility's budget was on the same order of magnitude as that of the KWI for Biochemistry.[115] This budget paid for the salaries of Melchers, Danneel, Schramm, and Bergold, two scientific assistants (for Melchers and Schramm), ten technical assistants, and a dozen other employees, including secretaries. Additional research money on the order of around 90,000 RM for the period 1940–45 was provided by the DFG or Reichsforschungsrat.[116] Both von Wettstein (Genetics and Cytology) and Kühn (Developmental Physiology) presided over work groups in the Agriculture and General Biology Department [Fachgliederung] of the Reichsforschungsrat.[117]

A further change came in the wake of the move of Melchers, Schramm, and Bergold from Dahlem and Oppau to Tübingen: Butenandt took over the sole direction of the virus group. In a September 1943 letter to Kühn, who had already moved his laboratory to Hechingen, Butenandt wrote:

> Mr. Melchers has, after a fairly lengthy conversation with me, taken due note of the fact that you and Mr. von Wettstein have made an irrevocable decision to resign from the leadership of the Virus Research Facility, and that I alone am to direct it in the months ahead. Like the other three gentlemen, he is willing, under these conditions, to continue with his work, as long as no official steps are taken with the President at present. For the sake of maintaining the continuity of the virus work and ensuring the Facility's smooth transition to Tübingen, Mr. von Wettstein and I consider it opportune to leave matters there for the time being—conditional, of course, on your consent.[118]

The others were in agreement. A few days later, Butenandt informed the "gentlemen" by means of a circular bearing Telschow's signature that "the Virus Research Facility of the Kaiser Wilhelm Institutes for Biochemistry and Biology is not an independent Kaiser Wilhelm Institute. All correspondence, therefore, including personal correspondence with the heads of the different sections, must pass through Professor Butenandt's hands."[119] Although Kühn had expressed reservations, Butenandt decided to remain in Dahlem for the time being.[120] Von Wettstein remained as well, to the extent that he was not away on official business. Late in July 1944 Butenandt arranged for more members of his group to be transferred to Tübingen. "A few days ago," Kühn wrote, "Butenandt was here. He now wants to transfer everything to Tübingen after all, with the exception of a minor post."[121]

In Tübingen the virus group was broken up and fanned out over several university institutes. As I have said Melchers was able to find himself a place in the Botanical Institute, but had practically no working facilities there. Schramm and his equipment found a haven in the Institute of Hygiene, whose director, Otto Stickl, was Rector of the University of Tübingen at the time. Bergold had made the move to Tübingen in August 1943, settling along with his group in a small summer house belonging to the Pharmacological Institute. He had brought his Phywe centrifuge with him, which he had technically adapted in such a way as to be able to measure both diffusion and sedimentation constants of particles with the same optical system.[122] He was "a very capable organizer,"[123] as Melchers put it, in addition to being a talented *bricoleur*. In Tübingen, he began analyzing together with Schramm the structure of ribonucleic acid, while at the same time developing more sensitive methods for determining the molecular weight of TMV.[124] The measurements balanced out around a molecular weight of 40.7 x 10^6. (Schramm had earlier reported a value of 23 x 10^6, identifying a "heavy" component of 46 x 10^6 as the result of the bonding of two particles.)[125] As in the past the polyhedral viruses and polyhedral protein crystals derived from *Porthetria dispar*, *Lymantria monacha*, and *Bombyx mori* remained Bergold's experimental object of choice. The paper in which he presented the final summary of his findings on this question was based on a total of 434 sedimentation and ninety-two diffusion measurements, most of which had been carried out and evaluated by his assistants Martha Vialon and Ursula John.[126] Bergold also measured the diffusion and sedimentation constants of other biological macromolecules; among the clients for whom he made these measurements in Tübingen were Behring-Werke Marburg (serum albumin from horses), Boehringer Mannheim (prolactin), and several academic institutes and KWIs.[127] He also continued to work for his former hosts in Oppau. "Because I had the only still functioning UC, I.G. [Farben] people often came to see me in Tübingen in search of high polymer investigations, IGAMID, Neopren etc."[128]

In fall 1944 Schramm and his coworker Hans Müller were ordered to go to the island of Riems. In March 1943 the research station on the island had been put under the command of the Interior Ministry and officially designated as a center for the production of vaccines against animal virus diseases, especially foot and mouth disease.[129] "On the basis of intelligence reports, German military leaders feared that the enemy would utilize biological weapons, among them cattle-plague virus. . . . Riems was accordingly charged with developing a vaccine against it."[130] However, since the scientists there were unable to procure viruses capable of reproduction, the work ground to a halt. Another project, based this time in the Tübingen area, also ran aground. Bergold was in contact with members of the German "Uranium Club," on retreat in Haigerloch, who were contemplating construction of big rotors for uranium separation. "We wanted to build titanium rotors," he recalls. "But we did not receive any titanium. So we said that we'd make them with carbon fibers and make a sort of U-turn around the two holes. But nothing came of all this."[131] In the last months of the war in January 1945 Butenandt asked Kühn for a few slides on their joint work on insects. "Since I don't have the right to publish anything on the rest of our work at the moment," he added, "I'm thinking of telling the Academy in Berlin something about the 'bar' problem. The slides would of course be very useful for that."[132]

As the war was coming to an end the Tübingen KWIS were first taken over by a commission of the American T Force and then put under the supervision of the French Centre National de la Recherche Scientifique.[133] Officers of the French Liberation Army, among them lieutenant-colonel André Lwoff, chief of the French *mission scientifique*, allowed the institutes to continue their work virtually uninterrupted. "Our operation was only interrupted for half a day!" Kühn reported to Danneel.[134] Bergold too remarked that he "did not lose a single hour of work."[135] Neither Kühn, von Wettstein, Melchers, nor Friedrich-Freksa had ever been members of the National Socialist Party. Others, such as Butenandt,[136] Schramm,[137] and Danneel,[138] had been. While the University of Tübingen's electron microscope was being dismantled by the French, Butenandt managed to contact the American liaison officer and as he wrote Kühn on May 16, 1945, "by showing the officer [his] correspondence with the Rockefeller Foundation and the many instruments that the Foundation had given [him]," to secure protection for the institute as an American sphere of interest.[139] The instruments were tagged with plastic labels marked "donated by the Rocke-feller Foundation";[140] Bergold's ultracentrifuge too remained in the sum-mer house of the Tübingen Pharmacological Institute in Wilhelmstraße where it continued to spin. On the subject of his personal fate, Butenandt remarked: "I hope that my way is the right one for the future as well." His recommendation to Kühn ran, "If you get into trouble and conclude that I have chosen the right path, take the same steps yourself!"[141]

In February Wettstein, who had already been showing signs of exhaus-tion, died of septic pneumonia in his home in Trins/Tirol.[142] Melchers strove to detach his research team from the virus research group and have it incorporated into the KWI for Biology; this organizational reshuffle took place in the fall. Kühn received a job offer from Paris, but chose to remain in Hechingen.[143]

The Virus Research Facility was dissolved on January 1, 1946. Schramm's and Bergold's groups were integrated into Butenandt's Institute for Bio-chemistry as an autonomous "Department for Virus Research." Shortly thereafter Friedrich-Freksa joined the same department and the past col-laboration was restored.[144] Bergold left Tübingen in 1948 for the Labora-tory of Insect Pathology of the Canadian Ministry of Agriculture in Sault St. Marie. When Butenandt intimated that he should choose a successor, Bergold suggested the animal virologist Werner Schäfer, whom he had met in 1939 during his mission in East Africa. Friedrich-Freksa became the founding director of the Max Planck Institute for Virus Research in Tübin-gen in 1954. Two years later Schramm and Schäfer were also appointed to posts as directors.

In retrospect we can say about this episode from the history of the Kaiser Wilhelm Institutes for Biology and Biochemistry under National Socialism that the two institute directors von Wettstein and Butenandt had managed to come as close to the center of political and industrial power as humanly possible for people continuing to work as scientists. In comparison Kühn was more hesitant and cautious, remaining in the background.[145] Wettstein and Butenandt maintained close contact with the Reichsforschungsrat and its departments for the various disciplines (*Fachgliederungen*); here Backe, state secretary in the Food and Agricul-ture Ministry and later minister, played an important role as an intermedi-ary.[146] Through middlemen like Hörlein and Krauch, Wettstein and Buten-andt also maintained multiple connections with industry whose interests in virus research had been well served by the working group at I.G. Farben.

I do not wish to and indeed cannot make a judgment about the personal and political convictions of the two directors, von Wettstein and Butenandt. What is certain is that they could not lose in some sense no matter what happened. If National Socialist Germany won the war, they had already placed their own followers among the next generation of scientists in strategic positions: cultivated plant research, virus research, and cancer research. If Germany lost, the scientific work of the KW Institutes for Biology and Biochemistry as well as the Virus Research Facility would rank as so fundamental that the two directors would not be denied the opportunity to play leading roles in postwar German academic research. As soon as the war ended, Butenandt contacted his predecessor Neuberg again to let him know how his old institute was faring. Neuberg wrote back from the United States: "I never doubted that somebody of your qualities and youthful capabilities would hold his own in all circumstances."[147]

PART III

Concepts and Instruments

STUDIES IN THE HISTORY OF MOLECULAR BIOLOGY

Each of the three chapters in this section centers on the role of a particular concept (chapters eight and ten) or instrument (chapter nine) in the experimental development of molecular biology. In examining these developments, I take the perspective of both a *longue durée* and a micro-study. Chapter eight assesses from a bird's eye perspective how the notion of the gene changed in the twentieth century from classical to molecular genetics by mid-century and from "classical" molecular genetics to genomics by the end of the century. My main argument is that the productivity of a concept such as that of an elementary unit of heredity lies in its epistemic and operational plasticity in the search for unprecedented knowledge rather than in its rigorous and unambiguous definition. Chapter nine focuses not on a concept but an instrument. The role that the uses of radioactive tracing played in addressing basic features of life on a molecular level in the second half of the twentieth century has long been neglected and underestimated by historians of molecular biology. The chapter on the Liquid Scintillation Counter (LSC) is the first detailed account of the rise of an instrument that became emblematic for molecular biology laboratories in the decade between the elucidation of the double helix structure of DNA and the deciphering of the genetic code. The LSC not only opened new possibilities for measurement, but in combination with a multiplicity of radioactive tracer molecules allowed experimental designs of hitherto unconceivable structure and scope. In chapter ten, I apply a fine-grained textual microanalysis to the examination of the papers of the scientist François Jacob at the Pasteur Institute in Paris in the narrow time span between 1958 and 1970. Specifically, I examine one cluster of concepts concerning language and information. I argue that the new discursive regime did not simply replace an older one, but became superposed on existing layers of biophysical and biochemical terminology to form a hybrid discourse of pervasive heuristic power.

8. The Concept of the Gene

Molecular Biological Perspectives

"[It is] the vague, the unknown that moves the world."
—Claude Bernard, *Philosophie: Manuscrit inédit*

"The real core of gene theory still appears to lie in the deep unknown. That is, we have as yet no actual knowledge of the mechanism underlying that unique property which makes a gene a gene—its ability to cause the synthesis of another structure like itself, in which even the mutations of the original gene are copied. . . . We do not know of such things yet in chemistry."[1] These remarkable lines were written in 1950 on the fiftieth anniversary of the rediscovery of Gregor Mendel's experiments with plant hybrids and three years before publication of a model of the double helix structure of deoxyribonucleic acid (DNA) by James Watson and Francis Crick. Ten years later, no molecular geneticist would have echoed Herman J. Muller's chemical profession of ignorance. Yet molecular biology has not solved the riddle of the gene once and for all: it has by no means finally torn it from the "deep unknown." Rather, it has redefined its properties and boundaries, succeeding over the course of the past half century in changing our conception of this enigmatic epistemic thing again and again—almost beyond recognition every time.

In what follows I attempt to shed some light on the various epistemic and experimental arrangements by means of which molecular biology approached genes in those fifty years. In doing so, my aim is to offer neither a systematic assessment of the way molecular biology has appropriated the concept of the gene nor an exhaustive presentation of the complicated history of the gene as one of twentieth-century biology's foremost experimental objects.[2] Rather, I ask two questions: since the 1950s, where has molecular biology taken our conception of the "unique property which makes a gene a gene"? And what has it taught us in the process?

The following considerations are divided into three parts. The first part argues for an epistemology of the imprecise: it depicts the historical and disciplinary evolution of representations of the gene as illustrative of the trajectory of a fuzzy concept. In the second part, we see early molecular biology arriving at its simple solution to the gene problem; in this part I also retrace in the subsequent development of molecular biology various events that nullified this simple solution. The third part suggests certain conclusions that might be drawn from this history; here I take up the idea of the "integron" that François Jacob has elaborated with a view to bringing genomes and phenomes into a symmetrical relation.[3]

Epistemology: Fluctuating Objects and Fuzzy Concepts

The specific experimental practices observable in modern research fields give rise to concepts that are bound up closely with the objects of that research. As such, they constitute attractors that despite their imprecision—even one suspects because of it—acquire to one degree or another the power to guide people's thinking and advance the world of research. Occasionally entire disciplines are built up around one or a few of these imprecisely defined epistemic objects, which thus regulate exchanges across the borders of neighboring disciplines and facilitate transitions between them.[4] The atom was long such an object in physics as was the molecule in chemistry and the species in evolutionary biology. In classical genetics the gene took on this function. Such objects derive their specific historical contours from variable epistemic practices. In classical genetics, the gene unquestionably served as a formal entity that made it possible to explain in the context of ever more ingenious experiments in cross-breeding, the emergence or disappearance of certain characters in subsequent generations. Classical genetics was distinguished from nineteenth-century investigations of heredity by its new practices: the production of pure lines and cross-breeding. It was thanks to them that the notion of character discreteness rooted in the Darwinian and early De Vriesian traditions could be combined with August Weismann's distinction between germ plasm and body substance. The result read back into Mendel's experiments and laws in the wake of their "rediscovery" was a clear line of demarcation between genetic entities—*Anlagen* in the terminology of Carl Correns[5]—and the characters they conditioned; or taking each as an ensemble between genotype and phenotype.

Let us now cast an equally rapid glance at molecular genetics with its biophysical and biochemical practices and possibilities for genetic manipulation. At the experimental level molecular genetics was characterized by a switch in model organisms—from higher plants and animals to bacteria and phages—and a transition to *in vitro* systems. Molecular genetics transformed the object it had inherited, the formal gene, into a physico-chemical substrate or substance. It then transformed this object into an entity with informational properties. The first transformation solved the problem posed to classical genetics by the stability—and mutability—of its basic elements by claiming that genes are made of a class of metastable macromolecules: the nucleic acids. The second transformation solved the problem posed to classical genetics by the way its basic elements reproduced themselves and by the relationship between genotype and phenotype. The new claim was that nucleotide sequences, particularly DNA, can be specifically and faithfully replicated by virtue of the complementary stereochemical properties of their basic building blocks; moreover, thanks to their ordered sequence of nucleotides, which can be translated with the help of a complex cytoplasmic apparatus into the corresponding sequences of amino acids, stretches of DNA specify structural proteins and enzymes. These enzymes in turn catalyze the many different metabolic reactions that occur in the cell.

The ex post facto clarity of this extremely schematic overview is altogether misleading. For it contains no indication that the special fertility of boundary objects in research consists in the fact that they can be assigned no exact, systematically fixed meaning from the outset. On the contrary, premature attempts to set the conceptual boundaries of research objects that are still "in flux" may even be counterproductive. As long as such objects remain imprecise their conceptualization must also remain "in flux."[6] In other words, imprecise objects engender—positively—imprecise concepts. The fertility of the latter depends on their operational potential; or as Petter Portin says, "all definitions of the gene require operational criteria."[7] Such criteria alone determine the aptness and possible effectiveness of different definitions.

The spectacular ascendancy of molecular biology began without the help of a comprehensive, exact definition of what "makes a gene a gene." As I shall show in greater detail in the second, historical section of this chapter this holds for both stages of the process that distinguished the gene of molecular biology from the gene of classical genetics, the stage in which the gene was represented as a material entity and later that in which it was depicted as an information-carrier.[8] The meaning of both of these notions remained blurred, and could in no case be isolated from the various experimental contexts in which the new biology had been evolving since the mid-twentieth century from the identification of DNA as the hereditary material in bacteria in 1944 to the genome sequencing projects of the late 1980s. Indeed, everything indicates that efforts to produce overly precise definitions have tended to function as epistemological obstacles[9] in this history, or at best as theoretical artifacts: the early attempts to arrive at a strictly quantitative definition of biological "information" in the sense of contemporary information theory are a case in point.[10] Consider another example: on the basis of his bacteriophage mapping experiments Seymour Benzer tried in 1955 to bring a degree of order into this bewilderingly complex field by breaking the gene down into three entities, one bearing on expression or function, the second on recombination, and the third on mutation. He christened them "cistron," "recon," and "muton," respectively.[11] These distinctions were no doubt theoretically justified, based as they were on the most advanced experiments in phage genetics. A cistron could be broken down into recons, and a recon into mutons; mutons in turn would presumably lead to the simplest possible unit, a DNA base pair. Yet, despite their great clarity and terminological precision, these distinctions eventually proved in the estimation of the variegated community of molecular experimenters to be too restrictive and partially redundant. They sowed confusion and in the long run failed to gain acceptance.

In my view, it is not the task of epistemologists to criticize imprecise scientific concepts or put forward more precise definitions with, say, the well-meaning intention of helping scientists clarify their reasoning and do more precise science with more precise ideas. What is crucial for both epistemologists and scientists is how and why fuzzy concepts, half-baked definitions, or definitions that overshoot the mark can have positive effects in science.[12] As long as epistemic objects and their concepts remain blurred, they generate a productive tension: they reach out into the unknown and as a result they become research tools. I call this tension "contained excess." In a similar context François Jacob speaks of a "play of possibilities."[13] In writings by leading contemporary molecular biologists we often come across very loose definitions of the "gene"—if we find definitions at all. Manifestly the coherence of molecular genetics as a whole does not depend on any such definition. We ought to try to learn something about the particular dynamics of science from this, rather than complain that scientists handle the basic conceptual building blocks of their research carelessly. It is quite instructive to note, for example, that the glossary to be found in *Les secrets du gène*, by the French molecular biologist and former director of the Pasteur Institute, François Gros, contains no entry for "gene," although it does have one for "genome."[14] This is by no means a quirk of more recent literature on the subject. In the glossary to Leslie Dunn's classic *A Short History of Genetics*, first published in 1965, we find at the end of the entry for "gene" this caveat: "at present, discussions of properties to be explained are more useful than attempts at rigid definition."[15]

Do molecular biologists need a unified, general concept of the gene? As we have seen, examination of the pertinent literature suggests that there is no unified, unambiguous, rigorously determined usage of the term. What we find instead is context-dependence. Molecular biology, notwithstanding its claim to general validity, is a mosaic of many different contexts—contributions from different disciplines, diverse experimental systems, and divergent views of the genome. Imprecise epistemic objects and concepts work because they are malleable and can be integrated into different contexts in accordance with changing needs. To ensure that the enterprise as a whole preserves a minimal coherence, however, such adaptation must be to some extent reversible. Exaggerating for the sake of clarity we might say that two tendencies are at work here. The first seeks to give imprecise objects the sharper contours mandated by specific experimental contexts. Paradoxically this often leads to the exclusion of the more precisely defined objects from the field under definition. For example, the attempt to use *ribosomal* ribonucleic acid (RNA) as a template for bacterial protein synthesis led Marshall Nirenberg and Heinrich Matthaei to the characterization of a *non-ribosomal* template RNA (messenger RNA) and correlatively a noncoding ribosomal RNA.[16] In lucky cases such as this one, shifts in the "reference potential" of scientific expressions[17] go hand in hand with the unintended emergence of new objects that are at least as imprecise as those from which the search initially set out. The second tendency seeks to render the conceptual framework immune to the fluctuations produced by the first. In the instance just cited microsomal templates remained a possibility for all the experimenters whose work was based on eukaryotic, not bacterial, cells. Such conceptual flexibility at the frontiers of research implies a certain level of imprecision as well as mobile boundaries.

Molecular biology is a hybrid science combining experimental systems from biophysics, biochemistry, and genetics, among others. It uses a great diversity of model organisms in its search for biological functions at the molecular level. It is no wonder that its concepts too are hybrids. But it does not follow that it is inconsistent: the expansive potential of its discourse resides precisely in the hybrid nature of its concepts. Today that discourse pervades biology as a whole, evolutionary biology included. Molecular genetics did not emerge directly from either the experimental regime of classical genetics or the assumptions of synthetic evolutionary theory *and yet* it completely transformed the conditions under which we now think about heredity and evolution. This is reason enough to take a closer look at the forms of such "hybrid consistencies": how they surge up, what effects they have, and the way they function and develop. Here, this will be done in a short overview; in chapter ten, more in-depth on the example of the notion of information.

Let us consider a few options for possible fragmentary definitions of the gene as they are determined by experimental systems in the several fields just mentioned. Biophysicists working with a crystalline DNA fiber and an X-ray machine will argue that a gene may be characterized adequately by a particular conformation of a double helix, defined in terms of the atomic coordinates of the nucleic acid bases. Biochemists working with isolated DNA fragments in the test tube may define genes adequately as nucleotide polymers exhibiting certain stereochemical features and recurrent sequence patterns. They will perhaps try to give a macromolecular definition of the gene based on the unique chemical properties of DNA. Molecular geneticists for their part will describe genes as informational elements of chromosomes that give rise to specific functional or structural products: transfer RNA, ribosomal RNA, enzymes, and proteins serving other purposes. They will be inclined to approach the question of the gene mainly in terms of the replication, transcription, and translation of informational elements; they would argue that translation products of hereditary units must be examined in any discussion of genes. For anyone interested in evolution genes will be products of mutated, reshuffled, duplicated, transposed, and rearranged bits of DNA in a complex chromosomal environment that has evolved thanks to differential reproduction, selection, or other evolutionary mechanisms. Evolutionists will consequently appeal to concepts such as transmission, lineage, and historical contingency. For developmental geneticists finally genes can be described sufficiently in two ways: as hierarchically ordered switches that when turned on or off provoke differentiation and as batches of instructions synchronically activated by these switches. Thus developmental biologists are likely to refer to the regulatory features of genetic circuits when defining a gene or a larger transcriptional unit such as, say, an operon in a fashion they consider to be relevant to their work. The list could be extended.

The question is whether we really need a unified concept of the gene. Is it necessary, is it desirable? Should an attempt be made to coordinate developments in these different subdisciplinary specializations with the help of a unified concept and to bind the various niche-solutions together in a larger whole? In the half century since molecular biology came into existence, that has plainly not happened de facto. Nor do I think that a "central concept of the gene" would have contributed significantly to the development of the whole field. Indeed, even today the result would be in the best of cases an exercise in scientific rhetoric. The coherence of molecular biology depends neither on an axiomatics nor on an algorithm; it resembles a landscape made up of complex, more or less loosely interconnected experimental systems. Out of each of these systems there have arisen over time particular epistemic practices that invalidated previous affirmations but also spawned fresh ambiguities. Genes as we know them today are still imprecise objects. They owe their existence more than to any theory to the practices and instruments that helped bring the new biology into being.

History: Early Ripples, Recent Waves[18]

As we have seen, the development of the concept of the gene appears in hindsight to be quite straightforward. With enough historical distance, we may be struck by surprising convergences. However, it is doubtful that from such a distance the history of genetics can be reconstructed meaningfully as a history of gene concepts: unless the succession of the concrete embedding of these concepts is reconstructed in sufficient detail at the same time the end result could only be a deceptive epistemological artifact. The sharply condensed account that follows can merely suggest what such a two-track reconstruction might look like. It offers an inevitably foreshortened account of the meandering experimental explorations inspired by a wide range of research questions that were sometimes even quite unrelated to genetic problems at the outset.[19]

In contrast to classical genetics, the search for the material substrate of the elementary units of heredity stands at the beginning of molecular genetics. In the late 1920s both Muller and Lewis Stadler showed that X-rays could be used to provoke mutations in genes.[20] Yet most classical geneticists were for at least the first three decades of the twentieth century busy selecting spontaneous mutations and performing extensive crossing and breeding experiments; in their experimental regimes the material composition of the gene was not a problem that required an urgent solution.

This was not the case for the physicist Max Delbrück, who turned to biology in the 1930s precisely in order to examine the gene's material composition. Delbrück reasoned that genes should be characterized as autocatalytic proteins that could be permanently altered by physical influences. After a few early attempts undertaken in Berlin with Nikolai Timofeev-

Ressovskii and Karl Zimmer to define the gene as an elementary physical unit[21] Delbrück went to the United States and began to study phages, which he took to be the smallest independent equivalents of genes occurring in nature.[22] It is one of the many ironies of the history of molecular biology that the experimental systems of phage research eventually proved to be as formalistic as the systems of classical genetics. The contribution of physics was at first confined to measuring techniques and statistics. Only in the context of other developments and after long detours did phage research at last begin to contribute to the elucidation of the physical nature of genes. Timofeev pursuing a different path in Berlin attempted to combine experimental analyses of mutations in *Drosophila* with population genetics in the wild and research on evolution. The irony here lies in the fact that this "synthesizer" among the experimental geneticists in Germany, whose declared objective was to "discover, from an evolutionary standpoint, possible gaps in our knowledge of variability in order to stimulate the search for new facts and mechanisms," was unable to translate that objective into *a particular* research program that would have gone beyond classical genetics.[23] He shared this fate with the majority of his fellow evolutionary synthesizers in America and England.

George Beadle's and Edward Tatum's "one gene-one enzyme" hypothesis, in contrast, has been a milestone on the road leading from classical to molecular genetics. This hypothesis originated in biochemical genetics, a hybrid discipline that came to be organized around the model organism *Neurospora crassa* (a filamentous fungus).[24] The one gene-one enzyme hypothesis should not be confused with the "one gene-one character" postulate of early classical genetics. As a rule, enzymes intervene in the metabolic production of characters or phenes. One particular character can depend on a whole cascade of enzymatic reactions. As we have seen, the zoologist Alfred Kühn spoke in this connection of "gene-action chains" that intervene in "substrate chains."[25] And a single mutable entity can affect many different characters if the corresponding enzyme (or gene product) is active at the basis of metabolic bifurcations, a circumstance that explained the long-observed phenomenon of pleiotropy. Yet neither Beadle and Tatum in the United States nor Kühn in Germany came any closer to understanding the genes they were studying by experimentally tracing their biochemical products. Consequently none of them succeeded in elucidating the physical structure and chemical mechanisms of gene activity.

The new genetics would have to go through several further stages—examples are the transformation experiments that the group around Oswald Avery carried out with various types and strains of *Pneumococcus*[26] or Erwin Chargaff's chemical analysis of the specific base composition of DNA—before it became clear in the late 1940s that genes are made of DNA and cannot be represented as autocatalytic proteins.[27] The one gene-one enzyme hypothesis became the "one segment of chromosomal DNA-one protein" hypothesis.

DNA now moved to center stage and it became desirable and even urgent to determine its molecular structure. Since the closing decades of the nineteenth century the molecule was known to be a major component of the cell nucleus; until the late 1940s, however, its long biochemical and biophysical history[28] had had very few points of contact with the discourse of genetics. DNA was, rather, part of the history of structural chemistry, and at times an object of textile fiber research. When James Watson, Francis Crick, Maurice Wilkins, and Rosalind Franklin presented their double helix model of DNA in 1953 it became easy to envisage a simple, straightforward mode of replication for this specific class of biological macromolecules. A solution to the enigma of *auto-catalysis* began to fall into place: the replication of genes could now be conceived of as the dissolution and subsequent formation of specific base pairs joined by hydrogen bonds.[29] Yet *heterocatalysis*, the mechanism by which genes are converted into gene products, would keep molecular biologists, geneticists, and especially biochemists busy for another twenty years.

Another direction of inquiry, until then completely independent of genetics—biochemical analysis of the cellular fabrication of proteins—gradually began to insinuate itself into molecular biology. In spite of numerous attempts to find a theoretical solution for the "problem of coding,"[30] it was in the context of biochemical experimental systems that the concept of genetic information and genetic information transfer acquired a specific biological meaning in the late 1950s. Eventually, it corroborated the formulation of the "central dogma" of molecular biology: DNA is transcribed into RNA and RNA is translated into proteins.[31] The "information flow" was deemed irreversible according to Crick's central dogma.

By around 1960 the principle of colinearity between DNA sequences and corresponding sequences of amino acids had been established. It enabled a definition of what a gene was supposed to be on the molecular level: a finite, linear sequence of nucleotides containing the instruction for the fabrication of a corresponding finite, linear amino-acid sequence of a polypeptide. The translation was granted by a code that attributed particular base triplets to each amino acid. The establishment of a strict colinearity obtained between gene (polynucleotide) and gene product (polypeptide)[32] was crucial: it served as the foundation of an experimental regime without which the "code" would have remained an empty concept. Colinearity could be operationalized — if only for a while; that is, until the unsuspected complexity of the eukaryotic genome gradually came into view.

When in 1961 Marshall Nirenberg and Heinrich Matthaei succeeded in translating in an *in vitro* protein synthesis system a nucleic acid homopolymer (polyuridylic acid) into a homopolymeric protein (polyphenylalanine) the characteristic penchant of the new molecular genetics for hasty oversimplification attained its peak.[33] The experimental strategies of molecular biology seemed to have solved the problem altogether. The gene had first become a physico-chemical *molecule* and then a carrier of sequence *information*. But that same year the plot thickened once again. François Jacob and Jacques Monod presented their operon model for the regulation of lactose degradation in the bacterium *Escherichia coli*.[34] Their new finding was that genes had to be divided into two classes, structural and regulatory genes, and that it would be necessary to imagine on parts of the chromosomal DNA operator regions that did not code for polypeptides, yet played a determining role in regulating gene expression.

Since then non-coding DNA elements involved in gene regulation have proliferated almost past counting. There are promoter and terminator sequences; upstream and downstream activating elements in transcribed or non-transcribed, translated or untranslated regions; leader sequences; externally and internally transcribed spacers before, between, and after structural genes; interspersed repetitive elements and tandemly repeated sequences such as satellites, LINES (long interspersed sequences), and SINES (short interspersed sequences) of diverse classes and sizes. The details are bewildering and the functions of these and other elements are still far from having been studied exhaustively.[35] Should these batteries of elements be considered integral parts of genes or not?

On the level of post-transcriptional modification, the picture has become equally cloudy and complex. It was soon understood that DNA transcripts such as transfer RNA and ribosomal RNA first have to be trimmed in a complex enzymatic process and then have to "mature" in order to become functional molecules. It had to be understood as well that the messenger RNAs of eukaryotes undergo extensive post-transcriptional modification at both their 5'-ends (capping) and their 3'-ends (polyadenylation) before entering the translation machinery. In the late 1970s to everyone's surprise, molecular biologists had to get used to the idea that eukaryotic genes consisted of modules and that after transcription "introns" had to be cut out and "exons" spliced together if a functional message was to emerge. This insight was one of the first significant scientific by-products of recombinant DNA technology. As a result the colinearity postulate that had had so decisive an impact on the early experimental history of the genetic code faded into the background. What was a gene? Was it the whole stretch of DNA from which the primary transcript was derived or was it the spliced messenger RNA that sometimes contained less than ten percent of the primary transcript? Since the late 1970s we have become familiar with several kinds of RNA splicing: autocatalytic self-splicing, alternative splicing of a single transcript for the production of different mRNAs, and even trans-splicing of different primary transcripts for the production of a hybrid message. One of the surprising findings of the human genome project was that only about one-third of the expected roughly one hundred thousand coding sequences that may be described as genes were actually found in the human genome. "Genic diversity," it follows, is due mainly to splicing.

One last mechanism that was initially deemed to be a curiosity to be found only in a few unicellular trypanosomes merits mention. It turned out that "messenger RNA editing" can occur at the level of RNA transcripts.[36] Here the nucleotide sequence is systematically altered after transcription. The original transcript is not just cut up and pasted back together; rather, various guide RNAs and enzymes excise nucleotides and insert new ones before translation begins so that the product is no longer even complementary to the DNA stretches from whose transcripts it was composed. What, then, is a gene?

The problem of the gene in molecular biology made itself felt at the level of translation as well. It was shown that the translation process can begin at different start codons on one and the same messenger RNA. There can be "obligatory" frameshifting within a given messenger; without such frameshifting a nonfunctional polypeptide would result. There are also instances of protein modification after translation, involving, for example, the excision of individual amino acids or chemical modifications. Moreover, a process known as "protein splicing" has been observed in which portions of a precursor polypeptide have to be catalytically or autocatalytically split and reassembled in a new order to produce a functional protein.[37] In the snail *Aplysia*, for example, one and the same stretch of DNA gives rise to eleven such protein products with a bearing on the reproductive behavior of this mollusk.[38] Finally according to yet another baffling report from the field of translation a ribosome is capable of translating a single polypeptide by reading two different messenger RNAs: an instance of "*trans*-translation," to use Raymond Gesteland's term.[39]

How shall we define "the gene"? The fact that it is autocatalytic used to be regarded as one of its defining features, but autocatalysis has long been identified as a property of the DNA structure as such: not all DNA consists of genes, but all genomic DNA is self-replicating. What properties distinguish heterocatalytic entities? Which sequence elements must be included and which left out? In the middle ground of transcription ambiguities are proliferating wildly; and as we have seen they are hardly decreasing on the other side of the divide between genes and phenes, that is, proteins. To be sure the fact that the final protein products have to function after all does justify a conclusion of sorts. After a long career, the molecular biologist François Gros has arrived at the paradoxical, if not heretical thesis that the gene can be defined (if at all) only by "the products resulting from its activity," that is, the functioning RNA molecules and proteins whose emergence it has triggered.[40] But is that satisfactory? What do we do with all the nonfunctional products resulting from either DNA mutation or errors of transcription and translation? And where do we situate regulatory elements?

Let me make one final point here. Eukaryotic model organisms have come to the fore over the past three decades; as a result, the genome is increasingly conceived as a flexible, dynamic configuration. It is not just the mobile genetic elements observed by Barbara McClintock in her experiments with maize over sixty years ago that have returned to be both regularly and irregularly excised and reinserted across the genome in the form of transposons; at the DNA level we know of still other forms of gene shuffling. The immune responses of higher organisms—hence potentially the production of distinct antibodies by the millions—involve at the DNA level extensive, extremely complex somatic gene tinkering. No genome would be big enough to handle this task without somatic parceling of genes and the permutation of their component parts. Gene families also contain silenced genes ("pseudogenes"), jumping genes are to be found, and there is polymorphism at the DNA level, that is, multiple genes and isoforms. In short there exists a wealth of mechanisms and entities that together accomplish a kind of hereditary "respiration" or breathing.

Molecular biology has barely scratched the surface of this flexible genetic apparatus. Only recently has the attempt been made to understand its place in evolution and in embryonic development. To become adult organisms or produce viable offspring these gene-phene or phene-gene complexes must be reproduced in their own context, in their genomic as well as their cellular and inter-cellular environments. Molecular biology—often reproached for flattening biology out with the bulldozer of its peremptory reductionism—has in the course of its development made it impossible to conceive of genes as simple stretches of DNA associated in colinear fashion with amino acid strings, defined by ready-made instructions deposited in their nucleotide sequences, and confined by precise starting and end points. Today it seems more appropriate and may ultimately even prove sufficient to speak of genomes or simply "genetic material"[41] rather than genes, whether it is a question of organismic function, development, or evolution. For by now it is evident that the genome represents a dynamic ensemble of endlessly assembling and reassembling fragments and manifold forms of genetic iteration. Biologists are currently attempting to illuminate this structure through genome sequencing and intelligent sequence comparison programs.

If evolution is ever to be understood in terms other than those of the classical synthesis, it can only be from the standpoint of this dynamic configuration. The presumed central component of this complex machinery, simple point mutation, is by no means the essential moment of the genetic process, but only one of many other elements in the immensely varied DNA "tinkering" of which François Jacob has spoken.[42] A comparatively simple arrangement of genes in the bacterial chromosome may not reflect a primitive simplicity, but may be the result of a billion years of streamlining. Thanks to molecular biology, we have traveled a long road from the classical gene to today's genome. But the long road from genome to organism still lies ahead and calls for the efforts of a new generation of developmental molecular biologists. The road from there to populations and communities—and back again—will quite likely be no shorter, and is doubtless reserved for yet another generation of scientists.[43]

The process of putting "the gene" back into proper perspective is well under way.[44] Jürgen Brosius and Stephen Jay Gould proposed to abandon the concept altogether. They suggested a new terminology according to which every segment of DNA with a recognizable structure and/or function (such as a coding segment, repetitive element, or regulatory element) should be called a "nuon"—that is, an entity that has "meaning." By duplication, amplification, recombination, retroposition, and similar mechanisms, nuons can give rise to "potonuons," that is, entities with the potential to serve as nuons in their turn. Potonuons can be transformed into either "naptonuons," meaning that they would lose the non-adaptive information they possessed earlier without acquiring new information, or "xaptonuons," that is, elements "exapted" to a new function.[45] However tempting this evolutionary genome-terminology may appear, it remains to be seen just how operational it will prove.

Integrons and a Conjecture or Two

We are farther than ever from being able to define "the gene" as a simple constituent of DNA and an information-carrier for a simple polypeptide chain. One thing, however, seems to have weathered the to-do around "le gène éclaté"[46] (the shattered gene) and to have lent some resistance to the genetic engineers' undiminished discourse on the gene with its naturalistic, often deterministic overtones. This is the core of the central dogma of molecular biology as defined by Crick some fifty years ago. It postulates that the flow of genetic information runs from DNA to proteins and not the other way around. Whatever may happen at the level of DNA, RNA, or protein en route, "once 'information' has passed into protein it cannot get out again."[47] Although it is clear that the genetic material cannot replicate and recombine at all without the help of a complicated apparatus of enzymes and DNA-binding proteins and although we now have indications of the existence of some sort of directed mutation,[48] no mechanism of an "intelligent," environmentally or somatically steered alteration of genetic material has been found so far. While it is known that thanks to epigenetic inheritance mechanisms gene activation or repression patterns such as DNA-methylations can be carried over into the future generation(s),[49] a functionally improved protein capable of bringing about a corresponding sequence change in its own stretch of DNA has yet to be discovered. And although some continue to dream this Lamarckian dream, no one has the faintest notion as to how such a mechanism might work in a biochemically specifiable manner. It would seem that a protein code for producing sequence-specific strings of DNA has not come about through evolution.

The hard core of Weismann's legacy lies here. Yet the time has come to treat genomes and phenomes as two halves of an embodied whole. With Evelyn Fox Keller we may say that "owing in large part to the reemergence of developmental biology, molecular biology may well be said to have 'discovered the organism'" again. But we should also give due attention to her admonition that "the subject of the new biology, however whole and embodied, is only a distant relative to the organism that had occupied an earlier generation of embryologists."[50] This new biological body is saturated with an assortment of linguistic tropes, of which the most important are information, code, message transmission, signaling, and communication.[51]

Taking up a suggestion that François Jacob made forty years ago, I propose that we speak of "integrons" rather than genes and gene products.[52] Consider a piece of DNA that produces a DNA-polymerizing enzyme and ignore for a moment that this process presupposes a complex machinery of transcription and translation; the resulting polymerase produces more DNA of the same kind. Any mutation in this DNA that increases the efficiency of the product (i.e., that results in a polymerase capable of making still greater quantities of DNA) will lead to the selection of this prototypical "integron" from among its competitors, at least if space and resources are limited. Given such a "hypercycle"[53] the sequence information of the DNA segment itself takes on a biological meaning. In its simplest form it is identical to the capacity of proliferation. Yet more has to be involved even in this simple case than information in the quantitative sense of information theory. There is no genetic information dissociated from its biological meaning. Genetic information has to "make sense" in terms of a biological function. Even if we reject the strictly hierarchical aspects of Michael Polanyi's theory of boundary conditions we can accept his thesis that there is a "semantic relation" between the genetic and organismic level within which the latter endows the former with "meaning."[54] When we postulate such a relation we enter the realm of the symbolic. Organisms may indeed be regarded as symbolic machines even if their only basic "signifying" activity consists in further signifying, that is, reproducing. Just as there is no text without a context, so as Howard Pattee says, "a molecule does not become a message because of any particular shape or structure. [A] molecule becomes a message," Pattee goes on, "only in the context of a larger system of physical constraints which I have called a 'language.'"[55] But we should remain on our guard. Biologists have always happily borrowed terminology from other fields (and have repeatedly been forced to go beyond the borrowed analogies). Neither natural languages nor technical information-processing systems generate and transmit "information" the way living systems do. These processes are not isomorphic and what ultimately counts here are the differences. If we can assume that meaning is something that is induced at the point of intersection between different strings of signifiers, then we must in looking for meaning in organisms attend not to their genes, but to the multiple interfaces between the genome and the body.[56]

To avoid the ambivalence inherent in the language of information some prefer to talk about "instructions"—or a "blueprint," to cite Manfred Eigen's term.[57] The potential advantage of this performative idiom is, however, lost if either of the terms "instruction" or "blueprint" is equated with the term "genetic program." As Jacob reminds us the so-called "program" of a living being not only needs the products of its messages in order to be read and executed, but also regulates the execution of the instructions.[58] Henri Atlan therefore is inclined to turn the metaphor around: the whole metabolic machinery of the organism is a "program" and the genes are the "data" with which the machine is fed.[59] Where this point of view might lead remains open. A great deal more remains open as well: the biosphere as a whole might well be conceived of as a set of biosemiotic regimes comprising countless different levels of information transmission, from biochemical signals to human communication.[60] In this perspective the hereditary process might be considered just one—very basic—way of (to play on the title of J. L. Austin's classic) "doing words with things."[61]

The reader who has followed me to the end of this brief history of the gene painted with broad strokes will hardly expect a culminating definition of a gene. From the perspective of the history of science there is none to be had, since the concept remains in flux. But an epistemological lesson can be drawn even if it is likely to disappoint non-deconstructionists. Paraphrasing the virologist André Lwoff, François Jacob's early mentor at the Paris Pasteur Institute and also one of Gunther Stent's teachers, we might say that a gene is a gene is a gene.[62] This is an affirmation with far-reaching implications.[63] Taken literally, it means that in science every presumed referent will turn into a future representation.

The epistemic entity known as a "gene" was and remains quite influential in the history of heredity, not least because it displays the imprecision typical of such entities. Yet this situation should not lead us to make blanket generalizations: not all fertile scientific concepts must be polysemic. My aim is not to strike precision from the list of scientific values. What people understand by precision is, however, historically variable. Moreover, arriving at a more precise understanding of what imprecision means *in science*—that is, treating it as an epistemic possibility rather than rejecting it out of hand—is one of the main objectives, for example, of current research in AI and the neighboring disciplines. Lofti Zadeh once declared that "there is a rapidly growing interest in inexact reasoning and the processing of knowledge that is imprecise, incomplete, or not totally reliable. And it is in this connection that it will become more and more widely recognized that classical logical systems are inadequate for dealing with uncertainty and that something like fuzzy logic is needed for that purpose."[64] Although the solution that the syntax of fuzzy logic has to offer is purely technical, it throws up a methodological question: do we need harder metaconcepts to come to terms with the perpetual conceptual tinkering characteristic of research—that is, concepts more rigid than the first order concepts that we epistemologists and historians analyze? I think not. Why should historians and epistemologists—at least if they consider their trade too a field of knowledge in the making—work in less imprecise, less practical, less opportunistic fashion than scientists? This does not mean that they should make their histories more incoherent than their scientific subjects! Rather, they must come to see that "there is an incompatibility between precision and complexity. As the complexity of a system increases, our ability to make precise and yet non-trivial assertions about its behavior diminishes."[65] The history of the gene has a contribution to make to our understanding of this epistemological uncertainty principle. It has something to teach us about how to handle complexity: not by setting out in quest of an all-encompassing theory, but by looking for patterns of "moderate compressibility," as the theoreticians of complexity put it.[66]

9. The Liquid Scintillation Counter

Traces of Radioactivity

Perhaps no other instrument has symbolized the techno-myth of ascendant molecular biology and biomedicine of the 1960s and 1970s better than the radioactivity measuring device known as the liquid scintillation counter (LSC). This apparatus, which embodied three key twentieth-century technologies at once—mechanical automation, electronic circuitry, and radioactive labeling—experienced a meteoric rise: in the span of a dozen years the LSC evolved from a greasy, unmanageable detection device into a generic technology that was by the 1970s ubiquitous in the laboratories for molecular biology and medicine, where it fundamentally altered molecular-biological work and especially the time needed to carry out experiments. Of the early models it was above all Lyle E. Packard's Tri-Carb® Liquid Scintillation Spectrometer that made its way into universities, national laboratories, state hospitals, and the research departments of private industry. A Packard Tri-Carb, or simply a "Packard," with its calculator data output connected to an IBM typewriter-printer, became the icon of a modern molecular-biological laboratory (fig. 21). Yet in contrast to other instruments and techniques characteristic of molecular biology, such as electrophoresis, ultracentrifugation, electron microscopy, NMR (nuclear magnetic resonance), and PCR (polymerase chain reaction),[1] the LSC has so far attracted little interest from historians of science.

In order to chart the unprecedented career of the LSC, I begin in what follows with an account of the introduction of emitters of low energy β-particles (electrons) such as ^{35}S (sulfur), ^{14}C (carbon), and ^3H (tritium) into physiological research in the 1940s and 1950s. Measuring technologies initially available to follow these isotopic markers in biochemical reactions were relatively inefficient. After World War II the conjunction of a number of different epistemological, technological, and cultural factors established liquid scintillation counting as an effective alternative to traditional methods of solid sample counting or gas counting based on ionization. The first commercial LSC, which became the prototype of a production series, was built for the University of Chicago by Lyle E. Packard in 1953. I have chosen the history of this prototype to illustrate the par-

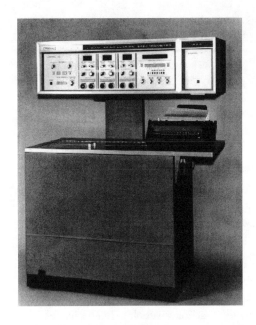

21. The first LSC with an absolute activity analyzer, mid-1960s. Collection Edward F. Polic, 702 Glenn Court, Milpitas, CA 95035–3330.

ticular trajectory of one research technology: between 1953 and 1970 the design of the instrument underwent a series of mutations that enabled its generic application. This transformation opened up new epistemic dimensions for radioactive experimentation in biology and medicine. I trace the main stages in the development of sample preparation, data processing, and the construction of the instrument—a process of development that was fostered in part by the commercialization of nuclear energy in postwar Europe and the United States. Finally, I look at the way the relations between producers and customers took shape as Packard's firm grew from a one-man company into an international corporation.

Radioactive tracing is a "research enabling technology," to use the term of Gerhard Kremer, a former president of the International Bureau of the Packard Instrument Company in Zurich, to describe technologies that open up new fields of investigation.[2] The application of this technology to the analysis of biodynamic processes has provided answers to questions that previously could not even be posed. Three components of this technology must be distinguished. The first is the production of suitable radioactive isotopes (^{3}H, ^{14}C, ^{32}P, ^{35}S) and their incorporation into organic molecules; it is characteristic of this process that the isotopes do not perceptibly alter the molecules' chemical or biological properties unless they decay. The second is the development of experimental systems, especially in vitro systems, in which these molecules can be used as tracers.[3] The third is the development of the corresponding counting devices. My presentation here focuses on the third aspect of tracer technology.

Radioactive Markers in Biological and Medical Research

During World War II and especially in its immediate aftermath, the emergence of artificial radioactive isotopes revolutionized the field of biological and medical research, rather than merely transforming it as in the late 1920s and the 1930s.[4] Of particular importance for our purposes are hydrogen, carbon, phosphorus, and sulfur, the main constituents of biological molecules. Radioactive phosphorus (^{32}P) was one of the first cyclotron-produced radioactive isotopes. In 1939 Luis Alvarez and Robert Cornog, using the cyclotron at the Radiation Laboratory of the Physics Department of the University of California at Berkeley, obtained radioactive hydrogen (^{3}H) by bombarding deuterium gas with heavy water.[5] Ernest Lawrence, the head of Berkeley's radiation library, immediately recognized the potential significance of this finding: "radioactively labeled hydrogen," he announced, "opens up a tremendously wide and fruitful field of investigation in all biology and chemistry."[6] He then entered into negotiations with the Rockefeller Foundation in hopes of obtaining substantial research funds. Before the year was out Samuel Ruben and Martin Kamen using Lawrence's apparatus obtained radioactive carbon (^{14}C) by bombarding graphite with heavy water.[7] With the construction of powerful particle accelerators and shortly thereafter the emergence of controlled fission the decisive conditions for the production of a large variety of new isotopes had been created.[8]

Like phosphorus and sulfur, hydrogen and carbon are ubiquitous constituents of organic matter. And like phosphorus and sulfur the radioactive isotopes of hydrogen and carbon emit electrons (β-particles) as they decay. These isotopes have a half-life that makes them excellent tracers in metabolic studies. Typically in such experiments the in vivo distribution or uptake of these atoms into biological molecules is monitored. Alternatively the metabolic fate of isotopically labeled molecules can be followed in vitro. Only in the case of phosphorus, however, was their radiation energy high enough to be measured reliably with conventional Geiger-Müller counting tubes. The counters recognized ^{14}C inadequately and tritium not at all, since its low energy β-particles could not penetrate the walls of the tubes.

The Manhattan Project, the vast atomic research program at the core of the massive American war effort, gave birth at the end of the war to a network of national laboratories, among them Los Alamos, Berkeley's "Rad Lab," Oak Ridge, and Argonne. The laboratories were supervised by a federal agency for atomic energy (the Atomic Energy Commission, AEC), created in 1947; the AEC's mission was to coordinate military, economic, political, and scientific interests in atomic energy. As byproducts of the development of atomic reactors, radioisotopes accumulated in large quantities. In 1947 alone, the isotopes produced in the reactor at Oak Ridge represented "the equivalent of thousands of years of cyclotron production."[9] The priority was to produce fissionable material as well as atomic warheads for military purposes. However, it was also incumbent on the AEC to endow atomic energy research "with a peaceful, civilian image"[10] and especially to foster studies in such fields as radiobiology and radiomedicine. A division for biology and medicine was accordingly added to the AEC in the first year of its existence. Thanks to the Isotope Distribution Program of the AEC conducted by the agency's director Paul Aebersold—who had written his doctoral dissertation at Berkeley under Ernest Lawrence's supervision[11]—American biological and medical institutions were in the following years supplied with massive quantities of ^{3}H, ^{14}C, ^{32}P, and ^{35}S.[12] In 1947 the AEC sold a millicurie of radioactive phosphorus (^{32}P) for $1.10, and the same amount of radioactive iodine (^{131}I), sulfur (^{35}S), and carbon (^{14}C) for $1.70, $35.00, and $50.00, respectively. From 1948 under the Atoms for Peace campaign biomedical research as well as cancer diagnostics and therapy were provided with isotopes at no charge.[13] Attempts to use ^{32}P and ^{131}I in the diagnostics and therapy of the disease had been underway for about a decade; ^{35}S and ^{14}C promised to become ideal tracers for biochemical assays.[14]

The impact of this policy can hardly be overestimated. Of the total number of papers published in the *Journal of Biological Chemistry* between 1945 and 1956 the proportion of those reporting use of radioactive isotopes rose from 1 percent to 39 percent.[15] The laboratory at Oak Ridge, which had been delivering radioisotopes to hospitals and labs since summer 1946, sent out more than 5000 shipments with a total radioactivity of 2.5 million curies in 1966.[16] The ubiquitous presence of radiation in military as well as civilian—environmental and medical-biological—contexts heightened the need for new, sensitive, reliable measuring and monitoring devices. Postwar declassification of studies of instruments for measuring radiation further stimulated the search for alternative counting methods.[17] Companies such as Radiation Counter Laboratories (Chicago), Instrument Development Laboratories (Chicago), the North American Philips Company (New York), the Victoreen Instrument Company (Cleveland), the

General Radio Company (Cambridge), the Cyclotron Specialties Company (Moraga, CA), the Engineering Laboratories (Tulsa), and the Geophysical Instrument Company (Arlington)[18] were soon turning out counters of all sorts and sizes and advertising them in scientific and technical journals. The beginnings of a civilian nuclear industry reinforced the trend.[19]

The First Steps toward Radiation Measurement

Radiation research began around the turn of the twentieth century. One of the first methods for measuring the activity of radioactive samples was developed in London. Sir William Crookes's method was based on the phenomenon of scintillation, the production of flashes of light that allowed Crookes to quantify the "emanations" of radium or in other words the "scintillation" that these emanations produced on a screen of zinc sulfide.[20] Crookes's observations were promptly confirmed by Julius Elster and Hans Geitel in Wolfenbüttel.[21] Five years later Erich Regener in Berlin used the scintillation method to record the α-particles of polonium.[22] The light flashes were counted visually using a simple microscope. This method was widely utilized in nuclear physics for about twenty years, although it had a major disadvantage—the "counters" were human beings: "Rapid fatigue of the observer and other subjective influences make it necessary to change observers frequently. They can only observe for thirty seconds to a minute, need to make long pauses between observations, and cannot make reliable observations for more than two hours a week. The efficiency of the method is poor; good, useful counts are to be had only at rates of 20 to 40 scintillations per minute."[23]

The scintillation method was relegated to oblivion when Geiger-Müller counters came into use in the late 1920s.[24] These instruments were based on the ionizing capacity of the emitted particles and the concomitant discharges produced in the electrical field of a gas-filled tube. Geiger-Müller counters proved useful for the detection of higher energy β-particles. They made it possible to measure γ-rays as well, albeit less efficiently, thanks to the secondary electrons whose emission they provoked when they penetrated the walls of the tube. Subsequent versions of the Geiger-Müller counting tubes had a thin mica end-window in front of which a solid sample could be mounted on aluminum planchets. This device, used well into the 1950s, allowed measurement of the β-particles emitted by radioactive carbon, for example, with an efficiency of around 10 percent. The weak β-emissions of tritium, however, remained beyond the scope of this technique. An alternative technique based on ionization consisted in producing a gaseous form of the sample: for example, by oxidizing ^{14}C-labeled compounds into water and radioactive carbon dioxide, and then using ionization chambers to measure the decay events of the gas. In theory this method of gas counting could be used for substances that produced weak emissions of β-particles as well. However, it had major disadvantages: the sample preparation procedure was laborious and it was hard to obtain quantitative measures of the samples.

Early in the 1940s as a result of developments in another field, photoelectricity, scintillation counting came back into use. Peter Galison, distinguishing an "image" tradition from a "logic" tradition in the history of registration methods, describes this event as follows: "What transformed the scintillator's flash and Cerenkov's glow into basic building blocks of the logic tradition was the electronic revolution begun during the war. When attached to the new high-gain photomultiplier tubes and strung into the array of amplifiers, pulse-height analyzers, and scalers that emerged from the Rad Lab and Los Alamos, then and only then did the scintillator and Cerenkov radiation become part of the material culture of postwar physics."[25] The physicist and biophysicist Adolf Theodor Krebs, then a staff member of the Kaiser Wilhelm Institute for Biophysics in Frankfurt and director of the Radiobiology Division of the U.S. Army Medical Research Laboratory at Fort Knox from 1947, was probably the first to develop an instrument in which the human eye was replaced by a highly sensitive photoelectric apparatus for detecting and counting scintillations.[26] Attempts to improve such combined scintillation and photoelectric devices—based mainly on constructing efficient, reliable photomultipliers— were undertaken more frequently toward the end of the war. Radio Corporation of America (RCA) and Electrical Music Industry (EMI) in Great Britain soon dominated the market for this technology, crucial in weapons control and guidance systems as well as civilian mass communication.

As a rule scintillation counters consist of a scintillating crystal and a photomultiplier.[27] New solid scintillation devices were constructed that were capable of counting α-particles[28] as well as β-particles and γ-rays.[29] An alternative to them, the Geiger-like photon tube counting device, soon became quite popular: it combined the classical scintillation arrangement with a photosensitive Geiger tube of special design and could be used for combined α-β-γ measurements, the detection of α-particles in the presence of a β-particle and γ-ray background, or for β-ray detection or γ-ray detection alone.[30] The main problem with instruments of the first type was the current produced by the photomultiplier, that is, the spontaneous activity of the device itself; the disadvantage of instruments of the second type lay in the time required for regeneration of the Geiger tube between discharges. Nevertheless the two devices were hailed by contemporaries as "the most important advances in devices for the detection of nuclear radiation since the invention of the Geiger-Müller counter."[31] Because of the resolution and efficiency of the counting process as well as the prospect of applications involving low specific activity they were celebrated as heralds of a "new era" of nuclear research.[32] As early as 1949 the first conference on scintillation counting was held in Oak Ridge.

Liquid Scintillation Counting

Counting technology took a new turn when Hartmut Kallmann of the New York University Physics Department[33] in collaboration with Milton Furst resumed work on his earlier observation that certain organic substances such as anthracene when dissolved in organic solvents such as toluene functioned as scintillators, and could be used in liquid scintillation counting in combination with electron-multiplier phototubes.[34] At Princeton George Reynolds and his colleagues were working on the same technology.[35] As in the case of solid scintillation the process was based on converting the energy of radioactive decay into photons and converting the photons into photoelectrons that could be amplified and then counted. The energy of the decay events was absorbed by the scintillator solvent, which transferred it to the scintillator, causing it to emit photons. The photons in turn could be collected in a photomultiplier tube and amplified. The energy transfer processes in the solvent system were hardly understood at the time; it took physicists years to elucidate the details of the process. Early work in this completely new field of liquid scintillation counting focused on the external measurement of high-energy radiation from sources such as radium (Kallmann) or cobalt (^{60}Co, Reynolds).

In 1951 Maury S. Raben of the New England Center Hospital and Tufts College Medical School in Boston and Nicolaas Bloembergen of the Harvard University Nuclear Laboratory suggested "that a simple and geometrically ideal counting system might be obtained by dissolving the material to be counted directly in [the] liquid. This method would facilitate particularly the counting of soluble compounds labeled with a weak β-emitter, such as ^{14}C."[36] Initial measurements showed that it might be possible to use this internal measuring procedure to count even nanocurie quantities of radioactive carbon. This finding promised a pronounced increase in the sensitivity, efficiency, and accuracy of the system, and brought the prospect of counting other weak β-emitters such as ^{35}S, and, for the first time, even ³H, closer. The reason was the homogeneous distribution of the radio-active sample and as a consequence the scintillator's nearly total absorption of the emitted energy.

Some early LSCs used to measure internal samples were basically improved versions of the crystal scintillation spectrometers that had become available by the late 1940s. Here the sample was contained in a glass bottle surrounded by a reflector and connected to the photomultiplier tube by an optical coupling fluid such as silicon oil, glycerin, or Canada balsam. This arrangement was completed by a preamplifier, an amplifier, a pulse-height analyzer, and a scalar element. The hope was that tritium (³H) could be measured by such an instrument with an efficiency of up to 20 percent. However, the "dark current"—spontaneous thermionic emissions from the photomultiplier cathode—became prohibitively strong at the high voltages required to attain such efficiency.[37] The noise could be reduced by the selection of appropriate multiplier tubes, pulse-height discrimination, and refrigeration, but it could not be eliminated. Refrigeration in turn restricted the scintillators' emission spectrum, since the solubility of the samples was temperature-dependent. Liquid scintillation spectrometers based on a single photomultiplier were in use for only a short time.

A decisive step toward establishing liquid scintillation counting as the preferred method for measuring low energy β-emitters was taken as a result of investigations conducted by the group working around Francis Newton Hayes and Richard Hiebert at Los Alamos Scientific Laboratory.[38] The group tried out various scintillator solutions with an eye to typical applications in biology and medicine and developed an improved, solid coincidence-counting technique.[39] Coincidence-counting can be traced back to Walther Bothe and Hans Geiger.[40] Engineers of the Radio Corporation of America at the Princeton Laboratory Division had adapted it for use in a solid scintillation counter (fig. 22).[41] Both Kallmann and Reynolds proceeded to adopt this method in their external liquid scintillation counters, while Raben and Bloembergen began using it in their internal liquid scintillation counters.[42] To simplify somewhat, in order to eliminate the noise generated by the electronic equipment, two photomultipliers were placed around a sample. After amplification only those pulses were counted that "coincided" at their arrival at the pulse-height analyzer and could therefore be assumed to arise from a scintillation event induced by one and the same decay electron, not from noise generated by the system. Under these conditions the noise rate of a single tube could be reduced from tens of thousands of counts per minute (cpm) to around ten.

Between 1952 and 1957, six internal LSC coincidence counters were built

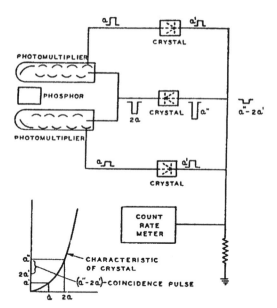

22. Coincidence circuit using crystal diodes. Morton and Robinson 1949.

for use in Wright Langham's Biomedical Research Group of the Health Division of Los Alamos Scientific Laboratory.[43] Other Los Alamos scientists experimented more adventurously with spectacular devices for external liquid scintillation counting. As a result of increasing military, industrial, and biomedical use of atomic energy, the risks of exposure to radioactivity and the means of preventing it were being discussed ever more vehemently. Ernest Anderson built a machine that could test a whole human body for naturally occurring radioactive potassium or heightened radioactivity after exposure to fallout; as a justification he pointed to the "sharp rise in public concern, in the past few years, about the effects of low intensity radiation on man." If present trends continued, he said sardonically, there might soon be "a legal prohibition of some of our most popular materials of construction, notably concrete and brick, on the basis of their high concentration of natural radioactivities such as radium and potassium."[44] One of Anderson's colleagues, Frederick Reines, was already working to build giant liquid scintillation detectors to measure neutrinos and neutrons.[45]

The constellation in Los Alamos is particularly interesting for what it reveals about the gradual development of generic, multi-purpose measuring devices. Initially the new technology of internal liquid scintillation counting tended to be reserved for special radiation measurement purposes; however, it rapidly evolved into a sophisticated assemblage of various physical, organic-chemical, and technical components into which biological and a variety of other materials could be inserted in order to be measured. The notion that it might ideally be possible to "prepar[e] counting samples by simply dissolving the substance in solvent in a bottle" had only just appeared on the horizon;[46] the prospect of constructing an instrument that could be operated by technically and scientifically inexperienced laboratory personnel carrying out routine laboratory work was still remote, as was generic use of the new device in the production of isotopes, the monitoring of radioactivity, or the disposal of waste in the fields of materials research, medical diagnostics and therapy, physiology, chemistry, or pharmacology. Among the other technical challenges to be met, aside from the technical core of the apparatus, were the geometry of the vials, the elimination of luminescence, the choice of the material to be used for the container, the optimization of photocathode sensitivity, the emission spectrum of the scintillator, and so on. The sample material too had to fulfill certain conditions. It had to be made soluble in the organic solvent of the scintillator, which was no small problem, and it had to be colored as little as possible so as to avoid "quenching," that is, reduction of the photon yield induced by the sample itself.

Lyle E. Packard

Although the first efficient LSCs were built in Los Alamos, the first commercial version of the apparatus originated in Chicago, another center of the development of American nuclear technology during the Second World War and thereafter. It was in Chicago in 1941 that Arthur Compton established the Metallurgical Laboratory (Met Lab), in which Enrico Fermi, Walter Zinn, and their colleagues built the world's first nuclear reactor—Chicago Pile-1—and produced a self-sustaining nuclear chain reaction late in 1942. Met Lab became part of the Argonne National Laboratory in 1946, a center for nuclear reactor technology and nuclear propulsion engines.[47]

Lyle E. Packard held a degree in mechanical engineering from the Illinois Institute of Technology in Chicago. During the war he served in the Navy, where he had his first experience of electronics and was soon working on radio, radar, and sonar projects. In spring 1946 he was hired as an engineer in the Institute of Radiobiology and Biophysics at the University of Chicago.[48] Together with the Institute for Nuclear Studies and the Institute for the Study of Metals this was the third new research facility established by the president of the university, Robert Maynard Hutchins, under the university's peace program, which had been launched immediately after the war.[49] The new institutes replaced parts of the Manhattan Project that had been operating under the aegis of the University of Chicago. Hutchins wanted these institutes "to advance knowledge and not primarily to develop the military or industrial applications of nuclear research." He observed that "for the past six years, the United States has abandoned both basic research and the training of a new generation of scientists. It is essential to our progress and our welfare that we overcome that deficiency."[50] Packard took on various assignments in the new institute. He set up laboratory space and also designed and built with a small group of technicians and engineers the equipment and special instrumentation needed by the various research groups: for example, apparatuses for studying nerve cells in squids for Kenneth Cole and George Marmont and a solid system for time-lapse photomicrography and various radiobiological instruments, including a Van de Graff generator, for the institute's director Raymond Zirkle and his colleague William Bloom.

Leo Szilard, who together with Aaron Novick was in the process of switching from biophysics to biology, had two special needs. He had thought up a "chemostat," an apparatus that would keep a bacterial population growing indefinitely, making it possible to study the long-term effects of mutations.[51] Szilard had joined Compton's Met Lab in 1942. After trying to no avail to convince President Truman not to use the atomic bomb against Japan, Szilard had taken an indefinite leave of absence from the Manhattan Project in Chicago in fall 1945. In 1946 University of Chicago President Hutchins appointed him Professor of Biophysics in the Institute of Radiobiology and Biophysics.[52] "When Szilard joined the University," Packard recalls, "he had a number of patentable ideas, things that he wanted to preserve for himself and he excluded those from his contract. It was on one of those things in particular that I have worked personally with him, after-hours, weekends and so on. So I got to know him a little bit. Very, very interesting experience."[53] Packard designed Szilard's laboratory, which was set up in the basement of what had once been the synagogue of a Jewish orphanage. The building had been taken over by the University of Chicago and was ready for use by January 1948.[54] The first chemostats were placed in a thermostatically controlled room held at a temperature of 37°C. "We did that at Szilard's request in a very inexpensive way by controlling banks of commercial 1500 Watt heaters."[55]

In 1948 Packard contacted a group of researchers at the Institute for Nuclear Studies who were measuring the decay of ^{14}C using mica end-window Geiger counters. These counters were inefficient; their yield was on the order of 10 percent. However, a staff member of the institute,

Nathan Sugarman, had recently invented an instrument without a window. Packard now set about constructing a workable windowless counter in which the sample could be inserted directly into the counting chamber. This was supposed to result in a higher counting yield, since it eliminated absorption of the β-particles by the window's mica plate.

The use of low energy radioisotopes had expanded rapidly as a result of the activities of the American Isotope Distribution Program. The electrons of weak β-emitters such as ^{14}C were difficult to monitor, because they had a very short range. They were used, nevertheless, for a simple reason: they made it possible to label a potentially unlimited number of different organic molecules so that they were suited for in vivo and in vitro metabolic studies of all kinds. In vitro systems based on these low energy radioisotopes were developed parallel to the tracer technology already in use; research on cell homogenates notably would have been inconceivable without them. The more widely these radioactively labeled compounds—nucleotides, amino acids, fatty acids, sugars, antibiotics, and so on—were disseminated, the more counting devices were needed. In addition the institutes were now confronted with the problem of radioactive contamination of the workplace.

Many visitors wanted to know, Packard recalls, where they too "'could get one of these'" instruments. "'Well, you can't get one,' they were invariably told; 'we make them here.'"[56] As a result of this demand Packard began to think in 1949 about founding a company of his own. With the permission of the university administration, he set up a firm that he named "Research Equipment and Service." Its first product—a windowless counter with a relatively slow transfer mechanism and a single position for inserting and removing samples—had been designed by Herbert Anker for the university's biochemistry department. It was soon followed by a version that had a considerably more efficient circular sample device with an internal counting position, a "pre-flushing" position, and an external loading and unloading position. Another of the firm's early products, an unspectacular fraction collector, had nothing to do with measuring radioactivity per se.

LSC—From Prototype to Product Line

At the Institute of Nuclear Studies Packard had befriended James Arnold, a former student and coworker of Willard Frank Libby, whose appointment to the General Advisory Committee of the AEC dates from this period. Libby was then using the naturally occurring carbon isotope ^{14}C for archeological dating: the method known as carbon-14 dating for which he received a Nobel Prize a decade later (1960).[57] He had obtained his first results with a solid sample Geiger-Müller counter, "a big thing about 4 inches in diameter, with all kinds of shielding around it."[58] Arnold then heard about the potential of liquid scintillation counting—the foundations for which had recently been laid by Kallmann and Furst in New York and Reynolds in Princeton—and had explored the prospects for applying internal liquid scintillation counting in the framework of the carbon-dating project. The instrument was a counter of the sort that Newton Hayes was building for Wright Langham's Biomedical Research Group at Los Alamos. Since Arnold was in close contact with both Hayes and Anderson, he had access to unpublished data and new scintillation materials. He was now planning to perfect the idea of internal sample measurement by converting low activity samples themselves into the solvent for the scintillator.[59] George LeRoy of the Argonne Cancer Research Hospital was also in contact with the physicians and physicists at Los Alamos and their work on a new counter had caught his eye. Packard was following Arnold's work with great interest as LeRoy knew. LeRoy also knew that Packard had turned his back on the university and had been running his own one-man company since 1952. He therefore inquired as to whether Packard might not be able to build him a liquid scintillation system.[60] The Argonne Cancer Research Hospital had been founded in 1948 with money from the AEC and was operated by the University of Chicago as part of the AEC program for fighting "America's number-two killer disease."[61]

Packard had hired his first coworker with an eye to constructing the prototype for a commercial LSC based on a coincidence system. The two of them worked in the front part of his apartment, which he had transformed into a workshop. It took the newly formed "Packard Instrument Company" approximately one year to build the first counter and put it into operation; it was installed in the Argonne Cancer Research Hospital in 1953 (fig. 23). None of the basic components of this machine was completely new, but thanks to his previous experience in the institute Packard knew what counted most in a research instrument and what users especially appreciated: versatility and ease of operation. LeRoy and his coworkers wanted to use the new instrument in evaluating double-label experiments with tritium (^{3}H) and radioactive carbon (^{14}C).[62] "That's when I came up with the idea of designing the production model especially for these two isotopes and naming it the Tri-Carb for ^{3}H and ^{14}C and gave it the model number 314." These requirements for the prototype determined

23. First commercial LSC made by Packard, sold to Argonne Cancer Research Hospital, circa 1953. Collection Edward F. Polic.

the way the electronic circuitry was laid out in the serial production that began soon thereafter.[63]

The price of the prototype, including the refrigeration unit, which basically consisted of an adapted commercial freezer, was $6,500 — at the time, about five times the cost of a Geiger counter equipped with a scaler. But in this context the price did not matter much. Eugene Goldwasser, a biochemist from LeRoy's laboratory, recalls that

> it was one of those peculiar times of history, at least from my perspective, when we had all the money we needed for research and George could go on and ask Lyle to build (an instrument) without worrying about where to get the money to pay for it. I came to Chicago from Copenhagen, and the first thing I had to do was sit in an unfinished room with stacks and stacks of catalogues and start a lab from nothing. And I figured roughly in 1952 I spent about a million dollars. [This] was all AEC money; there seemed to be no end to it. [That] was part of the original AEC charter from the Congress. They were to promote the use of radioactivity in research and therapy, and to promote the development of instrumentation for study of radioactivity. So that my work which had little to do with atomic energy or cancer research [was] funded under their umbrella because I used isotopes.[64]

In Denmark, Goldwasser had worked with radioactive adenine, which his mentor Herman Kalckar produced on his own.[65] Now he obtained his labeled compounds from Berkeley's Rad Lab.[66]

This episode was typical of the times. Yet a particular epistemic and technical constellation was involved: a mechanical engineer with experience in electronics who together with academic researchers was fiddling about with ultramodern technology that for the moment seemed to be applicable only to solving very specific archaeological questions of dating or to evaluating double-label experiments. Here is Packard's own assessment of this constellation:

> I don't think I can generalize my thoughts on the interactions of scientists and engineers. Based on nearly five decades of developing and manufacturing scientific instruments as well as five years at the University of Chicago functioning in somewhat varying roles, but all generally facilitating the requirements of scientists, I found extreme variations. To a certain extent, it depends on the field of science. As might be expected, physicists typically are more concerned with technical specifications and details of the mechanics and electronics of what they want from engineers. Biological and medical researchers, I have found, usually interact with engineers on the basis of the function they wish to accomplish—how easily, how fast, how precisely, etc.—and typically are not interested in details of the equipment as long as it performs what it is supposed to do reliably. There are exceptions and, particularly in earlier times, I have seen extreme cases. For example, I've seen biological scientists who like to play at engineering spend months of their laboratory time improvising something like a homemade fraction collector.[67]

Packard's company flourished as a result of these laboratory contacts. Most of his early products were purchased with money from the AEC and the federal Public Health Service, both of which were very well funded in the first years of the Cold War.

The Packard Instrument Company soon moved to the Chicago suburb of LaGrange, Illinois. Hardly had Packard delivered the Tri-Carb prototype to LeRoy than he received orders for two more scintillation counters. One went to Jack Davidson of Columbia University's Presbyterian Hospital. "That was really the first which I would call a production unit."[68] Davidson used his machine to carry out all sorts of optimizations of samples. He tested various sample sizes and different mixes of solvents containing 2.5-diphenyloxazole (PPO) and 1.4-di(2-[5-phenyloxazolyl])benzene (POPOP); these soon came to be called "cocktails" in laboratory jargon.[69] He considered liquid scintillation counting "a useful new technique," but "by no means the panacea for all counting problems."[70] Its most significant advantages were its great sensitivity to very weak decay electrons, its high precision, and its high absolute counting yield, as well as the fact that it was relatively easy to prepare samples for it. A survey of the early

LSC-literature leaves one with the impression that the first commercial machines were used mainly to test the new methods themselves: the application of the methodology then involved little more than its own optimization. It was a small, self-amplifying circuit.

I emphasize this point because in this period—the early 1950s—no one could foresee that the LSC would eventually become a kind of master-technique in biophysics, biochemistry, and biomedicine. Nuclear Chicago, one of the biggest and most experienced instrument builders in the field of commercial radiation technology, was still banking on solid sample counters and gas counters. According to the testimony of a contemporary user, Nuclear Chicago's D-47 Micromil gas flow counter, equipped with an ultra-thin Mylar window, counted ^{14}C with an efficiency of approximately 40 percent; the instrument had moreover a low counting background thanks to its anti-coincidence circuitry, while its sample changer was "nearly fail-safe."[71] Nuclear Chicago did not enter liquid scintillation counting until the early 1960s.[72] Yet even before it did Packard's Tri-Carb system had competitors. The Tracerlab company in Waltham, Massachusetts, which also produced radioactive biochemicals, was likewise working on building an LSC, as was the Technical Measurement Corporation (TMC) in New Haven, Connecticut.[73] Packard recalls a bitter comparison test that pitted one of his machines against one produced by the TMC:

> Our biggest competition in the early days came at the NIH. It seemed to us, in a much smaller way of course, like something they sometimes had in military procurement, where two companies would be requested to provide special-purpose airplanes for a fly-off to see which one was better. In this case, NIH requested one of our Tri-Carb systems and one of the TMC units for side-by-side comparisons. After extensive testing, our Tri-Carb system was selected and purchased as the first of dozens that NIH would acquire during the next few years.[74]

Edward Rapkin, who later joined the Packard Instrument Company, points in this connection to the peculiar, "unsymmetrical" system logic of the Packard machine. It required two photomultipliers with different functions. One was used for pulse height analysis, the other for monitoring coincidences (fig. 24). This arrangement necessitated only *one* good phototube and *one* good amplifier, an advantage that was not to be underestimated in the days when vacuum tubes were selected by hand.[75] By 1956 Packard's company already had twenty-five people on its payroll; it sold some twenty systems that year.[76]

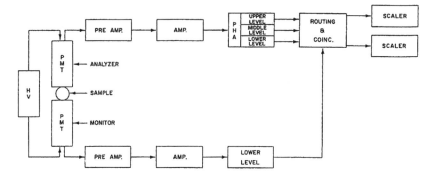

24. Packard Model 314 block diagram, 1954. Rapkin 1970.

Automation: A New Time-Management Regime

The early Packard Tri-Carb Spectrometer Model 314 was operated manu-
ally, the counting time was set mechanically, there was only one sample
position, and the electric circuitry was based entirely on vacuum tubes
(fig. 25). Although as Davidson notes, "the user of this apparatus [did] not
need to understand all of the electronics involved, any intelligent use of
the technique [called for] familiarity with the general principles of the
electronic equipment."[77] The first major change in design was introduced
in 1957 when an automatic 100-sample changer was incorporated into
the apparatus (fig. 26).[78] Until then changing a sample had been a time-
consuming business. Even the high voltage had to be switched off before
a sample could be removed and replaced with another. "Take it out, put
the next sample in, close the light-tight chamber, close the lead shield,
close the freezer door, turn the high voltage on. And then wait a little bit
and then start your count. And your count was manual. So, you sat there
and you watched the clock go around for a minute or two minutes or five
minutes or whatever."[79] The sample vial was a large glass vessel with a
capacity of 50 ml that was immersed in silicon oil to ensure a good optical
connection to the phototubes. The first prerequisite for automation was
a reduction in the size of the vials, which would also make the silicon-oil
connection unnecessary. Packard and his new partner Soderquist designed
the circular sample changing device. The former sample shield, which had
consisted of a horizontal cylinder, was replaced by a vertical iron cylinder, and the turntable was placed on top of it. This model sold exceedingly
well. In 1958 the firm, which now employed eighty people, produced and
delivered close to a hundred units. After TMC discontinued LSC production
Packard had a market share of almost 100 percent.

25. First serial production of a Tri-Carb LSC, a detector with a light-tight shutter—for 50-ml sealed ampoule samples or 20-ml vial samples, circa 1954. Collection Edward F. Polic.

26. First automated Liquid Scintillation Detector, steel shielding, dual elevators, 100 20-ml vial samples in four circular rows, 1957. Collection Edward F. Polic.

Its transformation into an automated machine made the LSC genuinely attractive in an "aesthetic" sense as well, perhaps—but it also made it an epistemically interesting device. With the conventional counters in use after the war each assay "even when efficiently organized . . . must have consumed an hour of professional time."[80] Now scientists confidently left their samples to the machine and to the machine's temporality. The possibility of making serial counts of hundreds of samples—unattended and overnight—converted experts at running LSCs into simple users with the most varied interests, who could now perform experiments of hitherto unsuspected dimensions and were also able for the first time to plan measurements in series. In kinetic experiments such as binding studies, for instance, samples were taken at different times and as often as possible. Series of experiments could now be carried out at different temperatures or different experiments could be combined and samples could be taken in duplicate or even triplicate in order to arrive at more stable results. In the case of big experiments this could easily mean testing a hundred samples or more. The automatic machine also made it possible to test large numbers of fractions obtained from chromatographic purification, separation columns, or preparative ultracentrifugation runs; furthermore, it allowed laboratory workers to include in the design of their experiments, not the minimum number of controls, but as many as they deemed necessary to make the differential signals as significant as possible. Not only did experiments become more extensive and efficient as a result; the anticipated capacity of the machines also altered their epistemic structure. Radiobiological experimental systems of a new type were born. Packard comments:

> Putting this turntable on, and making it automatic, just opened up the possibility for what then were massive studies. I mean, you would take a rat and sacrifice it, and take all the parts of the rat. [It] made it possible for people to design different types of experiments than they ever could have designed before. Previously you never would have designed an experiment for a thousand samples. [Well] I don't know how many times, but certainly *many* times I have had people come up to me at trade shows, really prominent names in the field and say, "you know, Lyle, all the work I have done in the last five years I could not have done if I hadn't had a Tri-Carb."[81]

If initially liquid scintillation counting had been nothing more than a potentially promising technology allowing scientists to add tritium labels to their tracer arsenal and exploit radioactive carbon more efficiently, the automated version of the device enabled them to scale up routine monitoring by orders of magnitude. At the same time it allowed them to set up hitherto inconceivable experimental systems. From an epistemic standpoint the operationality of the instrument took on new forms: what was originally only a locally relevant increase in the sensitivity of measurements became a pervading influence on the experimental regime of biomedical research in general. From the standpoint of routine applications the result was a manifold increase in monitoring capacity. With this massive improvement the automated liquid scintillation counter became an instrumental bridge between the purification of biomolecules on a preparative scale and analytical molecular biology. Wright Langham, a member of the Los Alamos Biomedical Research Group, put the challenges and possibilities that had appeared on the horizon by the late 1950s in a nutshell: "Biological and medical investigations by nature call for counting systems with the greatest of versatility. Among the requirements are (a) analyses of large numbers of samples with a minimum of processing; (b) high sensitivity; (c) wide adaptability as to variations in sample size; (d) accommodation of wide variations in the nature and chemical composition of the sample; [and] (f) dependable operation with a minimum of servicing."

Let us consider these points one by one. Large numbers of samples were necessary—the desire to perform serial experiments aside—because of the inherent variability of biological probes, which necessitate multiple measurements. A minimum of required processing steps between the experiment and the counting procedure reduced the risk of loss of activity attendant upon each of these steps. High sensitivity in measurement had the advantage among others of allowing the use of low doses of radioactivity: in the case of biological samples this made it possible to avoid damage caused by the tracer itself. Sample sizes varied in accordance with certain necessary components of the experimental design. Finally, dried pieces of filter paper were to be measured as well as substances dissolved in water or alcohol. The new automated LSC machines promised to satisfy all these desiderata. For Langham therefore "liquid scintillation counting is the most important recent development in the applications of radioisotopes to biology and medicine."[82]

Edward Rapkin, who joined Packard's company in 1957, recalls the enormous increase in sales figures after the automation of the Tri-Carb. "The thing . . . converted everything from an occasional sale of three or four a month to, some months, fifty."[83] By 1961, seven years after the first commercial unit had been delivered, more than 700 Tri-Carb Spectrometers had been installed around the world.[84] Direct sales to the United States government and its agencies now accounted for only about 15 percent of total sales. The rest were to universities, hospitals, and private industry.

In 1959 Tracerlab made a brief attempt to re-enter the market, this time with an automatic counter of its own and two pertinent innovations. Tracerlab was the first company to use the new EMI photomultiplier that soon became a standard feature of scintillation machines. Moreover, it was the first to introduce transistorized preamplifiers, which, however, could not yet match the performance of good vacuum tubes. But the Tracerlab machine lacked "important user requirements" such as high sample capacity or light-tight, trouble-free detectors.[85] The electronic improvements did not make up for these defects. A year later in 1960 Packard introduced the 314A Tri-Carb, which was entirely transistorized, except for the preamplifiers. The 314E series followed promptly in 1961. This model was now completely transistorized; the system logic had been so vastly improved that two different isotopes—for example, ^{14}C and ^3H—could be counted simultaneously in a sample with separation efficiencies that previously were achievable only by means of successive counts. That same year Nuclear Chicago also marketed an LSC. By 1962, it was offering an instrument with a serpentine sample transport and a sample capacity of 150 vials. This made it possible to perform repeated counting cycles one right after the other and also to arrange samples in separate groups so that different users could use the machine simultaneously. Moreover, a mechanical calculator facilitated data processing. The device competed successfully with the most recently introduced Tri-Carb, which had sample capacity of 200 vials. On Rapkin's estimate, Nuclear Chicago's market share rose to roughly 20 percent over the next five years, while Packard's fell from 85 percent to 63 percent.[86]

The market expanded very rapidly, as is shown by comparative statistics in specialized publications typical of the field, such as the *Journal of Biological Chemistry*: in 1959, only one in ten experiments using either ^{14}C or ^{3}H as a tracer was evaluated by means of liquid scintillation counting. Conventional Geiger-Müller counters with end-windows, vibrating reed electrometers, and gas flow counters widely dominated the scene, together with Nuclear Chicago's D-47 windowless gas flow counter. In 1963, a scant four years later, approximately one of every three comparable experiments involved an LSC (about 80 percent of the time a Packard Tri-Carb). By the end of the 1960s practically every sample in $^{14}C/^{3}H$-based experiments landed in an LSC. By that point the method was so widespread in experimental practice that it was no longer even mentioned in the technical sections of research papers.

The fierce competition that broke out at this time forced firms to make constant innovations in system logic and to offer ever more user-friendly machines and in particular more user-friendly counting devices.[87] In 1962 several former Packard employees founded the Vanguard Company, which marketed an automatic LSC with the advantage that it could be installed atop a laboratory bench. The business went bankrupt after delivering a few units. In March 1963, Rapkin left the Packard Instrument Company to open a small firm of his own, ANSitron. It produced the first serial device to incorporate automatic external standardization. Packard too had developed ideas about an automatic standardization that would correct for quenching, filing a patent for such a device in June 1965.[88] In 1966 ANSitron was bought out by the Picker Nuclear Company, which soon disappeared from the market. Also in 1966 Nuclear Chicago, Packard, Picker, and Beckman all brought out cheap LSC models. In 1967 both Packard and Beckman announced sophisticated systems whose automatic standardization equipment determined the effects of quenching before samples were counted.[89] Yet, however fierce the competition, the market was still expanding and could in principle accommodate a great many competitors (Table 2), since LSCs were just beginning to flood the laboratories. Ten times as many people were employed in Packard's company in 1966 (500) as had been in 1957 (50), and the net turnover had risen from half a million to fourteen million dollars, representing sales of about one thousand units per year.

Table 2: Estimated market shares of the main LSC producers between 1955 and 1970. Estimate Edward Rapkin 1996.

	% Dollar Market Shares (estimated)			
	1955	1960	1965	1970
Packard Instrument Company Inc.	95	85	63	45
Nuclear-Chicago; Searle (first sale, 1960)	—	10	20	15
Beckman (first sale, 1965)	—	—	5	20
Wallac; LKB	γ counters only until approximately 1973			
Ansitron (first sale, 1964)	—	—	10	6 (Picker)
Intertechnique (first sale, 1968)	—	—	—	15
Philips				
Picker Nuclear	—	—	2	acquired Ansitron 1966
Sharp Laboratories never sold LSCs (see Beckman)	—	—	—	—
TMC	5	discontinued		
Tracerlab	—	5	discontinued	
Vanguard	sold < 5 units total			
Nuclear Enterprises (Scotland)	very few units sold, none outside UK			
Berthold (Germany)	—	—	· —	3 (Germany only)
Aloka (Japan)	—	—	—	
Russia	no commercial counters, only home built			
Belin (France)	sold a few units to French AEC 1958–62			

With the optimized features that the liquid scintillation counter now boasted—it was equipped with an internal data processing system and it could be used to measure probes of different experiments simultaneously in bigger laboratories—the LSC could now be exploited for a wide variety of uses. In part because of the possibility of producing multipurpose scintillation cocktails it found applications not only in molecular biological research, but also in biomedical diagnostics, clinical settings, and the pharmaceutical industry. The production of increasingly more efficient machines between 1953 and 1967 had been enhanced by the exponential growth of biochemical and biophysical research, the expanding commercial production of radiochemicals, the huge programs for fighting cancer of the early post-War years,[90] and the development of nuclear medicine and the nuclear medical industry. Yet molecular biology was itself conditioned by the radiation technology whose development it stimulated. Without the massive spread of radioactive biomolecular tracers the molecularization of biology and medicine would have taken a different course; without the technology of liquid scintillation counting in particular the 1961–68 experiments that led to the full deciphering of the genetic code would have been inconceivable, to cite only one example. It is not that LSC technology itself delineated entirely new fields of research. It had a more diffuse and for that reason a more comprehensive impact on the planning of experiments and above all on the scope of metabolic analysis in the test tube, that is, on experimentation in vitro.

In this context the ever more diversified synthesis of substances labeled with ^{3}H and ^{14}C must also be mentioned. Initially these isotopes came mainly from the National Laboratory at Oak Ridge (fig. 27). Soon, however, commercial producers entered the business, among them Tracerlab, New England Nuclear Corporation, Isotopes Inc., and the Radiochemical Center in Amersham, England. Robert Loftfield recalls, "New England Nuclear set up a tritiation service: send us your compound, we will tritiate it over a one week period and return it to you for purification or experimental use."[91] But the tritiation process was complicated: it frequently resulted in highly labeled byproducts, and the exchange between incorporated ^{3}H and the solvent was difficult to monitor. The rigorous checks that were thus required ruled out routine application of the method. The development of LSCs and the production and purification of substances labeled with ^{3}H and ^{14}C proceeded parallel to one another for some two decades, requiring constant mutual adaptation. Had the "software" problems caused by the tracers not been resolved, the "counter hardware" could not have been introduced into molecular biology and biomedical

The Geiger counter measures the diffracted rays, whose intensities and angular position are characteristic of the material, and with the aid of suitable electronic circuits, translates the impulses into a hill-and-valley curve on the chart.

RADIOCHEMICAL SERVICES

Tracerlab, Inc., 55 Oliver St., Boston 10, Mass. Eight chemical services, which facilitate the use of radioisotopes, are available through the Radiochemical Division. These services are: purification of isotopes, dilution and standardization of isotopes, storage of radioactive materials, isotope sharing plan, synthesis of radioactive compounds, radiochemical analyses, radiochemical research, and consultation on special problems.

Tracerlab is approved to receive isotope shipments from Oak Ridge as an agent for the purchaser. These isotopes may be obtained for the following uses, in order of priority: (a) publishable researches in the fundamental sciences, including human experimentation, which require relatively small samples; (b) therapeutic, diagnostic and tracer applications in human beings and publishable researches in the fundamental sciences which require larger samples; (c) training and education by accredited institutions in the techniques and applications of radioisotopes; (d) publishable researches in the applied sciences.

LABELED COMPOUNDS

Isotopes Branch, Field Operations, U.S. Atomic Energy Commission, P. O. Box E, Oak Ridge, Tenn. Carbon-14-labeled compounds are being made available as a result of research on synthesis procedures by the Clinton Laboratories. Methanol, in which 3% of the carbon atoms present are radioactive, is the first such compound to be prepared. It will contain 10 to 20 millicuries of C^{14} and will cost $100 per mc of C^{14} in the form of CH_3OH. Application for the labeled methanol should be made on Form 313, "Request for Radioisotopes."

27. AEC advertisements for radioactive carbon. *Nucleonics*, October 1947, 85.

research. The instruments, the molecular probes, and the epistemic agendas into which they were inserted constantly determined and overdetermined one another.

The Industry and Its Customers

So far, we have seen how direct co-operation between researchers and engineers led to the development of LSC prototypes and later efficient and marketable machines. We have observed a direct reciprocal relation between users' needs and technical solutions (involving in the case to hand physics, chemistry, photoelectronics, and mechanics). Packard's coworker Leo Slattery, for instance, continually proposed new ways of applying his instrument to particular experiments; here manufacturing and research went hand in hand.[92]

As soon as the instruments began to be produced for commercial ends this symmetrical relationship changed. Packard himself insisted that from that point on most improvements of the instrument came from "inside"—that is, from within his own firm or competing firms that had marketed new models. But it had not therefore become less important to get the most encompassing information possible about how the users wished—or might soon wish—to apply the apparatus. Obtaining this feedback became the task of "combination people," to use Packard's term. Such "combination people" were both salespersons and service and repair personnel. According to Packard separating these two functions would have been not merely counterproductive, but disastrous. "If we had ever tried to do this through reps, it just would not have worked. [I] think our own people could give the best installation, the instructions and the theory of operation, and all of that necessary support."[93] Rapkin too has stressed that "the salesmen were all hired to be servicemen," adding, "one thing I think may have been the strength of the Packard Instrument Company in those days was its sales force . . . they were good about reporting back new requirements and problems."[94]

Gerhard Kremer has summed up his experiences in the LSC field with the remark that relations between research, technological refinement, customers, and marketing were decisive. Research and development engineers on the one side and customers on the other almost always had, he said, different ideas about how to perfect an instrument. The necessary mediation between the two sides had therefore to be assured by salesmen with extensive knowledge of the field, but also a feel for user needs. Paradoxically, the salesmen needed more knowledge and psychological competence the more the technology was adapted to routine use, since such adaptation made it also nontransparent.[95]

In one sector of the LSC business, however, knowledge unambiguously proceeded from the customers to the manufacturer. Refinements of sample preparation, including new recipes for scintillation cocktails, emerged through the efforts of the many different laboratories in which individual scientists were wrestling with their idiosyncratic experimental problems and trying to exploit the machines for their special purposes. Much of the early work of Hayes in Los Alamos, Arnold in Chicago, and Davidson in New York was deeply colored by such efforts. This is also the way plastic instead of glass vials eventually came to be used as sample containers. Rapkin reports, "A customer [Herbert Jacobsen, University of Chicago] told me that they were using plastic vials. And we tried it and it worked very well. . . . I think by having a wall that was diffusing the light, there was a better chance the photomultipliers would get the light. So for tritium counting, the improvement was significant as a per cent of the total count."[96] Although plastic had the disadvantage that the organic scintillation cocktail gradually permeated the walls of the containers, which therefore had to be disposed of after a short time, the cheap polyethylene vials at least partially replaced the more expensive glassware, which moreover could contain at least some naturally occurring radioactive isotopes.[97] Eugene Goldwasser has mentioned another example, the dual label counting he had to do while working with George LeRoy at the Argonne Cancer Research Hospital: It was, he said, "a two-directional kind of thing. Once it became known to experimenters that you could discriminate isotopes based on the magnitude of the pulse you get, then they would sort of talk to the people developing instrumentation saying: 'This is really what we would like to be able to do.'"[98] Although dual label counting seemed very promising from the moment it was introduced, it took years of tinkering and many setbacks before it could become routine.

The interaction between users and suppliers was vital. According to Robert Loftfield, Packard realized in 1958 or 1959 that many of his Tri-Carb machines had ended up in the hands of inexperienced customers and were either standing around in hospitals unused or at best turning out data of limited value. He therefore undertook a successful effort "to persuade the Atomic Energy Commission to set up an award sufficient to place Tri-Carbs in some 20 reputable laboratories where problems could be uncovered and applications developed that would increase the usefulness of the Tri-Carb for other hesitant scientists."[99] One of these machines — "a beautiful machine: coincidence counting, automatic sample changing, cooled

to about -10°C, automatic print-out, pre-selectable voltage gates, etc."—
went to the John Collins Warren laboratories of the Huntington Memorial Hospital in Boston, where Loftfield used it to explore the pitfalls associated with the direct counting of paper chromatography strips.[100]

Out of such epistemic-technical interaction new interfaces are constantly generated between industry and the lab. Highly developed "research-enabling" technologies such as liquid scintillation counting or ultracentrifugation with its multiplicity of rotor beakers and different types and sizes of rotors—"accessories" that nevertheless can enrich and expand existing experimental systems—require special forms of product management and marketing. Those who sell such technologies ascertain new needs in and for laboratories, as well as possible new applications; they bring about product modifications that influence a company's entire research and development program. A 1961 Common Shares Prospectus of Packard Instrument Company notes:

> The Company [maintains] a laboratory to study product applications, to handle trial samples for prospective customers and to devise and test techniques for utilizing both its existing products and new products under development. The Engineering Department devotes its efforts to the development of new products and improvement of existing models. During the year 1960 the Company had approximately fifteen employees engaged in research and development and spent approximately $210,000 for this purpose, exclusive of quality control and normal product testing. Consultants are utilized where special skills and knowledge can be more effectively obtained than with full-time staff members.[101]

Thus at that point Research and Development accounted for slightly less than 10 percent of net turnover ($2,964,161) and slightly more than 10 percent of the total number of employees (15 of 125).

Networks

Yet another aspect of this story of a "research-enabling technology" was—in addition to competition—the building of cooperative networks. As has already been mentioned, those who used the technology in biological research needed adequately labeled molecules. Those who developed it knew that the technical production of liquid scintillation counters—testing the electronic circuitry, calibrating the counting procedure, and establishing standards—presupposed pure, precisely quantified standard samples of the various isotopes for purposes of comparison. Packard therefore remained in close contact with Edward Shapiro and Seymour Rothchild of New England Nuclear Corporation, a company founded by former Tracerlab employees that produced and sold radioactively labeled compounds. Tracerlab itself had been one of the first private companies authorized to act as an agent for the distribution of isotope shipments from Oak Ridge and also to synthesize a variety of labeled molecules.[102]

Besides developing contacts with the isotope industry, Packard Instrument also intensified its public relations work. "When Edward Rapkin came to work with us, what we wanted was for him to become the leading liquid scintillation oracle and publish little newsletters for us, which he did. So from that time on, the person who knew everything that was being done, all the techniques, all those solvents, and all the cocktails, was Dr. Rapkin."[103] Rapkin had seen his first Packard Tri-Carb while doing his military service in the Argonne National Laboratory, where he operated a mass spectrometer. Before joining Packard he had investigated the alkaline digestion of proteins with hyamine for the firm. The goal of his research had been to make biological samples containing protein soluble in toluene-based scintillation mixtures. This was a critical aspect of the process of preparing samples,[104] since most biological samples contained protein in varying quantities.

Publishing and distributing a technical bulletin was only one facet of the necessary public relations effort.[105] A bigger challenge was setting up conferences to stimulate discussion between scientists and engineers, research institutes and laboratories. In this way the company itself became part of a communications network that brought together all interested parties, academic and commercial. To this end Packard and Seymour Rothchild of New England Nuclear and Atomic Associates sponsored a long series of conferences beginning in 1957 on "Advances in Tracer Methodology"; they were later edited by Rothchild and published in four volumes.[106] Packard himself occasionally gave papers, as he did, for example, at a conference on liquid scintillation counting held at Chicago's Northwestern University in August 1957 and another on organic scintillation detectors that brought more than two hundred participants to the University of New Mexico in August 1960. Both events were sponsored by the National Science Foundation; the second was co-sponsored by the AEC.[107]

The conference at Northwestern University reflected the rapid increase in the demand of liquid scintillation counting observable toward the end of the 1950s. The list of participating institutions was impressive: it included research institutes of renowned universities such as Princeton, Columbia, New York University, the University of Chicago, and the University of California at San Francisco; national laboratories such as Los Alamos, Argonne, and Brookhaven; the National Bureau of Standards and the National Institutes of Health; hospitals such as the Veterans Administration Research Hospital; international laboratories such as Saclay in France, the Atomic Energy Research Establishment in England, and the Weizmann Institute in Israel; and companies such as the Packard Instrument Company, Technical Measurement Corporation, Shell Oil Company, and Tracerlab. Interest in the new research technology was widespread and the range of potential users reached from its epistemic core—research in physics, chemistry, molecular biology, and biomedicine—to medical diagnostics, precision measurement, standardization, and radiation control. It thus brought together experimental physicists, chemists, radiologists, biochemists, archaeologists, medical researchers, electronics engineers, mechanical engineers, and instrument builders. All either contributed to the physical, chemical, or engineering parts of the apparatus and/or applied it in their research.[108]

The LSC was on its way to becoming the fulcrum of an informal transdisciplinary community of researchers, engineers, and industrialists. James Arnold, who was among the participants at the Northwestern University conference, remarked shortly thereafter: "As one of the early workers in the field, I am made rather complacent by the fact that it has ramified in so many unexpected directions. This is actually a rather good case history of the unexpected applications which result from 'pure' research. I do not think that either the people at Los Alamos or the others who were working in the field at that time would have been much more successful in predicting all the applications represented at this conference."[109] Yet to cast the story as one of pure research and its application is at least questionable. For what the story of liquid scintillation counting shows is that separating "pure" research from "mere" application is not a good conceptual choice for understanding and presenting the origin and development of research instruments. At least in this particular history the question as to where basic research leaves off and application commences must remain open.

Let us recapitulate. In the exploratory phase between 1950 and 1955, liquid scintillation counting was one option among others; indeed, there existed much more firmly established methods of radioactive counting. What made liquid scintillation operative and finally capable of competing with other methods was a conjunction of heterogeneous factors. Among them were research into a physico-chemical principle (liquid scintillation) of potentially generic applicability in bioassays; promotion of the photoelectronic industry as a result of intensified arms production and mass communication; the pervasive use of biologically pertinent weak β-emitters in the aftermath of atomic fission; and mechanical automation, which revolutionized laboratory work. Another, overarching factor was the basic predilection for "wet" experimentation in biochemistry and molecular biology, coupled with the fact that a "liquid" interface between the measuring device and the probe proved to be extremely adaptable and widely applicable.

Rapkin remembers that "it used to be a very active field for discussions of technique . . . and there were many, many conferences . . . about the best counting solutions, how . . . you measure steroids, all that kind of thing."[110] Journals such as *Nucleonics*, founded in 1947, aggressively promoted radioactive counting and its industrialization. The name "nucleonics" alluded to the "Prospectus on Nucleonics," a secret, thirty-eight-page strategy paper containing guidelines for postwar nuclear policies that had been drawn up by scientists of the Met Lab in Chicago and submitted to Arthur H. Compton in November 1944.[111] The journal billed itself at its launching as a "medium for the cross-fertilization of technical advances in all phases of nuclear technology, [a] meeting place for the exchange of ideas between engineers, physicists, chemists, life scientists and teachers."[112] There followed in the 1950s *International Journal of Applied Radiation and Isotopes* (1956) and the *Journal of Nuclear Materials* (1959). Over the years countless smaller conferences on scintillation counting were held, such as the Symposia on Tritium in Tracer Applications, the Northwestern University Conference, a good dozen big conferences, including the Scintillation and Semiconductor Counter Symposia in Washington, the Annual Symposia on Advances in Tracer Methodology, the 1960 University of New Mexico Conference on Organic Scintillation Detectors in Albuquerque, the 1961 I.A.E.A. Symposium on the Detection and Use of Tritium in the Physical and Biological Sciences, the 1958 and 1961 I.A.E.A. Conferences on Nuclear Electronics, a 1969 conference on Liquid Scintillation Counting at the Massachusetts Institute of Technology,[113] and a 1970 international conference on Organic Scintillators and Liquid Scintillation Counting at the University of California.[114] Before long systematic monographs complementing the conference proceedings also began to be published.[115]

One of the conferences jointly financed by the Packard Instrument Company and New England Nuclear was organized in Switzerland and took place in Zurich's Kunsthaus. "The biggest [conference] I think we ever put on was in Zurich. God, they made a real deal out of that. [The] Bürgermeister came and talked, and we had one of the great big halls, right downtown in Zurich. It was a two or three day seminar, and then a big dinner. [People] from all over Europe came to that."[116] By this time, Packard's business had expanded abroad. The first international sale was made in 1957 to Saclay, the French atomic energy research site in Gif-sur-Yvette. Sales throughout Europe followed, including Yugoslavia, Hungary, and the USSR. A chemical plant in the Netherlands that produced scintillation substances was acquired. The company even constructed a few whole body counters for monitoring radioactivity in living human beings, one of which went to the University of Hamburg. Together with Niilo Kaartinen of the University of Turku, Finland, Packard built an oxidizer machine.[117] With this "Packard-Kaartinen combustion machine," otherwise insoluble material could be converted into gaseous and liquid form; the transformation of organic materials into CO_2 and H_2O through combustion provided as a physical side effect an elegant means of separating radioactive hydrogen from radioactive carbon. This separation was above all essential in critical dual label experiments that operated with low ^3H and high ^{14}C content. The machine was widely used, although its mechanical functioning was susceptible to breakdown due to frequent soot deposits and the delicacy of the mechanics involved in closing and opening the combustion chamber.

To promote international sales Packard set up Packard Instrument Sales Corporation in 1957 and a subsidiary named Packard Instruments International S.A. in 1959. Under the direction of James Kriner a widespread network of international offices centered in Zurich was brought into being.

> We had the most advanced sales and service arrangement in our field, by having a corporation in Sweden, a corporation in Italy, a corporation in Belgium, a corporation in Germany, in France, in England, in Israel. And each of these corporations had a local person who was the top man. Whether he was the president or the director general or whatever other title, it was always a local person, with local staff, and everybody was bilingual, at least bilingual: their own language and English. And, all the business was done in the local currency with local bank accounts and it worked out just really beautifully.[118]

This was unusual business practice at the time for local branches were usually run by general representatives. Furthermore, the then strong dollar helped hold down costs. In 1969 Packard's worldwide sales and service network included in addition to the countries just mentioned agencies in Australia, Canada, Denmark, Japan, Norway, and the Republic of South Africa.

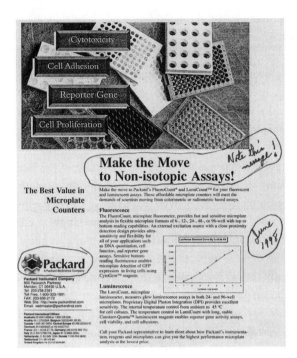

Cytotoxicity

Cell Adhesion

Reporter Gene

Cell Proliferation

Make the Move to Non-isotopic Assays!

Note this merger!

The Best Value in Microplate Counters

Make the move to Packard's FluoroCount® and LumiCount™ for your fluorescent and luminescent assays. These affordable microplate counters will meet the demands of scientists moving from colorimetric or radiometric based assays.

Fluorescence

The FluoroCount, microplate fluorometer, provides fast and sensitive microplate analysis in flexible microplate formats of 6-, 12-, 24-, 48-, or 96-well with top or bottom reading capabilities. An external excitation source with a close proximity detection design provides ultra-sensitivity and flexibility for all of your applications such as DNA quantitation, cell function, and reporter gene assays. Sensitive bottom-reading fluorescence enables microplate detection of GFP expression in living cells using CytoGem™ reagents.

June 1998

Luminescence

The LumiCount, microplate luminometer, measures glow luminescence assays in both 24- and 96-well microplates. Proprietary Digital Photon Integration (DPI) provides excellent sensitivity. The internal temperature control from ambient to 45 ℃ for cell cultures. The temperature control in LumiCount with long, stable Constant-Quanta™ luminescent reagents enables reporter gene activity assays, cell viability, and cell adhesions.

Call your Packard representative to learn more about how Packard's instrumentation, reagents and microplates can give you the highest performance microplate analysis at the lowest price.

Packard
A Packard BioScience Company

Packard Instrument Company
800 Research Parkway
Meriden, CT 06450 U.S.A.
Tel: 203-238-2351
Toll Free: 1-800-323-1891
FAX: 203-639-2172
Web Site: http://www.packardinst.com
Email: webmaster@packardinst.com

Packard International Offices:
Australia 03-9543-4956 or 1-800-335-636;
Austria 43-1 2702604; Belgium 32(0)2-481-85-30;
Canada 1-800-347-9633; Central Europe 42-456-32(00)-6;
Denmark 45-42905825 or 45-43527751;
France (33-1) 46.46.27.75; Germany (49) 6103-365-151;
Italy 39-2-33415796/74; Japan 81-3-3966-5690;
Netherlands 31-53-549 1295; Russia 7-095-250-9650;
Switzerland (41) 481-69-44;
United Kingdom 44-(5)118-9844481.

28. Packard Company advertisements for a microplate fluorometer and a microplate luminometer. [1998].

Eclipse

In 1967 Packard sold his firm to the American Bosch Arma Corporation. Here is how he explained his decision to sell:

> Basically, the reason for selling the company was that we needed additional financial support to continue to expand, compete and maintain our dominant market position. An undesirable alternative would have been to retrench both technical and marketing development and concede the leading market share to the bigger companies, Searle and Beckman, that had entered our market with acceptable products by acquiring our smaller competitors, Nuclear Chicago and Sharp Laboratories. [The] merger (with American Bosch Arma Corporation) did provide us with the necessary financial support to compete with the larger companies we were facing in 1967. [It] is very tough to compete when big companies buy their way into a business and are willing to lose money until they win or drop out.[119]

Packard nevertheless remained in the reorganized company for three and a half more years. Then he sailed around the world and bought two islands in the Caribbean. Later, he invested in several smaller companies. When

I was researching this chapter he was the President and Chief Executive Officer of Advanced Instrument Development, Inc. (AID), based in Chicago. AID produces X-ray equipment for medical diagnostics. Packard Instrument Company, taken over by the Packard Bioscience Company (formerly Canberra Industries, Inc.), continued to focus on liquid scintillation counting. It has preserved its strong market position in the field of counters and sample preparation solutions to the present day.

The once overwhelming significance of radioactive labeling in molecular biology and biomedicine has nevertheless been waning in the late twentieth century. Concomitantly the days of almost unlimited growth were over in nuclear research technology as well. Alternative tracer methods based on fluorescence and luminescence are now increasingly used. They have made the use of radioactivity superfluous for many applications. Research technologies emerge in particular historical constellations; their producers survive only if they keep pace with epistemic shifts in research technology. Lyle Packard's former company has accordingly taken a new tack: "Make the move to non-isotopic assays!" (fig. 28). Although it is quite possible that in certain domains radioactive tracing will remain indispensable for some time to come, non-radioactive visualization techniques using molecular tracers have today become a viable alternative.

10. The Concept of Information

The Writings of François Jacob

In the fall of 1965 French molecular biologist François Jacob won the Nobel Prize for Medicine or Physiology together with André Lwoff and Jacques Monod. The prize was awarded in recognition of their work on the regulation of gene expression in bacteria. Jacob had joined the Paris Pasteur Institute in the early 1950s at the "right moment" as he said at the beginning of his Nobel Lecture in Stockholm. For one thing biology had just appropriated relatively simple new organisms for experimentation — microbes and viruses; for another it was forging new ties with physics and chemistry. And not least it "was changing its ways of thinking."[1] In the following sketch I look at how this change in "ways of thinking" came about in Jacob's case in the short span between 1958 and 1970,[2] paying particular attention to the emergence and development of a richly connoted concept: biological information. With the help of one concrete example I chart the decisive stretch of the convoluted path on which the molecular-biological language of information crystallized and then acquired the very considerable heuristic significance it came to enjoy.[3] In keeping with this limited purpose I essentially ignore both the 1950-58 period during which Jacob co-operated closely with Elie Wollman at the Pasteur Institute, publishing work on lysogenic phages and bacterial conjugation, as well as the 1970-90 period when switching experimental objects, he undertook the study of embryogenesis in a mammal, the mouse. My account zeroes in on the years in which Jacob and Monod joined forces, developing a common experimental system based on their model organism, the bacterium *Escherichia coli*. It was a hybrid system that combined analytic practices of biochemical genetics with techniques for the induction of temperate phages in bacteria. Monod had developed the former over a period of many years using the induction of lactose metabolism in *E. coli*; credit for the latter went to Jacob, who had conducted parallel studies of gene transfer induced in the same bacterium by both sexual conjugation and phages. The language of biological information, which made its entry at the Pasteur Institute in this period, became a mediator between these different kinds of molecular biological research. As we shall see it also contributed substantially to the productive handling of the epistemic objects that took shape in Jacob's and Monod's joint experimental system.

The PaJaMo Experiments

The second main stage of Jacob's work at the Pasteur Institute began in 1958 with what is known as the PaJaMo or "pajama"-experiments,[4] which Jacob carried out with Monod and Arthur Pardee, a biochemist who had come to Paris on a sabbatical from Wendell Stanley's virus laboratory in Berkeley. The story of this collaboration has been often told by historians as well as contemporaries who took part in it.[5] Therefore I shall not repeat it here, but rather take a detailed look at the texts and lexicon of the three consecutive papers published by Pardee, Jacob, and Monod in 1958, 1959, and 1960.[6] How did the terminology they used to describe the results of the PaJaMo experiments evolve? Jacob describes this series of experiments as "conceptually simple, but difficult to perform technically."[7] Monod had succeeded in characterizing *E. coli* mutants (called z^-) that were no longer able to synthesize the enzyme β-galactosidase even after being supplied with lactose, the sugar degraded by β-galactosidase. He had furthermore identified a factor i, a mutant form of which, i^-, caused the bacterial cells to synthesize β-galactosidase even in the absence of lactose induction—on their own power or "constitutively." Jacob had developed a technique for transferring genes from one bacterial cell to another. The idea behind the experiments Jacob and Monod now planned to carry out ran as follows: a male bacterium, to whose fertility factor wild type genes of β-galactosidase and the i-factor had been attached, would transmit its genetic material by conjugation to a female bacterium that was mutant in both genes and consequently no longer able to produce active enzyme on its own, although it had been converted to constitutivity with i^-. The assumption was that if the process could be technically controlled it would be possible to observe the behavior of the now partially diploid cells, each possessing one mutant and one wild type allele of each of the two genes. Jacob and Monod were especially interested in finding out whether the gene for intact β-galactosidase transmitted to the female bacterium was activated. If so, they wanted to know how long it took to activate it and under what conditions synthesis occurred. And what would happen, they wondered, if the experiment were reversed, that is, if mutant genes were injected into a wild type bacterium?

The first set of experiments yielded a surprisingly clear result. After injection of z^-i^- genes into z^+i^+ bacteria no enzyme was synthesized even after a prolonged incubation period. However, when a combination of z^+i^+ genes was injected into z^-i^- bacteria synthesis of β-galactosidase occurred a few minutes after the male genes had reached the cytoplasm of the female bacterium. From this behavioral difference Pardee, Jacob, and Monod concluded that their system fulfilled an important technical prerequisite: after gene transfer no appreciable quantities of cytoplasm—and thus of already synthesized enzymes—were transferred from one bacterium to another.[8] But the experimental results also contained a real surprise. Contrary to what Monod had long assumed the "constitutive" i^- mutant turned out to be not dominant, but, oddly, recessive. Whenever a wild type allele of the i-factor was present in a bacterium that had been made partially diploid, the bacterium behaved normally, blocking its enzyme system in the absence of the inducing sugar.

Nowhere in reporting on these experiments did the researchers at the Pasteur Institute ever use the term "biological information." In their 1958 paper Jacob, Monod, and Pardee presented the relations between genes and gene products exclusively in terms of "expression" and "determination" (the same process was referred to as either expression of the gene or determination of the gene product). And the i gene did not control the production of a substance that activated enzyme synthesis, but "co-operated" by way of the cytoplasm in producing the enzyme; the i gene, according to the authors' hypothesis in this period, was responsible for the production of an enzyme which in turn produced a "repressor" whose chemical composition was still unknown. The repressor molecule "blocked" synthesis of β-galactosidase, but could be inactivated by a metabolic "inducer" such as lactose. This inactivation was effective, however, only when it affected both alleles in a cell diploid for the i gene.

Thus it can be seen readily that to describe gene action the scientists at the Pasteur Institute combined a general language of *control* with a specific language of *regulation* that mobilized expressions such as "repression," "co-operation," "blockage," and "induction." The terms "control" and "regulation" themselves were not, however, explicitly employed in this first paper.

In contrast the next paper published in 1959 and following in principle the same experimental strategy includes the phrase "genetic control" in its very title. Pardee, Jacob, and Monod present the subject of the paper as "the control mechanisms involved in induction."[9] The i gene is now said to "control" the synthesis of a cytoplasmic product, the presence of which prevents synthesis of β-galactosidase. The product itself is now identified as a "specific cytoplasmic messenger."[10] The i factor, the authors write, summing up their findings on this point, "sends out a cytoplasmic message which is picked up by the z gene, or gene products. Postulating, as we must, that this message is borne by a specific compound synthesized under the control of the i gene, we may further assume that one of the alleles of the i gene provokes the synthesis of the message, while the other one is inactive in this respect."[11] The term "cytoplasmic messenger" (or "cytoplasmic message") appears for the first time in this second paper. However, the term by no means designates as it would later a substance serving as an intermediary between the DNA of the genes and proteins as gene products. It signifies only a "specific compound synthesized under the control of the i gene," that is, a product of the i gene that in its active form represses synthesis of β-galactosidase. The authors thus speak of a "message" here in order to indicate that this product functions as a metabolic signal. The paper also distinguishes for the first time between "'structural' genes" and "a repressor-making gene" whose function is to "regulate the expression" of neighboring structural genes. The repressor-making gene is now expressly situated in the context of—in the authors' words—a "generalizable picture of the regulation of protein synthesis."[12]

The "repressor-making gene" is not yet designated by the generic term "regulatory gene"—although Jacob and Monod would introduce the distinction between "structural" and "regulatory" genes that same year in a brief contribution to the *Comptes rendus de l'Académie des Sciences de Paris*.[13] The question of the chemical composition of the repressor and the level at which it acts (whether at that of the structural gene or the cytoplasmic gene product) is explicitly left open. Interestingly the observation that β-galactosidase is synthesized in the recipient cell immediately following transfer of a wild type copy of the gene to the recipient cell inspires no conclusions other than that transduction is manifestly followed without delay by *de novo* enzyme synthesis, further decreasing the likelihood that cytoplasmic products are transferred from one cell to another during mating. The notion of information does not appear in the 1959 paper any more than it had the previous year.

Let us now take a look at the third paper in this series, which appeared in 1960. Its title alone—"On the Expression of a Structural Gene"—testifies that the authors' attention has now shifted from the cytoplasmic "inducibility" of an enzyme and genetic "control" over its induction to the processes surrounding the "expression of a structural gene."[14] The problem they set themselves is cast in new terms as well: they now state that the general aim of their experiments is to elucidate "the mechanism by which structural specificity is transferred from gene to protein."[15] New significance accrues to an earlier observation that was initially interpreted as a mere index of the technical feasibilty of their experiments: "the z^+ gene which determines the structure of β-galactosidase in *E. coli* functions without significant delay, at maximal rate, when it is transferred by sexual recombination into the cytoplasm of a cell which possessed only an inactive allele of this gene."[16] This kinetic behavior is now said to authorize certain conclusions about the way genes are expressed. The rapidity with which the gene becomes fully functional, says the 1960 text, militates against the prevailing model of protein synthesis built on the assumption that cytoplasmic RNA plays the role of a "stable intermediate template." A stable RNA such as ribosomal RNA, the authors go on, is consequently not a possible candidate for the job; their observations indicate, rather, that the gene itself acts directly as template, or at the very least points to the formation of a functionally "*unstable* intermediate."[17] Inasmuch as all the evidence that had been accumulated by this time suggested that cytoplasmic particles, that is, ribosomes rather than chromosomes, were the seats of protein fabrication, the second of the two alternatives—the hypothetical unstable intermediate—seemed the more promising choice.

In this context the term "transfer of information" makes its debut and the authors also introduce the term "information-bearing RNA," albeit only diffidently and negatively: by affirming the unlikelihood that in the transfer experiment an "information-bearing RNA" could have penetrated the receptor cell bundled together with DNA.[18] Both Denis Thieffry and Lily Kay have described how averse Monod was for quite some time to ascribing such an information-bearing role to RNA.[19] Hubert Chantrenne, the biochemist in the Brussels group of molecular biologists around Jean Brachet, recalls the animated discussions the group had with the Pasteur Institute molecular biologists between 1956 and 1958: "Monod (who would believe it today!)," he writes, "was for a long time staunchly opposed to the idea that RNA played a part in protein synthesis and enzyme adaptation. He took a sharply critical stance toward the results we had arrived at in Brussels, firmly maintaining that while we perhaps had good clues, we had no proof (and he was right about that)."[20] Interestingly attention to the lexicon of the 1960 paper reveals that the authors fall back on stock biophysical terms such as "structural specificity" and even "template," and use widely current biochemical expressions such as "intermediate product" to identify the hypothetical entity or emerging new object of epistemic interest—which they simultaneously assume is the vector for the "transfer of information" between genes and proteins. Thus their terminology itself remains hybrid. Still more remarkably the term "messenger," which would later serve as a generic designation for any molecule responsible for such "transfer of information," remains here reserved for the product of the regulatory gene, which it identifies as a signal of metabolic regulation. In a paper published the same year, in which Jacob and Monod introduce the concept of the "operon" (to which I shall return), the ambiguous formulation "cytoplasmic replica" (of the "genetic material") is used to refer to this gradually emerging new intermediate entity.[21]

The concept of information itself as introduced in the 1960 paper has a clear-cut definition. It is employed as a strict synonym for the term "structural specificity" in the sense that what is "transferred" from the gene to the protein is a precise specification or "determination" of the protein's structure—to pinpoint a term in the first PaJaMo paper that would now yield to another.

The Papers Published in 1961

The situation changed with the May 1961 publication in *Nature* of a paper by Jacob, Sidney Brenner (Cambridge), and Matthew Meselson (Pasadena).[22] In summer 1960 the authors had come together in Meselson's laboratory at the California Institute of Technology[23] to prove experimentally that during protein synthesis a newly synthesized RNA was picked up by the ribosomes already present in the cell plasm. As their report indicates they prepared different batches of bacteria labeled with both heavy and radioactive isotopes for this purpose. Using density gradient centrifugation they were able to demonstrate that after conjugation a newly synthesized radioactive RNA attached itself for the purpose of protein synthesis to pre-existing ribosomes labeled with heavy nitrogen.

This 1961 paper exhibits new and subtle shifts in meaning. We are now told that the intermediate molecule described as "unstable" "carries information" or is a "carrier of information" from genes to ribosomes and as such acts as a "messenger."[24] Thus the messenger previously treated as a cytoplasmic signal has become an—unstable—courier transporting information from the locus of the genes to that of protein synthesis. Brenner, Jacob, and Meselson were able to observe this courier en route. The observation led them to the assumption that "the messenger RNA" must be "a simple copy of the gene," a hypothesis which if correct would have "interesting implications for coding mechanisms."[25] This is the first appearance of the concept of "coding" in Jacob's writings: the molecular object is now "a simple copy of the gene" as well as a molecule taking a direct part in the fabrication of a protein. The molecule could be represented as a discrete entity, making it possible to envisage experiments in which the "nucleotide composition" of RNA could be correlated with the amino-acid composition of the corresponding protein.

In December 1960 Jacob and Monod submitted a long article to the *Journal of Molecular Biology*. Published in June 1961 it summarizes the work of the preceding three years with its accordant conceptual shifts.[26] Jacob later described the composition of this text as a replacement of "the real order of events and discoveries by what appears as the logical order, the one that should have been followed if the conclusion were known from the start." Monod "was a past master at this game."[27] The paper picks up on the concept of "operon" that Jacob and Monod had introduced the year before in the *Comptes rendus* of the French Academy of Sciences. They had coined the term to characterize an "intermediary genetic organization" for the coordinated expression of a group of structural genes regulated by a new "genetic element" on the DNA, the "operator." They had deduced the existence of this element from the behavior of a set of specific mutations involving a stretch of DNA with which they suspected the repressor interacted.[28] This "intermediary genetic organization" insinuated a structured space between the chromosome as a whole and the individual gene; the latter was conceived of—classically—as an "independent entity [for the specification of a] biochemical function."[29] A radically transformed description of genetic organization resulted: the concept of "messenger" was severed from that of "message." DNA is now said to contain a "message" that "defines the structure of a protein" through an intermediate molecule christened the "structural messenger" in reference to its presumed complementarity with the gene.[30] The messages or information on the DNA appear to be embedded in an integral "system of information transfer" in control of the "flow of information" between chromosome and cytoplasm, determining which messages would and would not be "translated" at a given moment.[31] The protein-synthesizing centers, the ribosomes, are now said to receive by way of messenger RNA "specific instructions" from the genes for the production of defined proteins.[32]

Against this background Jacob and Monod consider a "fundamental problem of chemical physiology and of embryology" in their conclusion. As they describe it the problem is to explain "why tissue cells do not all express, all the time, all the potentialities inherent in their genome."[33] The use of the Aristotelian category of "potentiality" to grasp developmental processes and processes of differentiation recalls Jacob's "On the Genetic Control of Viral Functions," the 1958 Harvey Lecture in which he had already hinted at the possibility that the "genetic material might express its potentialities by group[s] of genes operating as units of activity which could be switched on or off under the action of cytoplasmic substances."[34] By 1961 Jacob and Monod were probably not using the term "potentiality" innocently. The concept of the "program," too, appears at the end of their 1961 article for the first time in a context of this kind. By introducing it the two researchers from the Pasteur Institute distanced themselves from classic "bean-bag" genetics. They concluded their text in this spirit: "the genome contains not only a series of blue-prints, but a coordinated program of protein synthesis and the means of controlling its execution."[35]

In the history of molecular biology the operon model of gene expression and gene regulation represents a kind of turning point. To begin with it brilliantly confirmed the comparatively simple, not to say reductive, notion of the gene as a linear segment on an equally linear (or circular, in the case of bacteria) chromosome,[36] rounding out this view of matters with the notion of "messenger RNA," said to mediate between structural genetic information and proteins with their biochemical function. At the same time, however, the operon model evoked with the distinction between structural and regulatory genes the notion of a hierarchy of genetic information and thus the possible existence of feedback loops through which the genome could interact with an organism's metabolism. This brought the phenomena of development and differentiation into view. The operon model might accordingly be said to have provided the common denominator of a *single* epistemic object for two central categories of the new biology: information and control.

That same year Monod and Jacob presented their findings once more at a symposium on cellular regulatory mechanisms in Cold Spring Harbor. Asked to sum up the work of the conference they declared in their closing remarks that "microbial systems" had proven to be key to research in molecular genetics. Such systems also provided excellent "models for the interpretation of differentiation,"[37] which would have to be studied in the cells of higher organisms some day. This was a long-term perspective. Two other problems, however, could be tackled immediately as they saw it: "the nature of the code" and "mechanisms of information transfer from DNA to enzyme-synthesizing centers."[38] Over the next five years other research teams solved both problems.

Jacob's 1965 Inaugural Lecture at the Collège de France

In 1965 François Jacob not only won the Nobel Prize, but was also elected to the Chair in Cellular Genetics at the Collège de France. His inaugural lecture contained certain basic elements of his now famous history of heredity, *La logique du vivant*, published five years later.[39] In his lecture Jacob expanded on the distinction between "plan" and "program" adumbrated in the closing sentence of the great 1961 essay that he had co-authored with Monod. The genetic message, he declared, not only contained "the building plans for cells"—called "blue-prints" in the 1961 paper—but also a "program coordinating the syntheses and the means of ensuring their execution."[40] The cellular building plans consisted of specific "instructions" spelled out in the genetic message,[41] while the program contained the rules governing their realization. The program itself was at least in part—for instance, in the guise of operon structures—embedded in and part of the organization of the genome.

With regard to the earlier publications the most conspicuous feature of this lecture is yet another two-fold terminological shift. First the circuits regulating cytoplasmic expressions of genome structures are now explicitly described in the language of "communication" and signaling systems—indeed as "communications networks": "chemical activities are commanded by specific regulatory circuits that all seem to be constructed on the same principle: a large molecule that is determined by a gene recognizes, on the one hand, certain metabolites that function as specific chemical signals; on the other hand, there are certain cellular components to which it attaches itself in order to modify their activity in accordance with the signals it receives."[42] The other, still more conspicuous terminological shift pervades the whole text: the mechanisms that ensure the transfer of genetic information in the narrow sense of the word are now described expressly in terms of *language* and *writing*—as "l'écriture de l'hérédité."[43] The real "surprise" for Jacob resides in the fact "that genetic specificity is written not in ideograms, like Chinese, but with an alphabet, like French, or, more exactly, Morse code. The meaning of the message derives from the combination of signs in words and the arrangement of words in sentences. The gene is thus a sentence several thousand signs long, with punctuation marks at the beginning and end."[44] The gene is now identified as a "chemical text"[45] that is first transcribed into a messenger RNA and then translated into a protein. The latter step, translation, is carried out by a "translation machine," we are now told—that is, by a class of molecules known as "adaptors" capable of "reading the two alphabets," the nucleotides' alphabet and the amino acids. These molecules are therefore said to "hold the key to the code."[46] Jacob was certainly not the first to describe the expression of genetic information (which incidentally he also equated with an "order" in his inaugural lecture at the Collège de France—"an order, or as one says today 'information'")[47] in terms of transcription and translation; in the years preceding his lecture language of that sort had become the veritable hallmark of molecular genetics. Yet for Jacob the image of a "chemical text" was more than a convenient analogy or a metaphor calculated to appeal to the general public. Almost a decade later he would again explain, in detail this time, what he meant by the "linguistic model in biology."[48]

The Logic of Life

In *La logique du vivant* (1970) Jacob returns to the distinction between *plan* and *program*. But now he charges the concept of "program" with a significance unique within the history of biological ideas. For Jacob the concept of the "program" could abolish or at least go a long way toward reducing fundamental oppositions at work in the life sciences such as those between "finality and mechanism, necessity and contingency, stability and variation." The concept of program, he says, has two complementary parts, "memory" and "project" (*projet*). Memory stands for heredity, the evolutionary accumulation of information passed down from ancestors to descendants: in other words, for the phylogenetic process. Project means the "plan which controls the formation of an organism down to the last detail," and thus ontogeny, the unfolding of development and differentiation.[49] Jacob sees these two components of the biological program embedded in two "systems of symbols": one serves to "maintain [the] information" accumulated in the course of generations, while the other "unfold[s] the structures" in each generation.[50] The fundamental mechanism underlying both is *reproduction*. Reproduction "acts as the main operative factor for the living world. On the one hand, it provides an aim for each organism; on the other, it gives a direction to the aimless history of organisms."[51] Jacob is aware that the notion of program is drawn from the field of electronic data processing and that it "equates the genetic material of an egg with the magnetic tape of a computer." For that very reason he readily acknowledges a basic *difference* between the two modes of reproduction: unlike the program of a machine, that of an organism controls the production and reproduction of its own component parts and thus also of the very "organs which are to execute the programme."[52]

But the differences run still deeper, for an organism's program not only governs the production of its own constituents, it also presupposes them. Jacob affirms expressly that a cell and its genome always already form an integrated whole—in his terms an "integron," or more precisely a hierarchy of integrons.[53] Integration is the condition for the emergence of new properties in a system. "Able to function only within the cell, the genetic message can do nothing by itself. It can only guide what is being done. To produce machines from plans, there have to be machines. None of the substances that can be extracted from the cell can reproduce. Only the bacterium, the intact cell, can grow and reproduce, because only the cell possesses both the programme and the directions for use, the plans and the means of carrying them out."[54] When today Jacob's concept of "genetic program" is criticized for being the very incarnation of genetic reductionism[55] it should be recalled that he employed the concept to exactly the opposite end: that of opening up a non-atomistic, trans-reductionistic perspective on genetics. He begins *The Logic of Life* with the affirmation that heredity is currently conceived in terms of "information, messages, and code."[56] He concludes it by taking a step back and asking his readers to consider that from the perspective of history of science there is no ontological reason that this should continue to be the case. "Today the world is messages, codes, and information"; the discourse of heredity, its present-day categories and concepts, is shaped by this social and technological context accordingly. Jacob consequently brings his guided tour of the history of heredity to a close by raising the question of the historicity of his own vision: "tomorrow what analysis will break down our objects to reconstitute them in a new space?"[57]

It should be evident at the end of this brief historical reconstruction that the emergence and development of the terminology of molecular genetics can be traced in exemplary fashion virtually step by step through Jacob's writings. We were able to follow the evolution of a hybrid discourse that successively introduced new levels of description in an attempt to characterize gene expression and regulation. Initially the discourse of specificity and determination borrowed from biological chemistry was combined with performative concepts of prescription, execution, and control drawn for the most part from Jacob's own study of lysogeny and Monod's analysis of the lactose metabolism of *E. coli*. There followed a series of further significant linguistic shifts: from control to information, message, and code, thence to communication and signaling systems, and finally to language itself, to the text, written and read. No neat caesurae separated the individual shifts, which were, rather, superposed in the form of distinct registers that allow for nuanced modulations and changes of tack in interpreting old experiments and designing new ones.[58] All these experiments revolved around one central point: the kinetics of enzyme expression following a gene transfer.

The language of an experimental science originates as Jacob might say in ongoing *bricolage*: it operates in a framework established by the technical means and media available for the development and realization of experimental systems to which it refers and that lend it its force. In the story recounted here the concept of information is the nodal point in a far-flung discursive network. But not accidentally the concept of "information" has never in molecular biology assumed the purely quantitative meaning that information theory conferred upon it earlier. Instead, "information" in molecular biology serves as a stand-in for older definitions of biological specificity, composing the structural repository of a biological function that directly or indirectly continues to refer to differential reproduction. This is the reason that Jacob could undertake to bring reproduction and development or memory and project into reciprocal relation as two different symbolic orders.

PART IV

Epistemic Configurations

The short chapters of this last section lead back to the initial reflections on historical epistemology and epistemological history, and draw on materials of the case studies in the two middle sections by generalizing them and looking at them from new perspectives. They provide three different looks at the material culture of the laboratory as so many aspects of an epistemology of the concrete.

The first chapter revolves around the interface between instrument and scientific object in research. It shows for a number of research technologies characteristic of the life sciences of the past two centuries that examination of the shaping of their interfaces and interceptions is essential to understanding modern research's "phenomenotechnique"—to pick up on Bachelard's notion here. The perspective taken in this chapter is primarily that of the instrument part. The second chapter spins out the thread of the first by positioning itself from the object part. In so doing it brings into relief a peculiar kind of epistemic object of particular importance in the life sciences, the "preparation." I scan a number of biological fields from the eighteenth century to the present where preparations of widely different kinds came to dominance and left their mark on the scientific culture built around them. The last chapter of the book looks at the laboratory from the point of view of its "economy of the scribble." It highlights the generative function of laboratory writing in scrutinizing the intermediary space between the experiment with the traces produced in it and the printed product of research. Like the interface between the instrument and the epistemic object this other interface between the experiment and the epistemic subject is of crucial importance if one wants to understand science as a creative process.

11. Intersections

Since the seventeenth century scientific instruments have been regarded as the emblems of experimental science. Time and again they have been lauded as ideally transparent media that either help to prolong, reinforce, and heighten the powers of our senses or to isolate, purify, and quantify experimental observations. The historical literature on scientific instruments, which has grown considerably over the past few decades and will not be reviewed here in detail, has fundamentally questioned their much vaunted transparency. In particular many case studies have shown that as a rule instruments neither work nor produce insights by themselves. Rather, they are embedded in historical and local contexts, and their success depends on the more or less informed manner in which they are applied. If we ignore these contexts, which also determine the limits of the circulation of instruments, we can understand neither their history nor their efficacy.

Preliminary Epistemological Remark

This chapter is not another case study or reconstruction of a local historical context. Quite the contrary: it sets out to examine the fundamental relationship between instruments and experimental objects. What configurations can this relationship assume? How do experimental systems and instruments interact? How can we assess an instrument's specific epistemic value? I propose to show in what follows, taking various apparatuses used especially in the nineteenth and twentieth centuries as my examples, that instruments in the biological sciences generate opaque, very diversely configured *intersections* between themselves and the objects they are used to investigate. These intersections are the surfaces on which apparatus and object make contact. How are these planes and points of contact between the animate and the inanimate as it were, between organisms and technical apparatuses formed? The investigative value of an instrument depends on intersections; it is at the intersection that we can assess whether a given instrument and a given object can be brought together in a fertile analytical constellation. Hence such intersections also often command the kind of attention from the producers and users of instruments in which considerations of craftsmanship count for a great deal. At the intersection the instrument-maker's hand gives way to the experimenter's and even in the age of the industrial production of research technologies often enough remains the locus of joint handicraft work. In the development of new research technologies the exploration of such intersections between instrument and object is often at least as important as the realization of the technical principle embodied in the instrument, although it need stand in no direct practical or theoretical relation to that principle.

Unfortunately the work carried out at the intersection between instrument and object—supposedly a narrowly circumscribed field of investigation—has not always received adequate historiographical attention. I would like to combine an account of this work with a more broadly conceived epistemological question about the reciprocal relation between epistemic things and the technical conditions of their manipulation in experimental systems. The productivity of such systems depends on that relation. Technical things set the boundary conditions of experimental systems and in the process create the space in which an epistemic object can unfold. Scientific instruments as well—this is my general contention—acquire their epistemic meaning only against the backdrop of particular experimental systems or as a result of the fact that they have been tailored to determinate experimental systems. In and of themselves they are epistemically neutral even if they can be regarded as material embodiments of concepts or theories.[1] As a general rule instruments enter as technical things into an experimental arrangement and to that extent are among its identity conditions; within that arrangement, however, they themselves can become epistemic things or help produce them—if for example they elicit unexpected questions in their use. In the development of research technologies we often observe that the process by which a research instrument gradually takes shape meshes with that by which an epistemic object is gradually built up: the two go hand in glove.[2] The variable intersections between instruments and objects consequently form the critical zone in any experimental system.

In biological experiments, to speak of the intersection between the object of investigation and the techniques utilized to represent or measure it is to speak very concretely of the boundary between an organic and an inorganic entity. Here life and technology meet; since in general the living entity is wet and soft and the technological one is dry and hard special precautions have to be taken to ensure their compatibility. The scientific and cultural—indeed the aesthetic and ethical—significance of experimental systems turns on what is achieved at this intersection.

My concern here is not with boundaries between disciplines or the relationship between sciences and other major cultural formations, but with these manifold, often unspectacular intersections and interfaces in their own right. My epistemological questions arise from the specific materiality of each particular object of the biological sciences carved out in this manner. Pertinent as well are the many small, provisional borders between what may still be considered biological nature and what may have to be conceived of as an artifact. Ever since the life sciences took it upon themselves to turn organisms inside out it has been incumbent on them to define the limits of life. And ever since the beginnings of this empirical adventure they have been haunted by the question of the legitimacy of drawing these "minor" lines of demarcation. These minutiae are more dramatic and of greater consequence for the formation of the major frontiers between domains of knowledge than one might be ready to admit.

The Light Microscope

I begin with microscopy. The development and dissemination of the microscope from the latter half of the seventeenth century gave a new form to the question of what the object of observation was in natural history. The technology of magnification had its pendant in new, extremely small specimens—a good example of the way the forms of observation tied to new instruments determine the form in which the objects of interest have to be cast. As a rule things tailored to the lens cannot themselves be seen during the process of preparation, at least not in the detail that is the key to the success of the enterprise. The process of their production escapes the eye; only the gaze through the microscope decides after the fact whether it was successfully carried out. This requires that the preparer focus on the *regularities* of the production process, which must so to speak function blindly. It is therefore no accident that the scientific literature abounds in ever more circumstantial descriptions of preparation techniques. It is a truism that fresh specimens be prepared before each observation. But what guarantees that the procedure can be repeated in exactly the same way? Conscientious description of the procedure is supposed to ensure exact repetition. In his debate with Franz Ferdinand Meyen about how fertilization occurs in plants Matthias Jacob Schleiden lavished a great deal of attention on such detailed description. The second part of his *Botany as Inductive Science* (1846) says,

I will only add a few words on the preparation [*Darstellung*] of such dissections. If the seed-buds are not very closely enveloped, and immovable in the germen, I extract them, take them between the thumb and fore-finger in such a manner that with a sharp razor I can cut them accurately in half. . . . The two halves thus obtained I place one behind the other, the cut surfaces toward the thumb, between the thumb and finger again, and cut with the razor the thinnest possible slice from the surfaces of the sections. I then bring these two slices under the simple microscope, and dissect the respective parts with fine needles and knives, if they are not already exposed by the section, which is of course always the best.[3]

The ability to have specimens "exposed by the section"—that is to say, with a minimum of manipulation—is the mark of the master of fresh botanical preparations (fig. 29).[4] Soon enough, however, attempts were undertaken to make specimens durable since ultimately only permanent preparations could guarantee the persistence of the singular occurrence and consequently the possibility of iteration and comparison with new preparations.[5] With permanent preparations differential reproduction of specimens became possible, but at the same time their capacity to produce surprises diminished. Moreover, since additional manipulation of the object was required to make preparations durable, the question as to what was nature and what was artifact in the preparation took on particular epistemological urgency. For here in contrast to macroscopic observation there could be no examination of the "living" counterpart for purposes of control. Microscopic preparations were therefore scientific things par excellence—epistemically highly charged knowledge things. It stands to reason that most methodological critiques of the knowledge practices of the nineteenth-century life sciences should have flared up around them.

Another peculiarity of microscopic preparations is that they reduce the objects fixed in them to two dimensions, that is, flatten them out, a transformation conditioned by the way the optical apparatus works: its focus always lies on a planar surface, which does not necessarily match the surface of the object. The flat object of the microscopic preparation is realized in the "cut." In close step with the development of cell theory the tissue section became a distinctive feature of animal and plant microscopic morphology in the nineteenth century. The zoologists, botanists, anatomists, and physiologists of the second half of the century revolutionized the craft of the microscopic cut with acids and dyes, that is, with the help of inorganic and organic chemistry. The procedures of fixing, dying, and hardening rendered preparations durable and made things that were all

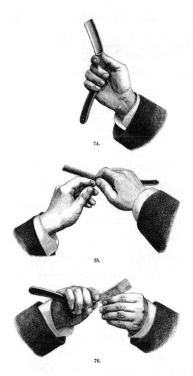

29. Making botanical cuts with a straight razor.
Behrens 1883, 150–52, Figs. 74–76.

too soft firm enough to cut, bringing out new contours. These procedures yielded literally inter-sections: that is, intervening, formative interfaces between the organic body and the optical apparatus. Epistemic object and instrument were interlocked and the new joints and sutures between the organic and the technical were materialized in a great number of vastly different microscopic experimental systems. The form in which the microscopic preparation became an image was decided here at the locus of magnification.

It was only with the new procedures for manipulating objects that the microscopic lens technologies elaborated in the nineteenth century could develop their potential in full. Thanks to them the locus of magnification was shifted from the wet to the dry side. The fresh cut of the "sharp razor" was immersed in an alternating bath of chemical reactions. The microscopic objects themselves treated with fixatives, dyes, and crosslinkers, mounted and immobilized between object carrier and cover glass began filling the cases of microscopic archives—as a new form of epistemic things made durable. What was especially delicate to cut, what could not be taken "between the thumb and fore-finger," was embedded in resin. Microtomes with the capacity to produce slices of an unprecedented thinness were developed. As microscopists started to utilize the whole arsenal of new techniques of this kind the work of preparation increasingly took the form of a dialectic of fact and artifact—one that should not, however, simply be regarded as the root of all epistemic evil or a dangerous knowledge trap. For it is also the driving force behind a typical research dynamic: the fact that such objects hover constantly on the edge of the artificial also means that they can be left behind time and again. Epistemic objects are by their nature made to be surpassed.

It is thus no accident that microscopy is of special interest to a developing epistemology of error—witness Jutta Schickore's studies of fallacies and error in the discourse of the nineteenth-century life sciences.[6] The work of microscopy embodies in a sense the methodical consciousness of a science that operates constantly on the border between the visible and the invisible—as well as on the border between the living and the dead—a science that in order to displace this border has to subject its potential objects to ever new manipulations.

Physiological Apparatuses

We have by now an abundance of studies on the experimentalization of nineteenth-century physiology. Of sole interest here is the intersection between organism and instrument, which was very different in nineteenth-century apparative physiology as compared to microscopy. The kymographion (fig. 30) is an example: the intersection that resulted from its application took the form of a bodily wound. Invented by the Leipzig physiologist Carl Ludwig this instrument was supposed to make it possible to measure animals' blood pressure. The basic principle was simple: a "communicator" connected a blood vessel through an open wound in the animal with a recording apparatus that could plot a curve. As Sven Dierig's detailed description shows, whether the result was a usable curve depended crucially on the form and characteristics of this connecting device. Physiologists collaborated with toolmakers to try out every possible form of connection and type of connecting material. Ultimately they settled on a mercury pressure gauge: through a glass cannula inserted into an artery an animal's arterial blood pressure was transmitted to the liquid metal, which converted it into motion and thence into a graphic trace at the other end of a U-shaped tube.[7] Media that were supposed to transform organic movements into technical procedures had to be connected directly to the living phenomena under study. In Ludwig's blood-pressure gauge the connection was made through a pressure-sensitive liquid. In experiments with nerves

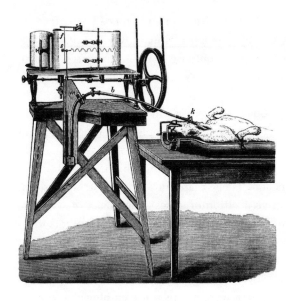

30. Blood pressure experiment with a kymographion. Langendorff 1891, 206, Fig. 169.

and muscles, in contrast, the conducting media were electric circuits. In Etienne-Jules Marey's apparatuses for measuring the gait, a compressible rubber ball formed the contact surface between the foot and the ground.[8]

In such experiments, typical of the technologically sophisticated physiology of the second half of the nineteenth century, an organism became an integral part of a seamlessly composed technical construction that not only measured certain of its life manifestations, but in limit cases also sustained them. Peter Geimer has drawn attention to the fact that the organisms whose life manifestations the apparatuses were originally supposed to measure and analyze evolved paradoxically into experimental hybrids that problematized the difference between animate and inanimate.[9] Where did nature end in these nineteenth-century Cyborgs, and where did technology begin? The major organic systems such as respiration, blood circulation, and the nervous system were objectified in a way that made the organism an organic element or an organic toggle-switch in a technical circuit. The individual organic functions were analyzed—and integrated— through corresponding mechanical elements of function.

Model Organisms

As a contrast to the hybrid living machines of nineteenth-century physiology, let us examine another, counter-intuitive "instrument" that has become a hallmark of the twentieth-century life sciences, the model organism. As Robert Kohler has forcefully argued by reference to the favorite of classical genetics, the fruit fly *Drosophila melanogaster*,[10] model organisms function not just as exemplars, but also always as research "instruments." But is that not just a metaphor? What does it mean epistemically speaking?

As an instrument the model organism is one of the technical conditions of an experimental system in which an epistemic object acquires its characteristic contours. To stick to the example of classical genetics *Drosophila* mutants were not themselves epistemic objects when it came, say, to drawing up chromosomal gene maps, but rather instruments that helped pin down the relative position of genes—the object of scientific interest—on chromosomes. Indeed many of the *Drosophila* mutants identified in Thomas Hunt Morgan's laboratory were not interesting in and of themselves—because of their specific, "monstrous" defects—but, rather, served as mapping markers. Eye pigmentation mutants and morphological mutants, for example (fig. 31), were in this sense employed as instruments quite early: all that mattered was the chromosomal position of their genes. They were transformed into epistemic objects only decades later with biochemical and developmental genetics when the genetic processes that led to the development of such features themselves became an object of study. In model organisms as instruments the organism itself has become an intersection. Model organisms are *organic* tools and for that very reason are all the more valuable since the instrument is made of the same organic stuff as the object under investigation.

Test Tube Experiments and the Ultracentrifuge

The question of the intersection takes on much more radical form with the kind of twentieth-century biochemical experimentation inaugurated shortly before the turn of the century by Eduard Buchner's attempts to obtain a cell-free alcoholic fermentation enzyme. What happens in the biological test tube experiment? Where does the boundary between biological nature and experimental culture lie in this newly created space?

At this point biologists began to distinguish between *in vitro* and *in vivo* experiments. As the name indicates in the case of in vitro experiments a transparent glass container replaces the skin of the organism or the cell membrane. What Claude Bernard called the *milieu intérieur*, the internal environment of organic processes,[11] is replaced by a chemo-technical environment: the *milieu intérieur* is turned inside out and in the process altered in such a way as to make new analytical and technical connections possible.

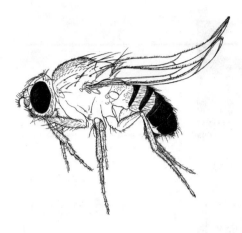

31. Fruit fly mutant with "ski" wings.
Morgan, Bridges, and Sturtevant 1925,
42, Fig. 16.

For biologists the in vitro experiment was for a long time counter-indicated because of their basic assumptions about the specificity of biological organization:[12] after all its essential function is to separate individual organic reactions and their substrates and produce and let them run in isolation from one another. In this case the intersections are cast between individual components of the organism itself and then vitrified. But what is measured here? Is it still a biological function or is it just a chemical process? Is it something that occurs in an organ or something created in the test tube? How can phenomena produced by an in vitro experiment be reinserted in the original internal environment of the organism? How can they be relocated in life? The legitimacy of a biochemistry that continued to see itself as biological chemistry, rather than simply organic chemistry or a chemistry of natural substances, depended on these questions.

The ultracentrifuge played a decisive role in this context. Developed in the 1920s by Theodor Svedberg, it was first used in the 1930s to separate cellular components and to represent viruses quantitatively.[13] A tube inserted in the rotor of a centrifuge and oriented in the gravitational field functions as a vessel or container in which the contents of homogenized cells are reorganized with respect to a single parameter, the molecular weight of their constituents. Thanks to centrifugation, researchers can decompose the cell sap into fractions separated by more or less sharp boundaries in the rotor tube. If the tissue is homogenized in the process, that is, if the cells are broken up, the morphological and functional relevance of the centrifugal bands must, with the help of supplementary testing systems based on intact cells, be connected back to and cross-checked against in vivo conditions. Such cross-checks may have a histological design as the especially elegant example of the centrifugation of intact cells shows

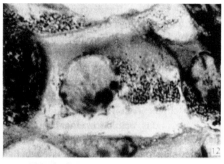

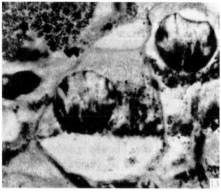

32. Intracellular segregation of the components of *Amphiuma* liver cells by high speed centrifugation. Claude 1947–48, 1950, 121–64, Figs. 12 and 13.

(fig. 32). Microscopic inspection of fractions is another possibility. But the supplementary tests may themselves be in vitro experiments: witness the example of fractional enzyme activity assessment. Thus centrifugal intersections too attest to the dilemma of modern experimental biology, which by definition must try to lay down the boundary between the still organic and the no longer organic after it has been violated. Every one of these violations is carried out in the expectation that it will be cross-checked with the help of another biological feedback whose success is never guaranteed.

X-Ray Crystallography

X-ray structure analysis, another molecular analytical technique, was first applied to macromolecules in the 1930s in connection with polymer fiber research. Here the intersection between the instrument and the biological object of investigation takes the form of a new physical entity, a crystal (fig. 33). What cannot be crystallized is not an epistemic object for this technology. Metaphoric descriptions of life in terms of "crystals" were already popular and widespread in the eighteenth century. The nineteenth and early twentieth centuries took up the theme in many different variations.[14] In the macromolecular biocrystals of the twentieth century it took on material form. The experimental crystallization of biological macro-molecules, especially nucleic acids and proteins, contributed decisively to the breakthrough of a new biological order. X-ray crystallography led to the image of the iterative, double helical structure of DNA in whose nucleotide sequence hereditary information is said to be stored; it also lent visible form to the three-dimensional structure of proteins as the translation products and functional correlates of nucleic acids. As Soraya de Chadarevian has pointed out the mathematical space engendered at the intersection between the crystallized organic molecule and the X-ray can only be translated back into the realm of biological form by way of three-dimensional macroscopic models.[15]

The Electron Microscope

The problems involved in preparing microscopic specimens as we have seen them became more acute with the advent of twentieth-century electron microscopy. This was a technology originally developed in and for the

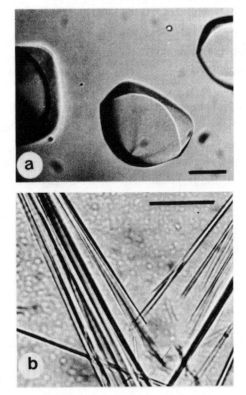

33. Crystals of ribosomes (70S) from the bacteria *Thermus thermophilus* (a) and *Bacillus stearothermophilus* (b). Yonath, Bennett, Weinstein, and Wittmann 1990, 134–47, Fig. 2 (a, b).

sciences of matter; biological applications of it were only later envisaged, and the instrument had to be specially adapted before it could be brought to bear on biological objects.[16]

Transmission electron microscopy required "ultra-thin sections," since specimens allowed passage of the electron beam only if their thickness was greatly reduced. New embedding procedures, microtomes with minimalized feed, and electron-dense "dyes" for contrast enhancement spawned new research enterprises that developed in step with the operational component of the instrument, the electron gun. The biological material itself had to be manipulated to withstand the harsh conditions to which it was exposed in the microscope's specimen chamber—a high vacuum and the high-energy bombardment of the electron beam—at least for as long as it took to produce and register a shade of scattered electrons. During the screening process and at the moment of image production in contrast to regular optical microscopy the interaction between the prepared specimen and the electron beam is so intense that the specimen is destroyed. Thus at this new highpoint of the technique of specimen preparation the object of investigation once again proves to be ephemeral like the fresh specimens of the nineteenth century: it vanishes in the act of being visualized.

The electron density and resistance of the material exposed to the electron beam are consequently the decisive parameters when it comes to shaping the intersection between the instrument and the epistemic object. Contrast enhancement plays a special role here. Thus specimens are stained, for example, with salts containing heavy metals; these salts accumulate in certain parts of the cell or the structure to be visualized. One of the most remarkable modulations of the electron-microscopic intersection between the electron gun and the biological material is produced when the organic cut is transformed into a metallic replica (fig. 34). The specimen is coated with metal that has been vapor-deposited at an angle from a metal source. The metal makes the object visible by throwing its contours into relief; for this purpose all organic remains of the original specimen must be macerated away from the metal copy. Thus the condition for the representability of the original specimen is its total elimination. The surface of contact between instrument and specimen, the intersection itself, is transformed into a resistant, virtually imperishable new object.

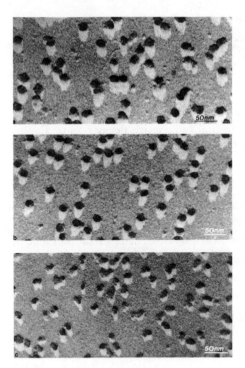

34. Freeze-dried and tungsten-shadowed ribosomes from *Escherichia coli*. Stöffler, Bald, Kastner, Lührmann, Stöffler-Meilicke, and Tischendorf 1980, 171–205, Fig. 3.

Radioactive Isotopes and Scintillation Machines

No review of the instruments that were of decisive importance to twentieth-century molecular biology can neglect the technique of radioactive labeling—an instrument of a special kind that revolutionized the analysis of metabolic processes after the Second World War.[17] It became possible to use radioactive labeling on a relatively extended scale once radioactive varieties of the most important atomic constituents of biomolecules, isotopes of hydrogen, carbon, sulfur, and phosphorus, became available in large quantities in the nuclear reactors of the Manhattan Project. As we have seen, at war's end the American National Laboratory in Oak Ridge became the center of a large-scale program of isotope production and distribution, carried out under the banner of "Atoms for Peace."[18] The practice of radioactive tracing accordingly made rapid headway in biological-chemical and biomedical research. It was perhaps owing precisely to this rapid spread of the new tool across the globe that the dramatic turn it initiated in the history of molecular biology has never been properly appreciated. Radioactive labeling did not come crashing onto the stage in the form of an imposing heavy machine; it produced its effects in capillary fashion.

In the span of two decades between 1945 and 1965 it wound its way through the biomolecular and biomedical sciences like a rhizome. Only recently have historians of the biological sciences begun to pay closer attention to this technology.[19]

Radioactive labeling consists essentially in replacing individual atoms in biomolecules with their radioactive isotopes: the fate of these molecules can then be traced by recording radioactive decay events. Radioactive isotopes serve as molecular tracing instruments. Lodged deep in various metabolic pathways they act like tracer ammunition to make biomolecules visible. This enabled the emergence of an entirely new kind of biological chemistry in the second half of the twentieth century: radioactive labeling makes it possible to measure concentrations so small as to be inaccessible to classical chemical analysis. The registration range of signals shifts from the micromolar to the picomolar level. As a result individual reactions or molecular components no longer have to be produced in microgram amounts and in the purest possible form in order to be measured for they can—in almost infinitely small concentrations—be visualized in mixtures that have been only very partially purified. Alternatively an autoradiograph makes the locus of a reaction visible in the tissue itself (fig. 35). The chemical composition of the molecules is not significantly altered by the introduction of the tracer. Hence the intersection between the tracer instrument and the epistemic object coincides with the epistemic object itself; yet the trace of this object originates and is observed entirely on the "wet" side of the borderline in contrast to, say, electron microscopy. Radioactive labeling exhibits the paradox of the production of traces as such, inasmuch as a radiogram makes something visible in situ that no longer exists: at the very moment that the trace is produced the marker by decaying irrevocably abolishes itself.

To obtain relevant data the radioactivity of the specimens under investigation has to be recorded and *measured* by technical means. In the case of the autoradiograph a sensitive photo plate is all that is needed. Other states of aggregation are more difficult to record. As discussed at length in chapter nine from the 1950s to the 1970s parallel to the massive use of radioactive carbon, hydrogen, and sulfur in physiology, medicine, pharmacology, and material science the liquid scintillation counter was developed as an alternative to traditional Geiger counters. It operates in the range of the weak β-rays of these isotopes. The procedure transforms radioactive decomposition events in an organic liquid into bursts of light that are subsequently amplified by photoelectronic means. The LSC swept triumphantly through molecular-biological laboratories within

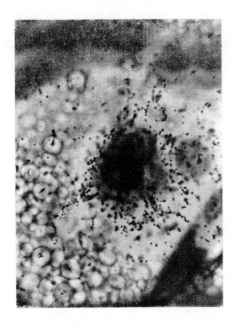

35. Autoradiograph showing extensive incorporation of amino acids (^{14}C-adenine) in the nucleolus of *Acetabularia*. Brachet 1960, 209, Fig. 32.

a decade, becoming together with the ultracentrifuge a symbol of modern molecular-biological lab technology. The development of an automated counting device for testing samples had a more than merely quantitative impact on molecular-biological experimentation. It made possible not only serial testing, but also qualitatively new experimental designs. Using the history of the liquid scintillation counter as an example, we can trace the way the incorporation of a new instrument in existing experimental systems—in this case the mere introduction of a new measuring device— can lend these systems completely new features. The advanced measurement technique realized the potential of the "fluid" intersection between the device and the radioactive sample, accommodating biochemistry's and molecular biology's predilection for "wet" experimentation. Notwithstanding the massive lead chamber and tangle of electric wires around the measuring point the probe finally remains in the liquid environment of a glass vial (fig. 36).

Outlook

The new kind of molecular-biological research that took shape around the middle of the twentieth century relied on physical and chemical instruments for the analysis of the organic—it was based on a consistent "extracellular" paradigm.[20] It mobilized a panoply of research technologies that

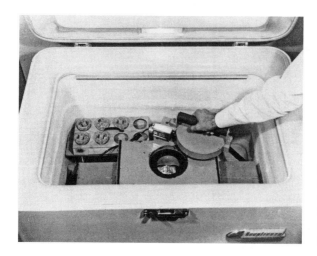

36. The first commercial liquid scintillation counter, with a lead chamber for 50-ml ampoules. Collection Edward F. Polic.

engendered a wide range of intersections between the central epistemic objects of molecular biology, biological macromolecules and molecular cellular structures composed of them, and each of these new technologies. The most diverse properties of macromolecules became measurable as a result. Ironically, consistent pursuit of this research program brought about a situation in which the objects of epistemic interest themselves, the macromolecules, were ultimately converted into an arsenal of molecular tools; the extracellular technology of early molecular biology has given rise to an arsenal of instruments that have found their way back into the cell, for example, in the genetic engineering of today. We do not know just how far these new tools will go: genetic engineers no longer build technologies up around organisms, but build them directly into cells and their instruments do their work from within, no longer analyzing the organism but recomposing and reformulating it. They are constructive and synthetic. As a result not only biological research, but the organism itself becomes for the first time the instrument of a new comprehensive project. As a consequence the fundamental distinction between instrument and object as supposed in this chapter, and between culture and nature for that matter, may begin to disappear altogether.

12.   Preparations

Recent history of science has made us aware of the extent to which scientific research itself—and by no means just its diverse technical products—is bound by culturally variable ways of handling material things. The cultural character of scientific things was for a long time of only incidental interest to historians of science, who generally preferred to study the history of concepts and the subdivisions and fusions of scientific disciplines. The *things* around which the knowledge process unfolds were relegated to the margins of this history. Yet they play a decisive role in the development of knowledge.

As a rule the primarily descriptive, systematizing sciences extract their objects from their traditional natural context and their "natural" ambiguity and incorporate them into an order specific to the sciences in question. The result is, say, a botanical garden, rock collection, or herbarium. The things thereby become part of a theoretical order and an object of determinate epistemic practices. Their previous meanings are modified—or perhaps "meaningful" aspects of them become perceptible only as a result of such modification. They become epistemic objects or scientific things in the first place thanks to this reordering.

The process is still more radical in the experimental sciences. Here generally speaking objects must be prepared before any actual observation can take place. The required manipulation of the things is often guided by the development of instruments of observation and measurement that intervene as third parties—on an equal footing as it were—between the manipulated objects and the instrument-based knowledge of them. Bachelard's term "phenomenotechnique" can be taken as an indicator as to how the modern natural sciences bring things into the realm of the visible.[1] In this chapter I will examine how we are to understand this process and how we might assess it. I will define epistemic things brought into existence by scientific procedures of various kinds as "epistemologica" whenever they have attained a certain stability. Epistemologica should be understood as all the material things rendered permanent in various ways that play a part in knowledge production by enabling facts to be exposed and elucidated.

"Preparations" represent a special class of epistemologica. They play a particular role in the biological sciences, the sciences that study living things. In these sciences diverse preparations and techniques for producing them have been developed, because special arrangements are required to fix and stabilize living things for gaining knowledge about them. The 1974 edition of *Duden*, an authoritative German dictionary, gives two definitions for the word *Präparat*, here translated "preparation." The first is "something that has been skillfully prepared (for instance, a medicine or chemical substance)"; the second is "a) a plant or the body of an animal that has been preserved (for didactic purposes); b) a section of tissue intended for examination with a microscope."[2] The first usage of the word comes from the field of pharmacology and medicine in the broadest sense. The second refers to organisms in the biological sciences, our focus here. The organisms in question are *prepared* organisms or parts of living things however small. In chemistry and in biology as well one still commonly speaks in German of a *Darstellung* when characterizing preparations. Correlatively in the case of a substance separated out of a mixture the term *Reindarstellung* is employed. Here the term *Darstellung* does not mean "representation," its primary meaning in German, but designates both the production (Herstellung) of a product and the product itself.

This concept of *Darstellung* also grounds Martin Heidegger's view of the essence of modern research and his elusive concept of the "world picture." According to Heidegger we live in the age of a materially and technically conditioned "planetary" configuration and reconfiguration of things in the sense of a representing production and a productive representation. "The fundamental event of modernity," Heidegger claims, "is the conquest of the world as picture."[3] However, as a result of such productions, things are not supposed to become distorted and artificial, but rather "come to themselves"—that is, become evident thanks precisely to the properties that the scientific process has brought out in them, thus rendering possible their ongoing definition and redefinition. This paradox of science becomes particularly clear in the case of scientific preparations: all the work that goes into manipulating them (to which the meaning of the Latin *praeparare* silently points) is deemed to have succeeded precisely when it has been made to disappear in its object. Only then is a preparation considered authentic. Whence the fundamental distinction between preparations and another class of epistemologica that can only be mentioned here: models.

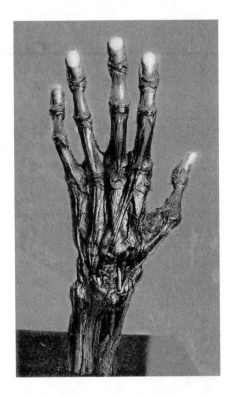

37. Dry preparation of a left hand showing the blood vessels. The arteries were injected with red dye. 1893. Loan from the Center for Anatomy of the Charité to the Berlin Medical Historical Museum of the Charité.

A model is defined as a limit-object or rather as the transposition of the model's object into another medium. Thus whereas a model can at best claim to "look deceptively like" the modeled, a preparation participates in the very materiality of the thing under investigation. It is made of the same stuff. It is a figuration that this stuff has been made to yield.

Anatomical Preparations

In the history of knowledge techniques the anatomical preparations familiar from natural history, medicine, and biology may well go back further than all others to be considered here.[4] Ingenious procedures for the production and preservation of preparations both wet and dry have been devised over the centuries. There exists a correspondingly broad range of such preparations, from mummies to skeletons, from individual bones or body parts that have been plasticized, injected with dyes, or stuffed with foam rubber, to normal or deformed embryos, organ systems, and organs that have been preserved in alcohol. Figure 37 is a dry preparation displaying the blood vessels of a healthy left hand; figure 38 shows a wet prepara-

38. Wet preparation
of gangrene of the
toes caused by
arteriosclerosis.
Inv. N° 799. Berlin
Medical Historical
Museum of the
Charité.

tion of a foot, both from the collection of the Charité Hospital in Berlin. The Dutch apothecary and physician Frederik Ruysch from The Hague assembled one of the first anatomical collections of organ preparations at the end of the seventeenth century. Together with the natural historian Jan Swammerdam of Amsterdam he developed an injection technique that none of his contemporaries was able to copy. It made it possible to exhibit for the first time even the network of blood vessels located deep in the body. Ruysch's preparations were meant to last; some have been preserved down to the present day. In the eighteenth century the French anatomist Honoré Fragonard took a different approach to overcoming the body's opacity: as his dissected, macerated mummifications show, the aim was to expose the inside of the body and bring its depths to light without sacrificing its surface contours in the process. The layered construction of the body was revealed by means of perforation and partial dismemberment. The result made it look as if the layers of the body illustrated in Vesalius's anatomical plates had been superposed one on the next and pasted together to form a new whole.

The preparers' original preoccupation was with the visible surface.[5] They were, however, faced with a new challenge as soon as the concealed organization of life forms became the central theme of biology toward the end of the eighteenth century. The example of Fragonard illustrates the means mobilized in the struggle to reveal depth and everything that does not appear on the surface. It was in the eighteenth and above all nineteenth centuries that anatomical preparation flourished. Universities everywhere acquired extensive anatomical and morphological collections often with a focus on pathology and teratology.

Anatomical preparations can serve one of two distinct epistemic ends. One is typological exaggeration: the individual exhibit becomes a prominent representative of a class of objects, an instantiation and a showpiece. What it is supposed to illustrate is highlighted in a particular way. Preparations of this kind are always hypostatizations. When the opposite end is pursued an item is exhibited not as the representative of a type, but as something rare or even unique, something that ultimately refers only to itself: a unicum embodying deviation or divergence from the type. The anatomical preparation oscillates between these two poles: representation of a class and singularity. Something similar holds for taxonomic collections. The tension between the two poles, which makes itself felt in any investigation of biological forms, may be traced back to the "logic" of evolution conceived as a play of similarities and·differences.

The authenticity or genuineness of anatomical preparations can likewise be addressed and staged in one of two different ways. The preparation can present itself as something untreated or unprepared: for example, a bleached intestinal loop in alcohol. If no attempt is made to prevent alterations such as the discoloration and deformation that preparations undergo because they have been detached from the body and conserved for a long time, they can seem especially authentic and untreated. On the other hand, if all correction is avoided, a preparation can under certain circumstances be altered beyond recognition so that it no longer fulfills any function: it becomes amorphous and no longer represents either a type or a singularity. Conversely one can do everything possible to convey the impression that an exhibit has just been plucked fresh from life. Here authenticity is achieved by retouching the preparation itself so that the exhibit comes as near as possible to what it has ceased to be: a living thing. The paradox of the preparation manifests itself in the two different forms of its authentication.

Herbaria

The history of the life sciences has seen another form of specimen collection. The *Herbarium vivum* emerged as the counterpart to the herbals of the sixteenth and seventeenth centuries: this new knowledge receptacle was supposed to contain living knowledge. In comparison with the botanical gardens that were springing up at the same time, however, it appeared lifeless. Herbaria can accordingly be regarded as the botanical counterparts to anatomical show cases. Like an anatomical preparation a plant in a herbarium can perform one of two epistemic functions: it can serve as a typical representative of or stand-in for a species,[6] or as a lucky find with a special aura, a *trouvaille* that reveals something about a territory or testifies to the unexpected occurrence of a species or variety in this particular region or country. Thus the Berlin Botanical Museum's *Silene chlorantha*, which Friedrich August Körnicke collected in 1854 on "sunny hills in Reinickendorf," functioned as such a "voucher," to use the botanists' technical term (fig. 39). In contrast to the forms used to present scientific objects in other media—models, illustrations, formulas, and so on—the plant in a herbarium is an object capable of becoming the subject of new studies at any moment. Like all other preparations it is a signifier, to borrow linguistics terminology, that participates in the material substrate of its signified. Indeed the essence of organic preparations qua knowledge objects resides in this material complicity, which ensures their duration and the permanent possibility of their epistemic recall. Herbarium plants lend themselves to such recuperation more easily than anatomical preparations, which as a rule have been subjected to more intensive preparative manipulation. This ease of scientific retrieval makes the herbarium plant the epistemologicum par excellence. Dried plants are knowledge things whose reversibility to and readiness for examination is maintained in a particular way. One of the consequences is that their taxonomic classification can be revoked at any time, clearing the way for new determinations.

Microscopic Preparations

Use of the microscope spread from the latter half of the seventeenth century. As we saw in some detail in the last chapter one of the inherent demands that this instrument put on microscopic preparation was that objects be presented in a flat, two-dimensional form, since microscopes produce a sharp image only in the focal plane.[7] Hence preparations had

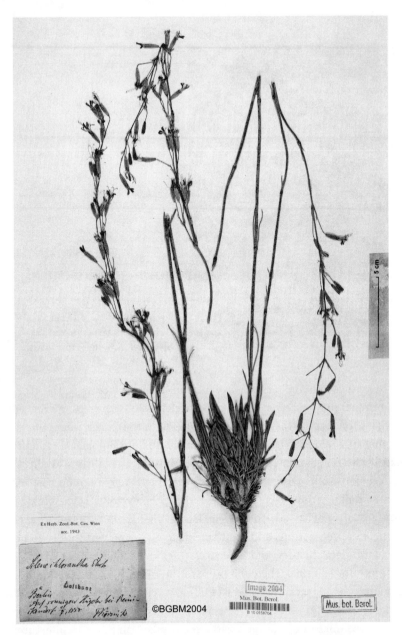

©BGBM2004

39. *Silene chlorantha* Ehrh., found on sunny hills in Reinickendorf by Friedrich August Körnicke (1828–1908). From the collection of the Botanical Museum Berlin-Dahlem.

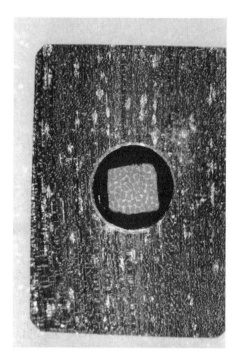

40. *Hepar cuniculi*. A cubic prepa-
ration from the case "Original prepara-
tions by Jos. Hyrtl, a gift for Virchow,"
Royal Pathological Museum, Berlin.
Collection of the Berlin Medical
Historical Museum of the Charité.

to be reduced to a *planar* surface, not an *outer* surface as in the case of
anatomical preparations. To this end various techniques of microscopic
preparation came into being. In close step with the development of cellular theory, zoologists, botanists, anatomists, and physiologists in the latter
half of the nineteenth century learned to produce thin films or slices of
watery consistency (or some other consistency): they became past mas-
ters at making tissue-sections and crush preparations. Microbiologists
and protozoologists enriched the existing arsenal with their impression
or *Stempel* preparations and their smears, giving a powerful fillip to both
developmental history and infectious medicine. The wealth of micro-
objects—for example, the cube preparation of a liver contained in a gift
case that Josef Hyrtl presented to Rudolf Virchow (fig. 40)—lent the late
nineteenth-century biological sciences and medicine their distinctive profile. These micro-objects even had a hand in shaping the architecture of
scientific spaces: laboratories were now fitted out with huge repositories
to store the many permanent preparations for repeated consultation as
well as display to large auditoriums with projectors. In addition traveling
cases of preparations became part of the standard equipment of scientists
on their lecture tours.

Preparations in Molecular Biology

A preparation of a different kind has become characteristic of twentieth-century molecular biology: chromatography. As the name indicates, chromatography involves making something visible by means of color. More precisely it is a graphic procedure that combines the production of a colored trace with the inscription of that trace in readable form. Like prior forms of preparation chromatographic preparations manipulate the objects under investigation in a special way. However, in one crucial respect chromatographic procedures differ from the classical graphic procedures of the nineteenth century, which have been much more thoroughly studied: chromatography does not simply register things—it does not function as a mere "pencil of nature"—but also graphically reconfigures them, inscribing itself in the preparation or specimen and penetrating the object under investigation. Chromatograms are thus on a continuum with microscopic preparations. Yet a conspicuous shift has taken place. As objects microscopic (even electron-microscopic) preparations remain miniatures. Although they stand in an intimate relation of dependence to the magnifying instrument, they remain external to the technology of the instrument. Matters differ in the case of chromatographic epistemologica. Although the prepared objects here are typically molecules and thus far beyond the resolution capabilities of optical microscopes, chromatograms represent these rearranged objects on a scale that can be assessed by the naked eye. That is to say that the chromatographic plate brings the technology itself into manifestation; it incorporates into the preparation itself the principle of magnification informing it.

Figure 41 shows a mini-chromatogram based on the procedure of electrophoresis. The objects here are the protein components of a cell organelle. The large subunit of a ribosome of the bacterium *Escherichia coli*—the molecular machine in charge of producing the proteins in the cell, including the proteins of which this machine itself is made—has been separated into its component parts, and these have been charted in a coordinate system. The plate is made of a porous gel consisting of synthetic, interwoven polyacrylamid fibers. The compound that is to be separated into its components, containing over thirty proteins of the large subunit of the cell organelle, is distributed in two different dimensions according to two physical parameters. In the first dimension (in our illustration, the horizontal dimension) the components drift toward either the cathode or the anode of an electric field; how far each one gets depends on the elec-

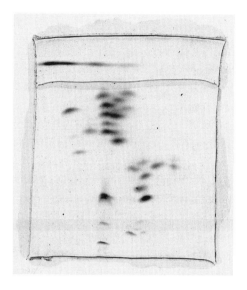

41. A two-dimensional miniature polyacrylamide gel with the proteins of the large ribosomal subunit (50S) of *Escherichia coli*. From the laboratory of Knud Nierhaus, produced by Detlef Kamp at the Max Planck Institute for Molecular Genetics, Berlin.

tric charge it carries. In the second dimension (from top to bottom in our illustration) a change in the acidity of the buffer solution has neutralized the charge so that all the components now drift in the same direction; how far each gets depends on its molecular weight or size. These components are then dyed with methylene blue so that they appear on the plate in the form of longer or shorter, smeary blue spots.

Chromatograms are perhaps best described as analytical preparations. In a chromatogram, substances in a mixture are arranged in a certain order in accordance with certain physical-chemical parameters: the mixture is separated out into its component parts, yielding distinct patterns. The same principle informs the production of the sequence gel of a DNA-specimen. This highly abstract form of digital presentation has become the second icon of the age of the genome after the double helix. In the base sequence of a sequence gel (fig. 42), the molecular-biological preparation par excellence, the "logic" of life seems to have come to itself at last.

: : :

The anatomical, botanic, microscopic, and molecular-biological epistemologica that have been passed briefly in review here constitute a very special type of scientific thing: their peculiarity resides roughly speaking in the fact that they consist of the same material as that for which they stand. In contrast other scientific things such as models are cut from a different cloth than the things whose models they are (to stick to the

42. Dust jacket of the German edition of François Jacob's *The Logic of Life*.

metaphor); they embody the representation of these things in a different medium. Analogical or digital images—simulations—might be regarded as a third kind of scientific thing and tentatively defined by the fact that they transport their objects into a different space.

We do not yet have an exhaustive typology of epistemologica. Yet deposited in their diverse configurations layer upon layer is a record of the modern research process. It can, moreover, be very precisely and vividly visualized in them; thus they make it possible to communicate to the public the idea that science is *research*—a comprehensive, far-reaching, manifold cultural process. As a rule we see only scientific instruments in scientific exhibitions organized by institutions specializing in the history of science and technology. The forms and techniques required to configure *objects* in such a way that these instruments can do their job are doubtless often less spectacular. They are, however, of critical importance when it comes to answering the question of how knowledge is thrown into relief.

13.   The Economy of the Scribble

Over the past three decades historians of science have been increasingly concerned with the subject of scientific writing and publishing, giving special attention to literary techniques such as rhetorical emphasis, persuasion, and dissimulation.[1] Here I address another aspect of scientific writing. The research laboratory is a place where new knowledge comes into being leaving behind it a trail of rough notes, scrips and scribblings, and revised write-ups that offer insight into concrete processes of knowledge formation. The past years have seen a growing number of microhistorical reconstructions dealing with laboratory notebooks and other research documents.[2] However, with a few exceptions[3] the generative function of such documents in the overall order of knowledge production has been neglected. In what follows I would like to suggest that we begin to observe and investigate in its epistemic positivity the "economy of the scribble" in the lab.

Laboratory Writing

A brief summary of how experimental systems work illustrates their active function in constructing epistemic things.[4] Experimental systems are material, functional units of modern knowledge production; they co-generate experimental phenomena and the corresponding concepts embodied in those phenomena. In this sense experimental systems are techno-epistemic processes that bring conceptual-phenomenal entities—epistemic things—into being. Thus as François Jacob once put it they are, "machines for making the future."[5]

Epistemic things themselves are situated at the interface between the material and conceptual aspects of science; hence we can also conceive of them as graphematic traces.[6] They are writing in the broad sense that Jacques Derrida gives to the word—they have the capacity to become detached not only from their initial referent, that to which they originally referred, but also from the one who writes, the one who produces the trace.[7] Notes and scribblings are the primary transposition of these traces on to paper, the beginning of their transformation into data. They are not yet of the order of printed communications addressed to the scientific community; they are still of the order of the experimental engagement and entanglement. They belong to the lab's knowledge regime.

Laboratories as spaces of knowledge production generate a wealth of such primary scriptural traces: excerpts from the scientific literature; notes on basic concepts, fragmentary ideas, or striking correlations; sketches of experimental setups; running data from a single set of experiments; tentative interpretations of experimental results, and corrections of such interpretations; provisional calculations; calibration protocols of existing instruments; designs for new equipment, and so on. These and many other traces of research activity are invaluable sources for historical analysis. They lie *between* the materialities of experimental systems and the conceptual constructs that leave the immediate laboratory context behind in the guise of sanctioned research reports.

Laboratory notes and write-ups in their primary form were long regarded as simple means for recording and storing data, as mere "temporary storage bins" along the way to some "final result." In this purely retrospective vision, however, only the definitive data counted; and these appeared as results painted directly by nature's brush, transparent and vanishing intermediaries between the material under investigation and its conceptualization. Textbook examples included products of "the curve method" of mid-nineteenth-century physiology or bacteriological microphotography of the late nineteenth century. But the fact that their availability always already depends on an experimental setup is just one feature of these graphic data; in their putatively "provisional" forms they are also part of a more comprehensive laboratory discourse.[8] The "containers" in which they are "temporarily stored" by no means compose a neutral scaffolding in an otherwise purely intellectual process of knowledge formation; these "temporary storage bins" are and remain—literally—*between* the epistemic objects and the knowledge processes bound up with them.

One eminent function of taking notes and writing up experiments at the laboratory bench is something I call the "redimensionalization" of the experimental dispositive. In the most banal sense this simply means that the organization of an experiment in time and space is projected onto a two-dimensional surface. This not only puts a laboratory in handy, transportable form.[9] Reduction to a surface facilitates exploration of new ways of ordering and arranging data: sequential events can be presented in synchronic form, temporal relations as spatial. At a somewhat more sophisticated level a laboratory protocol produces condensation effects. Data can go through different degrees of such condensation and at each stage of the itinerary—depending on the way they are arranged and the extent to which they have been compressed—new patterns can become perceptible. It is essential for a well-kept laboratory protocol that all deductions and reductions be reversible. A protocol becomes epistemically productive when it allows us to run through a chain of transformations in reverse and at certain points in the series to vary their order.[10] Hence laboratory protocols generally represent the memory of whole series or strata of experiments; they make it easier—indeed make it possible in the first place—to give meaning to the individual data they contain. The transformation of the lab into a writing surface, the transformation of actual experiments into a patchwork of signs consisting of icons, symbols, and indices, clearly has more than a merely reductive function. Quite the contrary: laboratory notes and protocols are themselves new resources and materials opening a space that alone gives research its distinctive contours and prevents it from being closed off prematurely.

Individual styles in the production of scientific novelty are also put to the test in the realm of such primary scientific writing. In contrast to what happens with standard scientific publications, which straightforwardly eliminate such traces, here the search process as such becomes visible—that is, research as it actually unfolds. As Jacob has remarked, scientists when they go public with their work "describe their own activity as a well-ordered series of ideas and experiments linked in strict logical sequence. In scientific articles, reason proceeds along a high road that leads from darkness to light with not the slightest error, not a hint of a bad decision, no confusion, nothing but perfect reasoning. Flawless."[11] Research notes on the other hand are residues of "night science," which Jacob distinguishes from well-ordered "day science:"

> By contrast, night science wanders blind. It hesitates, stumbles, recoils, sweats, wakes with a start. Doubting everything, it is forever trying to find itself, question itself, pull itself back together. Night science is a sort of workshop of the possible where what will become the building material of science is worked out. Where hypotheses remain in the form of vague presentiments and woolly impressions. Where phenomena are still no more than solitary events with no link between them. Where the design of experiments has barely taken shape. Where thought makes its way along meandering paths and twisting lanes, most often leading nowhere.[12]

Here we are in the domain of the pre-normative in which the opportunism of the process of acquiring knowledge can expose itself without restraint:

the domain of the experiment (in German *Versuch*, which also means attempt, assay, trial) in the profound sense of the word in which experiments are constitutive of research in the natural sciences[13]—as they are of literature and philosophy.[14] An experiment is not only a test or an examination, as it may be at times. More frequently, it is an exploratory movement, a game in which one plays with possible positions, an open arrangement. In research, such experiments transform the as yet unbound explanatory potential of experimental systems into a game of combinations still unrestricted by the rigorous limits of stringent compatibility.

The Protocol Bundles of Carl Correns

Let me illustrate these basic considerations of principle with an example taken from the history of research on heredity. As described at greater length in the first case study in this book,[15] the botanist Carl Correns began a series of crossing experiments with varieties of peas in spring 1896. For two years he had been seeking the solution to the following problem: can pollen of a different variety have a direct influence on the characteristics of the fruit and seeds of the mother plant? The xenia phenomenon had posed its riddle to Darwin before him. Correns had sifted through the literature for reports on plants that could be used to demonstrate such a direct influence. *Zea mays* and *Pisum sativum*—corn and peas—were among the species for which pertinent descriptions could be found. Correns planned first to produce experimentally an unambigouous, unimpeachable example of the phenomenon. Then by means of physiological-histological examination of the fructification process he hoped to find a definitive answer to the xenia question.

The protocols of his experiments with corn and peas have come down to us[16] and allow us to reconstruct step by step the process in which the original research question was not merely modified, but eliminated and replaced. In the end it was supplanted by the observation and explanation of regularities in character distribution in hybrids produced by crossing—a distribution that Gregor Mendel had already described in an 1866 paper published in the *Verhandlungen des Naturforschenden Vereines in Brünn*.[17] Mendel's paper had been to all intents and purposes passed over by his contemporaries. Only in the *annus mirabilis* 1900, which has gone down in history as the year of the "rediscovery" of Mendel's laws, did its broad implications gain public recognition. Correns had glossed over Mendel's work as he was beginning his experiments with the very different goal of laying the xenia controversy to rest in spring 1896. The spring 1900 publication of his findings was then precipitated by a paper of his colleague and contemporary Hugo de Vries from Amsterdam, who confirmed Mendel's laws shortly before Correns.[18]

What had happened in between? What do his protocol books reveal — and above all what did they accomplish? From 1894 Correns had been accumulating observations on the characteristics of the hybrid seeds of his two experimental plants and their varieties. He was obviously interested at the outset only in the xenia question. He began to experiment with various maize and pea varieties that he had selected for the characters of their fruits. His express aim was to produce a broad spectrum of hybrid fruits that could be used to study possible xenia. From the reciprocal crosses of these varieties he raised in each individual case a very small number of hybrid plants that he then carried on into the second and third generation together with a corresponding number of control plants and backcrosses whenever that seemed appropriate. His notes contain painstakingly exact descriptions for each cross: descriptions of the coloration and structure of the seeds as well as the embryos they contained. Correns conducted these experiments over a period of six years: the sheer volume of the sheets of paper he covered with his individual observations indicates how intensely he pursued his goal. For years he turned out virtually exhaustive descriptions of the arrangement of his experimental garden in Tübingen and — quite literally — the fruits of his experimental work. Thanks to these notes each *individual* seed and all the plants grown from these seeds could be traced back to the corresponding individual plants and seeds of the preceding generation.

This process of faithful reduplication notwithstanding, Correns's protocol notes were not mere passive databases, nor were they, most importantly for my argument, mere compilations of statistically relevant data — as they most likely would have been had he been following in Mendel's footsteps with a view to documenting the statistical regularity in the characters of offspring. Rather, the structural arrangement of his notes, complemented by seeds stored in boxes — a second, material protocol — facilitated the condensation and accretion of meanings. Correns was obviously looking for surprising seed characteristics, which he was hoping to see emerge at one or another point in his broadly conceived set of crossings. As a virtually complete inventory of the individual seed lineages his notes could be read forwards and backwards. This reflected his initial research project, the search for xenia. Yet his protocol notes were redundant enough (that is, they generated an excess of information) to enable Correns later retrospectively to reinterpret his entire research project in a new light.

Consider a page from Correns's protocol notes of 1897 (fig. 43). The *individualizing* style of this notation leaps to the eye from the numbering of the plants through that of the pods to the detailed description of each of the peas contained in them. On the left-hand side of the page Correns obviously left a broad margin for addenda. In the course of the year 1899 he added to the original 1897 protocol. One of the comments he appended in 1899 reads, "Σ 23, 20 are still in existence. Seven of them have green germs = 35.0 %—yellow germs, 13, = 65%."

Moving from the right to the left side of this protocol we pass from the descriptive, individualizing xenia regime to the arithmetical, statistical Mendel regime. The difference in the significance of the observations could hardly appear more clearly. The change came to Correns's thinking at a particular point in the course of his experiments—in all probability at the beginning of 1898. Examining the results of some experiments with corn plants that turned out to be inadvertent backcrosses (plants that were themselves doubtless descended of hybrids) Correns must gradually have come to understand that something important was going on in his experiments with corn and peas, something that he had so far ignored. At this point presumably Mendel's work acquired new and considerably deeper significance and urgency for him than it had had in April 1896 when he had read it with nothing but the xenia problem in mind. Now his reading was reorganized by the results of his own experiments. Only now did he realize that his peas, which had stubbornly refused to display the xenia phenomenon he was looking for, exhibited—somewhat more clearly than his corn—another regular pattern, a pattern that became manifest in the seed characters he had been observing: one of the alternating characters was dominant in the first generation of hybrids while in the second the suppressed character reappeared in a certain number of offspring. As the comments in the left margin of his protocol sheets show the ratio was about three to one, although it varied considerably due to the small number of plants that had to serve—the experiments having been designed for another purpose—as the basis for his calculations.

However, Correns did not need to run his experiments all over again at this point: what he had failed to do because he was concentrating on the direct action of the pollen on the mother plant—namely, attend to character transmission ratios—now became possible thanks to his extensive protocols. He could simply reread this material from the new standpoint, ignoring the various forms of the seeds as well as the coloration of the seed coats on which he had initially put the greatest weight, and focus on just one character pair, the coloration—yellow or green—of the germ. He

Bastart

„ gr + p B. 1. gelb "

V Pflanze I.   Hülse 1. Sa 3, scherbenfarbig (hellgrünlich orange) mit dunklen Punkten, sparsam, etwas faltig.

H. meist roth
überlaufen

≤ 2B, es sind noch da
20.
davon haben grüne Keime
7, = 35,0 %
gelbe Keime
13, = 65,0 %.

2. Sa 6 davon 2 stark faltig, klein (entwicklungsfähig?) schmutzig-gelbgrün, rothorange überlaufen.
   b) 4 voller, wenig faltig, grösser bläulichgrün bis hellbräunlichgrün, stark violett punktirt.

3. Sa 2, einer scherbenf., einer bläulich grün, beide (sparsam) viol. Punkt.

4. Sa. 2, gross, etwas faltig, sonst wie die von 2b,

Aussaat 3 gelb
≤ 7 grün = 30,4
16 gelb = 69,6

5. Sa. 4, davon 3 wie 2b einer wie 2a, aber etwas (ein Fleck auf beiden Seiten) punktirt

aussaat 2 gelb, 1grün
≤ 13, davon gelb 11 = 81,8
, grün 2 = 18,2 %

6. Sa 6, gelbscherbenf. bis bläulichgrün, fein punktirt, wenig faltig.

VI Pfl. II.   Hülse 1. Sa 2 scherbengelbgrünlich, schwach punkt.

≤ 13, davon sind noch
da 10. 3 wurden also
ausgesaet. 1 hat grünen
Keim (10%) die übrigen grün bis röthlich scherbengelb.
gelben, stark erds sonst
punktirt.

2. Sa. 2, gross, grobfaltig roethlichscherbengelb, stark punktirt, z. Th. fast fleckig

3. Sa 5, fastfleckig punktirt bis punktirt, fl. bis röthlichscherbengelb.

4. Sa. 4, 3 wie 3, 1 wie Pfl. I. 2b. rothgelb, klein etc.

43. Hybrid gr + p B. 1 yellow [gelb]. MPG Archive, Section 3, Folder 17, No. 115, File "Pisum Resultate 1897."

counted the seeds and identified those he had used to raise new plants. In this way he was able to reconstruct the original ratios. With the resulting information and after calculating the sum of all the individual plants and their progeny he was able to generate numbers big enough to perform a statistical analysis of the whole run of his experiments, something he had not at all intended to do in setting out. The calculations for the experiment that he published in spring 1900, which in fact represented a composite of all the crosses he had carried out over the years using the two pea varieties *Grüne Folger* and *Purpurschote*, were thus ultimately based on several hundred peas from a total of three successive hybrid generations.

Collective Forms of Laboratory Writing

Do there exist "collective" equivalents of such individual forms of scientific note-taking and write-ups—graphic forms of preserving traces that while not associated with just one experiment nevertheless do not constitute definitive statements either and as comments on the epistemic things under investigation still belong to the same field those things do? To use Michel Foucault's terms, how might we characterize these "discourse-objects" of an archaeology of the laboratory?[19]

In the gap between the laboratory bench and the organized public discourse of the scientific community we find different categories of collective writing and the preservation of traces that would repay closer examination. One such category comprises lists, tables, and other types of what might be called scientific bookkeeping or alternatively *counting techniques*.[20] In the research process these can be described on the one hand as laboratory records or archives, as systems of recall that make available information with which experiments can be set up or pursued. On the other hand, they serve as primary, constantly updated databases in which new research results are entered on a continuous basis, so that they can be made available for the collective work of entire networks of laboratories. Today this extended function is fulfilled by laboratory counting techniques in the form first and foremost of electronic data storage, retrieval, display, and communication. A conspicuous example is provided by the nucleic acid sequence databases that molecular geneticists and genetic engineers use in constructing their probes and comparing their results, feeding their own sequencing products into them in turn. Such digital information pools constitute one particular form of networking in the organization of the research process. They can also—thanks to their sheer size and the accelerated comparability of results—spawn new research questions.

A second category of collective write-up techniques consists of semi-standardized protocols and laboratory manuals. Let us call them *literal techniques*. In this rubric we may put, for example, written experimental procedures dependable enough to be applied in more or less routine fashion at least by initiates—in many cases by succeeding generations of experimenters. Literal techniques are means of handing down elements of the practice of a laboratory that have proven successful under certain local conditions yet cannot be generalized or are not supposed to be generalized because they contribute to the social identity of one particular laboratory, an identity in which all newcomers are trained up.

With an eye to such literal write-up techniques, which may involve repeated revision over extended periods of time, we might be tempted to treat laboratories themselves as scientific writing collectives or even collective authors representing more than the mere fact that a group cooperates to achieve certain ends. Laboratories preserve a particular tradition and an identifiable experimental style, which despite constant rotation of the participants and changing individual preferences are reworked in line with a protocol and so continue to be written, whether competitively or cooperatively. This laboratory feature has given rise to the long-standing discussion in science studies and history of science about research traditions and schools of research,[21] although the accent in this discussion is typically placed on sociological characteristics such as strong "pioneering thinkers," the idiosyncrasies of a local institution, or disciplinary alliances in a particular laboratory. It seems to me more important to recognize that notational techniques such as the counting and literal techniques just mentioned play a fundamental role in the emergence of these traditions and schools. Research traditions result from the material reproduction of local research cultures. A laboratory's scriptural idiosyncrasies—recipes, instructions about procedures, and log sheets as well as standardized experimental designs and even specially adapted software—are indispensable elements in this process.

These concluding remarks are intended as indications and stimuli. The laboratory "economy of the scribble," the economy of its scriptural forms, remains largely unexplored to this day, in particular in its contemporary electronic versions. For an epistemology of the concrete it is a promising source for the further elucidation of both historical and actual research processes.

ACKNOWLEDGMENTS

The following chapters in this book have been previously published in English. They have all been substantially reworked for the purposes of this book. Chapter one: "On the Historicity of Scientific Knowledge: Ludwik Fleck, Gaston Bachelard, Edmund Husserl" (David Hyder and Hans-Jörg Rheinberger, eds., *Science and the Life-World: Essays on Husserl's "Crisis of European Sciences."* Stanford, Calif.: Stanford University Press 2009, 164–170); chapter two: "Gaston Bachelard and the Notion of 'Phenomenotechnique'" (*Perspectives on Science* 13 [2005]: 313–28); chapter three: "Reassessing the Historical Epistemology of Georges Canguilhem" (Gary Gutting, ed., *Continental Philosophy of Science*. Oxford: Blackwell, 2005, 187–97); chapter four: "Carl Correns' Experiments with Pisum, 1896–99" (Frederic Holmes, Jürgen Renn, and Hans-Jörg Rheinberger, eds., *Reworking the Bench: Research Notebooks in the History of Science*. Dordrecht: Kluwer, 2003, 221–52); chapter six: "Ephestia: The Experimental Design of Alfred Kühn's Physiological Developmental Genetics" (*Journal of the History of Biology* 33 [2000]: 535–76); chapter eight: "Gene Concepts: Fragments from the Perspective of Molecular Biology" (Peter Beurton, Raphael Falk, and Hans-Jörg Rheinberger, eds., *Gene Concepts in Development and Evolution*. Cambridge: Cambridge University Press, 2000, 219–39); chapter nine: "Putting Isotopes to Work: Liquid Scintillation Counters, 1950–70" (Bernward Joerges and Terry Shinn, eds., *Instrumentation between Science, State and Industry*. Dordrecht: Reidel, 2001, 143–74); chapter ten: "The Notions of Regulation, Information, and Language in the Writings of François Jacob" (*Biological Theory* 1 [2006]: 261–67); chapter eleven: "Intersections: Some Thoughts on Instruments and Objects in the Experimental Context of the Life Sciences" (Helmar Schramm, Ludger Schwarte, and Jan Lazardzig, eds., *Instruments in Art and Science: On the Architectonics of Cultural Boundaries in the 17th Century*. Berlin/New York: De Gruyter, 2008, 1–19); and chapter thirteen: "Scrips and Scribbles" (*Modern Language Notes* 118 [2003]: 622–36).

ABBREVIATIONS

AEC	Atomic Energy Commission
AGM-UH	Archiv des Instituts für Geschichte der Medizin, Universität Heidelberg (Archives of the Institute for the History of Medicine, University of Heidelberg)
ARF	Rockefeller Foundation Archives
BA Koblenz	Bundesarchiv Koblenz (German Federal Archives Koblenz)
BLUC	Bancroft Library, University of California
Caltech Archive	Archive of the California Institute of Technology
DFG	Deutsche Forschungsgemeinschaft (German Research Foundation)
KWG	Kaiser-Wilhelm Gesellschaft (Kaiser Wilhelm Society)
LAPS	Library of the American Philosophical Society
LSC	Liquid Scintillation Counter
MPG-Archive	Archiv zur Geschichte der Max-Planck-Gesellschaft (Archives of the Max Planck Society)
RFR	Reichsforschungsrat (Reich Research Council)

Prologue

1. Rheinberger 1997.
2. Bachelard 1949, 103.
3. Daston 2000.
4. Latour 1993 (1991).
5. Cassirer 1950, 17 f.
6. Bachelard 1968 (1940).
7. Knorr Cetina 1999, 247.
8. Twenty-five years ago, Isabelle Stengers and Ilya Prigogine already declared, with respect to the sciences: "We no longer wish to concentrate on that which remains stable, but, rather, on that which changes: geological and climatic revolutions, the evolution of species, or the genesis and transformation of norms." Prigogine and Stengers 1979, 15.
9. Morange 1998 (1994), 1.
10. Morange 1998 (1994), 2.
11. Rabinow 2004, 63.
12. Rabinow 2000, 44.
13. See Joerges and Shinn 2001. Shinn calls apparatuses of this type "research technologies."
14. Lévi-Strauss 1966 (1962), 24.
15. Scholthof, Shaw, and Zaitlin 1999.
16. Canguilhem 1968a, 305–18.
17. Friedman 2000.
18. Cassirer 1961 (1942), 97 f.
19. Cassirer 1950, 19.
20. Forman 1997, 185, 189.

1. Ludwik Fleck, Edmund Husserl

1. Du Bois-Reymond 1912, 436.
2. Boltzmann 1905, 28.
3. Schiemann 1997, 156.
4. Hartmann 1927, 10 f. For a more detailed discussion of Hartmann's work, see chapter five of the present book.

5. Hartmann 1927, 4.

6. Schiemann 1997, 157.

7. Hartmann 1956, 155.

8. Ibid., 150.

9. See chapter two.

10. Bachelard 1987 (1928), 13, 284.

11. Hartmann 1956, 152.

12. Riezler 1928, 706.

13. Laubichler 2000, 298, 294.

14. Harrington 1996.

15. Hartmann 1927, 716.

16. Fleck 1929, 426.

17. Ibid., 426.

18. Ibid., 425.

19. Ibid., 426.

20. Ibid., 428.

21. Bachelard 1987 (1928), 297.

22. Ibid., 15.

23. Ibid., 298.

24. Fleck 1929, 429.

25. Ibid., 427.

26. Fleck 1979 (1935).

27. Fleck 1929, 430.

28. Fleck 1979 (1935).

29. See Latour 1990, 63.

30. Fleck 1929, 429.

31. Bachelard 1987 (1928), 300.

32. For a more detailed discussion, see chapter two.

33. Bachelard 2002 (1938).

34. Troeltsch 1922, 4 f.

35. Husserl 1970 (1954), 6 f., 12.

36. Ibid., 273, translation modified.

37. Ibid., 278.

38. Ibid., 355 f.

39. Ibid., 295.

40. Ibid., 17. Italics in the original.

41. Husserl 1962.

42. Husserl 1970 (1954), 370.

43. Ibid., 372 f.

44. Ibid., 373.

45. Canguilhem 1988 (1981), 5. See also chapter three in the present book.

46. Husserl 1970 (1954), 378.

47. Ibid., 360 f.

48. Derrida 1999.

2. Gaston Bachelard

1. Hyder 2003.
2. This aspect of Bachelard's work has so far been widely neglected. Notable exceptions are Gaukroger 1976; Castelão-Lawless 1995; and Gayon 1995.
3. Dominique Lecourt employs this concept in his *L'épistémologie historique de Gaston Bachelard*.
4. See esp. Latour 1993 (1991).
5. Bachelard 1972a (1951), 36.
6. Chimisso 2001, especially chapter three.
7. Gayon 1995, 3–11.
8. For a collection of excerpts from Bachelard's work that provides a useful overview of his major epistemological theses, see Lecourt 2001 (1971).
9. Dagognet 2003.
10. Gayon 2003.
11. Gayon 1995, 39.
12. Castelão-Lawless 1995, 45.
13. Bachelard 1970 (1931–32), 18 f.
14. Bachelard 1949, 103.
15. Ibid., 5.
16. Bachelard 1968 (1940), 11.
17. Gayon 1995, 4.
18. Bachelard 1968 (1940), 11.
19. Bachelard 1949, 9.
20. Ibid., 4.
21. Ibid., 132 f.
22. Bachelard 1984 (1934), 172, translation modified.
23. Ibid., 171, translation modified.
24. Bachelard 1972a (1951), 39.
25. Bachelard 1964 (1938), 1.
26. Bachelard 2002 (1938), 237.
27. Bachelard 1951, 25.
28. Bachelard 2002 (1938), 237.
29. Ibid., 217.
30. Bachelard 1949, 2 f.
31. Bachelard 1933, 140.
32. Bachelard 2002 (1938), 25, translation modified.
33. Ibid, 70.
34. Bachelard 1949, 56.
35. Bachelard 2002 (1938), 33.
36. Bachelard 1970, 14.
37. Bachelard 2002 (1938), 24, translation modified.
38. Ibid.
39. Bachelard 1951, 25.

40. Bachelard 1949, 55.

41. Ibid., 133.

42. Ibid.

43. Bachelard 1951, 6.

44. Ibid., 12.

45. Ibid., 11.

46. Ibid., 14, 12.

47. Ibid., 13.

48. Ibid., 9 f.

49. Gayon 1995, 38, citing Bachelard 1928, 160.

50. Bachelard 1949, 8.

51. Bachelard 2002 (1938), 69, translation modified.

52. Gaukroger 1976, 221.

53. Bachelard 1933, 141 f.

54. Bachelard uses this term as early as 1932 in the title of one of his books, *Le pluralisme cohérent de la chimie moderne*.

55. Bachelard 1972a (1951), 40.

56. Gayon 1995, 4.

57. Bachelard 1953, 215.

58. Gayon 1995, 36.

59. Bachelard 1951, 27.

60. Bachelard 1949, 9.

61. Bachelard 1951, 21–26.

62. Broglie 1947, 9, cited in Bachelard 1951, 21.

63. Bachelard 1951, 26.

64. Bachelard, 1972b (1951), 144.

65. Bachelard 1951, 13.

3. Georges Canguilhem

1. Grene 2000.

2. Gayon 1998, 307, note 8. See also Lecourt 1975 (1969). Lecourt was among the first to write on Canguilhem's "epistemological history"; see his "Georges Canguilhem's Historical Epistemology," in Lecourt 1975 (1972), 162–86. See also *Revue d'histoire des sciences* 53 (1) (2000).

3. Canguilhem 1991 (1966); Canguilhem 1952; Canguilhem 1955.

4. Canguilhem 1994. See also Braunstein et al. 1998.

5. Balibar 1993, 58.

6. Foucault 2006 (1961).

7. Foucault 1973 (1963).

8. Foucault 1998 (1994), 465 f., translation modified.

9. Georges 1965, 11, cited in Balibar 1993, 58 f.

10. Canguilhem 1968b, here 46, cited in Balibar 1993, 60.

11. Canguilhem 1968a, 9–23.

12. See Canguilhem 1979, 22, note 1.
13. Canguilhem 1988 (1977).
14. Canguilhem 1988 (1977). Introduction, 3, translation modified.
15. Ibid., 5.
16. Canguilhem 1994, 44, translation modified. (1968a, 13).
17. See Butterfield 1951.
18. See, for example, Andrew Pickering 1992; Latour 1993 (1991).
19. Canguilhem 1963, 35. See also chapter two in the present book.
20. Canguilhem 1994, 45, translation modified. (1968a, 14).
21. Canguilhem 1968a, 16 f.
22. Renard 1996, 33.
23. Canguilhem 1994, 30, translation modified. (1968a, 19).
24. Canguilhem, 1968a, 18.
25. Canguilhem 1994, 26, translation modified. (1968a, 17).
26. Kragh 1987, 74.
27. Canguilhem 1994, 51, translation modified. (1968a, 22).
28. Ibid., 49, emphasis added. (1968a, 20 f.).
29. Canguilhem 1975, 43–80.
30. Canguilhem 1988 (1971), 103–23.
31. Canguilhem 1994, 31, translation modified. (1968a, 20).
32. Jacob 1993 (1970).
33. Singer 1950.
34. Canguilhem 1988 (1977), 16 f., translation modified.
35. Canguilhem 1994, 31, translation modified. (1968a, 20). See also Bachelard 1972b (1951), 137–52.
36. Canguilhem 1988 (1977), 11, translation modified.
37. Canguilhem 1994, 29, translation modified. (1968a, 18 f.).
38. Ibid., 52. (1968a, 23).
39. Lecourt 1975 (1972). See also Gutting 2001, especially the section of chapter 8 entitled "Philosophy of the Concept: Cavaillès, Canguilhem, and Serres," 227–33.
40. Renard 1996, 7.
41. Balibar 1993, 68.
42. Jacob 1993 (1970), 13.
43. Bachelard 1984 (1934), 136, translation modified.
44. Canguilhem 1968a, 148.
45. Canguilhem 1994, 30, translation modified. (1968a, 19).
46. Canguilhem 1988 (1977), 81–102.
47. Ibid., 83, translation modified.
48. Driesch 1901.
49. Renard 1996, 77.
50. Bing and Braunstein 1998, 128.
51. Canguilhem 1988 (1981), 27–40.
52. Canguilhem 1968a, 226–73.

53. Balibar 1993, 66.

54. Ibid., 69.

55. Canguilhem 1981, 45. (The sentence is not translated in Canguilhem 1988.)

56. Foucault 1972 (1971), 224.

57. Canguilhem 1968a, 37–50.

58. Canguilhem 1988 (1977), 11, translation modified.

59. Renard 1996, 47.

60. Canguilhem 1968a, 140.

61. Bernard 1949 (1865), 18.

4. Pisum

1. Roberts 1929; Olby 1966; Orel 1996.

2. Jahn 1957–58.

3. Zirkle 1964; Meijer 1985.

4. MPG Archive, Section 3, Folder 17, No. 115; see also Rheinberger 1995.

5. Stein 1950, 457n.

6. Vries 1900a.

7. Tschermak-Seysenegg 1900.

8. Olby 1971, 422.

9. Rheinberger 1995.

10. Correns 1900a, 166.

11. Correns 1889.

12. Correns 1892.

13. Correns 1901.

14. Correns 1893.

15. Correns 1896.

16. Correns 1899a.

17. Correns 1897.

18. Correns 1895.

19. Selbstbiographie Correns (typescript). MPG Archive, Section 3, Folder 17, No. 1, 3.

20. Ibid., p. 4.

21. Rheinberger 2000b.

22. Navashin 1898; Navashin 1899; Guignard 1899a; Guignard 1899b.

23. MPG Archive, Section 3, Folder 17, No. 115, File "Pisum-Kreuzungen 1896–1900."

24. Gärtner 1849.

25. MPG-Archive, Section 3, Folder 17, No. 115, File "Pisum-Kreuzungen 1896–1900."

26. Ibid. See Rimpau 1884.

27. MPG Archive, Section 3, Folder 17, No. 115, File "Pisum-Kreuzungen 1896–1900." ♀ and ♂ denote female and male germ cells, respectively. The note refers to Mendel 1866.

28. MPG Archive, Section 3, Folder 17, No. 1, 4 f.

29. Roberts 1929, 335.

30. Meijer 1985, 194.

31. Focke 1881.

32. This off-print is now part of the Library of the Max Planck Institutes in Tübingen. Examination of it turned up nothing more in the way of either interesting marginalia attributable to readers or indications as to when or by whom it was read. I thank Heinz Schwarz for sending me photographic reproductions of all the pages bearing light pencil marks.

33. "Erbsenaussaat." MPG Archive, Section 3, Folder 17, No. 115, File "Pisum-Kreuzungen, Resultate 96."

34. Ibid.

35. It was four years before the corresponding paper was published. See Correns 1900b.

36. MPG Archive, Section 3, Folder 17, No. 80.

37. "1. VIII. 96 Pisum pm + gr." MPG Archive, Section 3, Folder 17, No. 115, File "Pisum-Kreuzungen, Resultate 96."

38. Correns 1900a, "Versuch I," 162.

39. "31. VII. Pisum p + gr." MPG Archive, Section 3, Folder 17, No. 115, File "Pisum-Kreuzungen, Resultate 96."

40. "Resultate 1897." Ibid.

41. Protocols "gr + p" and "p + gr." Ibid.

42. MPG Archive, Section 3, Folder 17, No. 115, File "Pisum-Kreuzungen 1896–1900."

43. MPG Archive, Section 3, Folder 17, No. 115, File "Pisum Resultate 1897."

44. "Ergebnisse 1897." Ibid.

45. "Pisum Bastarte." Ibid.

46. "Bastart gr + p ♂ A 1, (gelb)" and "gr + p B.1. gelb." Ibid.

47. "Bastart gr + p ♂ A 1, (gelb)." Ibid.

48. "I–III Topf, Bastart gr + p ♂ A1 gelb III"; "IV Topf, Bastart gr + p ♂ A1 IV"; "V Topf Bastart gr + p ♂ B1 (II)"; "VI Topf Bastart gr + p ♂ B1 (I)." MPG Archive, Section 3, Folder 17, No. 115, File "Pisum Resultate 1898."

49. "III Topf, Bastart gr + p ♂ A1 gelb III." Ibid.

50. "Topf XXIII–XXV, p + gr." Ibid.

51. For more on this, see chapter thirteen.

52. "III. Topf. Bastart gr + p ♂ a. gelb III." Ibid.

53. "VI. Topf. Bastart gr + p ♂ B1 (I)." Ibid.

54. "Topf XVI gr + p." Ibid.

55. "Topf XVI gr + p." Ibid.

56. MPG Archive, Section 3, Folder 17, No. 115, File "Erbsen 99."

57. "T. 21. a, BE + gr ("Xenien") Pfl. XI. 97." Ibid.

58. Correns 1900a, 159.

59. MPG Archive, Section 3, Folder 17, No. 85, File "Theoretisches etc."

60. "2. I. 98." Ibid.

61. "24. II. 98." Ibid.

62. "Resultate 1894." MPG Archive, Section 3, Folder 17, No. 84.

63. "Vers: Sommer 96, ♀ Pflanze alba, Pollen: caesia." MPG Archive, Section 3, Folder 17, No. 84, Notebook "Erndte 96."

64. "Vers: Sommer 96, ♀ Pflanze caesia 'λ', Pollen: caesia 'λ'." Ibid.

65. "Vers: Sommer 96, ♀ Pflanze *caesia*, Pollen: alba." Ibid.

66. "1897 a + c. alba ♀ + caesia ♂." MPG Archive, Section 3, Folder 17, No. 84, "Resultate 1897."

67. "1897 c + a. caesia ♀ + alba ♂." Ibid.

68. Undated, unbound sheet. MPG Archive, Section 3, Folder 17, No. 85, Notebook "Notizen."

69. Undated, unbound sheet. Ibid.

70. Undated, unbound sheet. MPG Archive, Section 3, Folder 17, No. 115.

71. Correns 1900a, 163 f.

72. Undated, unbound sheet. MPG Archive, Section 3, Folder 17, No. 115.

73. Correns 1902.

74. Roberts 1929, 337; de Vries 1900a.

75. MPG Archive, Section 3, Folder 17, No. 84.

76. Undated, unbound sheet. MPG Archive, Section 3, Folder 17, No. 115.

77. Correns 1900a, 158.

78. Correns 1900c; Correns 1900d.

79. Correns 1901.

80. Correns 1899b.

81. de Vries 1899; de Vries 1900b.

82. Correns 1901, 68.

83. See also chapter six.

5. Eudorina

1. Hartmann 1919.

2. MPG Archive, Section 3, Folder 47, No. 6. Curriculum vitae Max Hartmann.

3. Hertwig 1902/03.

4. Hertwig 1903, 61.

5. Bütschli 1876, 208.

6. Maupas 1888, 207.

7. Maupas 1889, 488.

8. Weismann 1882.

9. Goette 1883, 80.

10. Weismann 1892 (1884), 12.

11. Hertwig 1889.

12. Hertwig 1899, 146.

13. Hertwig 1903, 50.

14. Hartmann 1927, 495.

15. Klebs 1896, 432.

16. Klebs 1899, 213.

17. Hartmann 1919, 3.

18. Ibid.

19. MPG Archive, Section 3, Folder 47, No. 705. Curriculum vitae Victor Jollos.

20. Hartmann 1919, 5.

21. Hertwig 1903, 49.

22. Hartmann 1919, 5.

23. Hartmann 1917, 762.

24. Hartmann 1917, 763.

25. Ibid., 766.

26. Hartmann 1920, 557.

27. Hartmann 1921.

28. Woodruff 1911.

29. Woodruff and Erdmann 1914.

30. Hartmann 1917, 772.

31. Hartmann 1920, 557.

32. Hartmann 1921, 278 f.

33. Ibid., 281.

34. Ibid., 282 f.

35. Hartmann 1927, 1–18.

36. See Chen 2003.

37. Schaudinn 1905.

38. Hartmann 1927, 5–11.

6. Ephestia

1. On the work of Ephrussi and Beadle, see, for example, Sapp 1987, especially chapter five; Burian, Gayon, and Zallen 1988, especially 389–400; and Gayon 1994. On the work of Beadle and Tatum, see Olby 1994 (1974), 139–41; Kay 1989, and Kay 1993, especially chapters four and seven. Both stories are told by Kohler 1991 and Kohler 1994, especially chapter seven.

2. There are a few exceptions. See Olby 1994, 138 f. Also important are the studies of Jonathan Harwood on the style of Kühn's scientific thinking and its cultural context. See Harwood 1985, and Harwood 1993, especially chapters two and seven. See also Grasse 1972, Egelhaaf 1996, and Grossbach 1996. References to biographical memoirs and obituaries may be found in Egelhaaf 1996. For a detailed discussion, see Gausemeier 2005, especially chapter two, section 3.

3. Studienbuch Alfred Kühn. MPG-Archiv, Section 3, Folder 5, No. 2, Pt. 1.

4. Kühn 1908.

5. Kühn 1909; Kühn 1910.

6. Kühn and Schuckmann 1912; Kühn and Wasielewski 1914.

7. Kühn and Trendelenburg 1911.

8. Kühn 1917; 1919.

9. Kühn 1920.

10. Kühn, Letter of December 30, 1921, to Spemann. Hans Spemann Papers, Johann

Wolfgang Goethe-Universität, Frankfurt am Main, *Senckenbergische Bibliothek*, A1, No. 523.

11. Kühn 1964 (1922).

12. Harwood 1985; Harwood 1993.

13. Kühn published the results of these experiments only in 1926. See Kühn 1926.

14. MPG Archive, Section 3, Folder 5, No. 1/1. See also Kühn 1959, 277 f.

15. Harwood 1993, 57. For a summary of this work, see Goldschmidt 1934. See also Allen 1974, and Dietrich 1995.

16. Kühn and Henke 1929, 1.

17. Harwood 1984.

18. Henke 1924.

19. Seidel 1924.

20. Schlottke 1925. See also Kühn 1927.

21. Kaestner 1931.

22. Kühn 1959, 278.

23. Zeller 1879.

24. Whiting 1921.

25. Kühn and Henke 1929, 2. On the emergence of applied entomology in late nineteenth-century and early twentieth-century Germany and Albrecht Hase's role in this context, see Jansen 2003.

26. Szöllösi-Janze 1994.

27. Kühn and Henke 1929, 27–30.

28. Hase 1928, 265, 267.

29. Dr. Dietrich Stoltzenberg, Hamburg, personal communication of February 1999.

30. Kühn wrote from army headquarters in Falkenhausen to Weismann's successor in Freiburg, Franz Doflein, "The fight against parasites is a very gratifying task; there is always enough material to hand when troops are displaced. We have quite a few good delousing stations and we always work with small field contrivances so that we can quickly bring the evil and the 'eastern peril' under control everywhere." Kühn, Letter of January 5, 1916, to Doflein, MPG Archive, Section 3, Folder 5, No. 2.

31. To Phineas Whiting goes the credit for performing the first genetic studies on the wings of *Ephestia*. See Whiting 1919.

32. Kühn and Henke 1929, 23, 27.

33. Kühn and Henke 1929 (1932, 1936), 1.

34. Kühn and Henke 1929, 1.

35. Kühn, Letter of February 9, 1938, to Henke. MPG Archive, Section 3, Folder 58, No. 8. For an introduction to the plasmon concept, see Sapp 1987, chapter three. Kühn used the term throughout the first three editions of his *Grundriss der Vererbungslehre* in 1939, 1950, and 1961. He abandoned it only in the fourth edition in 1965.

36. Kühn, publication list to 1951 and dissertation list to 1936. MPG Archive, Section 3, Folder 5, No. 4, envelope 7.

37. Kühn and Henke 1930, 204.

38. Hermann Hartwig, personal communication of August 3, 1998.

39. Vermilion was observed by Morgan in November 1910, cinnabar by Clausen in September 1920. See Lindsley and Grell 1967. For a contemporaneous review, see Morgan, Bridges, and Sturtevant 1925.

40. Kühn and Henke 1929, 2.

41. Kühn, Letter of September 4, 1929, to Henke; Kühn, Letter of May 10, 1930, to Henke. MPG Archive, Section 3, Folder 58, Nos. 6 and 7.

42. Kühn 1932, 974.

43. This work subsequently led to a dissertation: Blaustein 1935.

44. Kühn, Letter of May 10, 1930, to Henke; MPG Archive, Section 3, Folder 58, No. 7.

45. Wagner 1931; Gierke 1932; Köhler 1932.

46. Kühn, Letter of August 9, 1930, to Henke. MPG Archive, Section 3, Folder 58, No. 7. Kühn never again mentioned these investigations. In all his future work he relied on spontaneous mutations, not artificially induced ones.

47. Karin Magnussen completed her dissertation with Kühn in 1933 and then left the university to become a school teacher. In 1941 she obtained a fellowship from the Deutsche Forschungsgemeinschaft to work at the KWI for Anthropology, Human Heredity, and Eugenics. In 1943 she became Otmar v. Verschuer's assistant. At the Institute she was involved in the histological investigation of heterochromatic eyes of murdered Sinti and Roma that had been sent from Auschwitz to the Institute by Josef Mengele. Kühn had nothing to do with her later abominable "career." See, for example, MPG Archive, Section III, Folder 75, Melchers, Correspondence, K2M, 1948–50. See also Lösch 1997, 408–15, Hesse 2001, and Weindling 2003.

48. Kühn, Letter of August 9, 1930, to Henke. MPG Archive, Section 3, Folder 58, No. 7.

49. Precise figures are available in Grasse 1972, 262.

50. Hartwig, personal communication of August 3, 1998.

51. Ephrussi and Beadle 1935.

52. Caspari documented the development of his technique in his doctoral dissertation. From the literature, he was aware of the studies of Alfred Sturtevant, who had repeatedly reported on the reciprocal influence of physically separated male and female parts of a gynandromorphic mosaic Drosophila. See Sturtevant 1920, 70 f., and Sturtevant 1927.

53. Caspari 1933, 354.

54. da Cunha 1935.

55. Caspari 1933, 373 f.

56. Ibid., 379.

57. Kühn 1959, 278.

58. See also Rheinberger 1997.

59. Tisdale, Wilbur E. "Natural Sciences: Log on Trip to Riga, Danzig and Leipzig." June 8–13, I. ARF, Record Group 12.1.

60. ARF, Record Group 1.1., Series 700 D, Box 21, Folder 150.

61. ARF, Record Group 1.1., Series 717, Box 13, Folder 123.

62. Kühn, Letter of July 3, 1934, to Dr. Wildhagen. BA Koblenz, Call number R 73, Vol. 12475.

63. Tisdale, Letter of July 18, 1934, to Weaver. ARF, Record Group 1.1., Series 717, Box 13, Folder 123.

64. Tisdale, Letter of August 1, 1934, to Weaver; and Weaver to Tisdale, August 13, 1934. Ibid.

65. Tisdale, Letter of August 28, 1934, to Weaver. Ibid.

66. As evinced by Kühn's letter of December 30, 1921, to Spemann: "Goldschmidt is cold in everything." Hans Spemann Papers, Johann Wolfgang Goethe-Universität, Frankfurt am Main, *Senckenbergische Bibliothek*, A1, No. 218.

67. On Rockefeller Foundation policy toward Nazi Germany in general, see Picard 1999, especially chapter eight.

68. Tisdale, Letter of August 10, 1934, to Weaver. ARF, Record Group 1.1., Series 717, Box 13, Folder 123.

69. Kühn, Letter of December 4, 1934, to the DFG. In consequence, something over 16,000 Reichsmarks were paid out in 1934. BA Koblenz, Call number R 73, Vol. 12475.

70. Kühn, Letter of January 31, 1935, to Dohrn. Archives of the Stazione Zoologica, Naples, Alfred Kühn's correspondence.

71. Dohrn, Letter of September 14, 1934, to Kühn; Kühn, Letter of October 9, 1934, to Dohrn. Ibid.

72. Kühn, Letter of January 31, 1935, to Dohrn. Ibid.

73. Kühn, Report of March 24, 1935, to the Rockefeller Foundation. ARF, Record Group 1.1., Series 717, Box 13, Folder 123.

74. Hartwig, personal communication of August 3, 1998.

75. Brandt 1934; Busselmann 1934; Umbach 1934.

76. Feldotto 1933; Eckhardt Hügel 1933; Behrends 1935.

77. Kühn and Engelhardt 1933.

78. Goldschmidt 1935.

79. Caspari, Letter of August 8, 1947, to Kühn. AGM-UH, Correspondence of Alfred Kühn. See also Kühn and Engelhardt 1937.

80. Gayon 1994.

81. Kühn, Caspari, and Plagge 1935.

82. Kühn, Report of July 2, 1935, to the Rockefeller Foundation. ARF, Record Group 1.1., Series 717, Box 13, Folder 123.

83. Rockefeller Foundation, Letter of July 22, 1935, to Kühn. Ibid.

84. Kühn, Caspari, and Plagge 1935, 25.

85. Boris Ephrussi was one of the few to note the potential significance of the distinction underpinning this conclusion, but he did not regard it as "proven." It was Ephrussi who brought Kühn's, Caspari's, and Plagge's approach to the attention of a broad audience of American readers interested in science. Ephrussi 1938, 20 f.

86. Kühn, Caspari, and Plagge 1935, 1.

87. Kühn, Report of December 26, 1935, to the Rockefeller Foundation. ARF, Record Group 1.1, Series 717, Box 13, Folder 123.

88. Hartwig, personal communication of August 3, 1998. See also Eicher 1987.

89. Ephrussi, Letter of September 2, 1937, to Beadle. CALTECH Archives, Collection Beadle, Folder No. 3.16.

90. Dunn, Letter of April 14, 1938, to Kühn; Caspari, Letter of October 31, 1939, to Kühn. AGM-UH, Correspondence of Alfred Kühn. Ephrussi, Letter of March 4, 1939, to Beadle. CALTECH Archives, Collection Beadle, Folder No. 3.16.

91. Caspari, Letter of May 27, 1944, to Beadle. CALTECH Archives, Collection Beadle, Folder no. 2.22.

92. Caspari, Letter of March 3, 1946, to Kühn. AGM-UH, Correspondence of Alfred Kühn.

93. Kühn, Report of December 26, 1935, to the Rockefeller Foundation. ARF, Record Group 1.1., Series 717, Box 13, Folder 123.

94. Kühn, Report of September 14, 1936, to the Rockefeller Foundation. Ibid.

95. Plagge 1936a.

96. Plagge 1935.

97. Kühn and Piepho 1936.

98. It eventually led to the identification of regular hormones in the metamorphotic cycle of insects. For a general overview, see Piepho 1972, and Karlson 1972. See also Rheinberger 2004.

99. Plagge 1936a, 251.

100. Kühn, Report of July 2, 1935, to the Rockefeller Foundation. ARF, Record Group 1.1., Series 717, Box 13, Folder 123.

101. Kühn, Report of December 26, 1935, to the Rockefeller Foundation. Ibid.

102. Kühn, Report of March 24, 1936, to the Rockefeller Foundation. Ibid.

103. Beadle and Ephrussi 1935; Beadle and Ephrussi 1936.

104. Plagge 1936b.

105. Plagge 1936c, 136.

106. Ephrussi, Clancy, and Beadle 1936, 545 f.

107. Tisdale, undated report to the Rockefeller Foundation. ARF, Record Group 1.1, Series 717, Box 13, Folder 123.

108. Kühn, Report of April 24, 1936, to the DFG. BA Koblenz, Call number R 73, Vol. 12475.

109. Kühn, Letter of April 24, 1936, to Tisdale. ARF, Record Group 1.1., Series 717, Box 13, Folder 123.

110. Kühn, Letter of May 15, 1947, to Caspari. Kühn Papers, AGM-UH. Melitta von Engelhardt recalls that Kühn was "desperate" when the first Professor, James Franck, left Göttingen, and he "immediately" accepted the offer to move to Berlin-Dahlem (Melitta von Engelhardt, communication of December 6, 1990, MPG-Archive). Kühn never explicitly stated his motives in public, not even to his students; indeed, he generally avoided declaring his opinion on political matters (Hartwig, personal communication of August 3, 1998).

111. Goldschmidt 1960, 300 f.

112. Kühn, Report of February 26, 1937, to the DFG. BA Koblenz, Call number R 73, Vol. 12475.

113. Kühn, Letter of November 25, 1936, to Tisdale. ARF, Record Group 1.1., Series 717, Box 13, Folder 123.

114. Kühn, Report of June 19, 1937, to the Rockefeller Foundation. Ibid.

115. Becker 1937, 507.

116. Becker and Plagge 1937, 809.

117. Kühn, Report of October 19, 1937, to the Rockefeller Foundation. ARF, Record Group 1.1, Series 717, Box 13, Folder 123.

118. DFG, Letter of May 18, 1938, to Alfred Kühn. BA Koblenz, Call number R 73, Vol. 12475.

119. Khouvine, Ephrussi, and Harnly 1936.

120. Khouvine and Ephrussi, 1937; Khouvine, Ephrussi, and Chevais 1938.

121. Thimann and Beadle 1937; Tatum and Beadle 1938.

122. Ephrussi, Letter of January 9, 1937, to Beadle. Caltech Archive, Collection Beadle, Folder no. 3.16.

123. Ephrussi, Letter of July 17, 1937, to Beadle. Ibid.

124. Ephrussi, Letter of August 4, 1937, to Beadle. Ibid.

125. Ephrussi, Letter of August 8, 1937, to Beadle. Ibid. See also Kühn 1936b.

126. Tisdale, Memorandum of October 8, 1937. ARF, Record Group 1.1, Series 100 D, Box 40, Folder 365.

127. Ephrussi, Letter of October 18, 1937, to Beadle. Caltech Archive, Collection Beadle, Folder no. 3.16.

128. Ephrussi, Letter of January 30, 1938, to Kühn; Kühn, Letter of February 8, 1938 to Ephrussi; Ephrussi, Letter of March 7, 1938, to Kühn; Kühn, Letter of March 29, 1938, to Ephrussi. AGM-UH, Kühn Correspondence.

129. Ephrussi, Letter of May 8, 1938, to Beadle. Caltech Archive, Collection Beadle, Folder no. 3.16.

130. Becker 1938.

131. Plagge for his part went to Göttingen in 1939 on Kühn's suggestion, becoming Henke's assistant. Kühn, Letter of March 3, 1939, to Henke. MPG Archive, Section 3, Folder 58, No. 9.

132. Neuberg was sent into retirement on September 30, 1934, but remained acting director of the Institute until September 30, 1936. For details, see Lohff and Conrads 2007.

133. DFG, Letter of March 6, 1939, to Kühn. BA Koblenz, Call number R 73, Vol. 12475.

134. Kühn, Letter of September 29, 1939, to the DFG. Ibid.

135. Kühn, Letter of December 16, 1939, to Henke. MPG Archive, Section 3, Folder 58, No. 9.

136. According to Helmut Risler, later Kühn's assistant in Tübingen, Kühn together with Theodor Heuss was close to the Naumann circle (Risler 1972, 49). After serving as an enthusiastic medical orderly in World War I (Letter to Franz Dof-

lein, MPG Archive, Section 3, Folder 5, No. 2), Kühn became a member of the German Democratic Party, remaining in its ranks until its dissolution. Denazification document, MPG Archive, Section 3, Folder 5, No. 4.

137. Kühn, Letter of December 16, 1939, to Henke. MPG Archive, Section 3, Folder 58, No. 9.

138. Kühn, Letter of June 7, 1941, to Spemann. Spemann Papers, Johann Wolfgang Goethe-Universität, Frankfurt am Main, *Senckenbergische Bibliothek*, A1, No. 218.

139. Spemann, Letter of July 12, 1941, to Kühn. MPG Archive, Section 3, Folder 5, No. 4, No. 21. Yet according to Georg Melchers, "Kühn was not daring" [*Kühn war nicht kühn*] (Interview of June 1997). In the early years of the Nazi regime, he had accepted to let one of his papers appear in a clearly racialist context, although the content of his paper did not accommodate to this environment. See Kühn 1934.

140. Ephrussi, Letter of March 4, 1939, to Beadle. Caltech Archive, Collection Beadle, Folder no. 3.16.

141. Ephrussi, Letter of July 21, 1939, to Beadle. Ibid.

142. Ephrussi, Letter of February 19, 1940, to Beadle. Ibid.

143. Tatum 1939.

144. Mamoli 1939.

145. Butenandt, Weidel, and Becker 1940a, 63 f. See also Butenandt, Weidel, and Becker 1940b, 447 f., as well as Wolfhard Weidel's dissertation (1940).

146. Danneel 1941a.

147. Tatum and Haagen-Smit 1941.

148. Kühn 1941.

149. Ibid., 257 f.

150. Kühn, Caspari, and Plagge 1935, 1, 23.

151. Becker 1939.

152. Kühn and Becker 1942.

153. Kühn 1941, 258 (emphasis added).

154. This was not stated expressly by Beadle and Tatum either when in 1941 they published their first paper on the participation of genes in the biochemical pathways of *Neurospora*. Beadle and Tatum 1941.

155. Kühn 1955, 25.

156. Erwin Baur (1875–1933) was the founder of the KWI für Züchtungsforschung, located in Müncheberg, near Berlin.

157. Kühn, Letter of February 9, 1938, to Henke. MPG Archive, Section 3, Folder 58, No. 8.

158. On this point, my interpretation differs from Harwood's. See Harwood 1985.

159. Kühn, Letter of March 9, 1941, to Henke. MPG Archive, Section 3, Folder 58, No. 11.

160. Kühn, report of December 9, 1940, to the RFR. BA Koblenz, Call number R 73, Vol. 12475.

161. Ibid.

162. DFG, Letter of May 2, 1941, to Kühn. Ibid.

163. Kühn, Letter of January 12, 1942, to RFR. BA Koblenz, Call number R 73, Vol. 12475.

164. Curriculum vitae of Wolfhard Weidel. MPG Archive, Section 3, Folder 32, No. 1.

165. Kühn and Becker 1942.

166. Kühn, Report of April 24, 1942, to the DFG. BA Koblenz, Call number R 73, Vol. 12475.

167. DFG, Letter of April 30, 1943, to Kühn. Ibid.

168. RFR, Letter of May 15, 1944, to Kühn. Ibid.

169. Kühn and Schwartz 1942.

170. Kühn, Report of January 12, 1942, to the RFR. BA Koblenz, Call number R 73, Vol. 12475.

171. Kühn, Report for 1943/44 of April 28, 1944, to the DFG. Ibid.

172. Kühn, Letter of August 13, 1944, to Stubbe. Kühn Papers, AGM-UH. See also Kühn, Letter of September 21, 1944, to Henke. MPG Archive, Section 3, Folder 58, No. 14. Georg Gottschewski joined Kühn in Dahlem after a 1936–37 stay as a Rockefeller Fellow at the California Institute of Technology. In 1939 he went to work with Otto Koehler in Königsberg (in exchange, Rolf Danneel came to Berlin-Dahlem). Gottschewski became a professor of genetics in Vienna in 1942.

173. Kühn, Application for 1944–45 of April 28, 1944. BA Koblenz, Call number R 73, Vol. 12475.

174. Kühn, Report of January 12, 1942, to the RFR. Ibid.

175. Kay 1989.

176. Kühn, Letter of November 27, 1944, to Renner. Kühn Papers, AGM-UH.

177. For greater detail, see chapter seven.

178. Kühn, Letter of September 3, 1945, to Danneel. Kühn Papers, AGM-UH.

179. Kühn, Letter of October 22, 1946, to Wigglesworth. Kühn Papers, AGM-UH. See also Kühn's two contributions to the FIAT Report on Science and Medicine in Germany between 1939 and 1946: Kühn 1948a and 1948b.

180. Kühn and Schwarz 1942.

181. Kühn 1944a, Kühn 1944b.

182. Kühn and Engelhardt 1940; Kühn and Engelhardt 1944.

183. Kühn and Engelhardt 1946.

184. Becker 1941; Piepho 1942.

185. Kühn and Engelhardt 1937; Kühn and Engelhardt 1940; Kühn and Engelhardt 1943a and 1943b; Kühn and Engelhardt 1944.

186. Caspari, Letter of August 8, 1947, to Kühn. Kühn Papers, AGM-UH.

187. Kühn, Undated letter to Ernst Caspari. Ibid.

188. Danneel 1941a. Kühn must have seen no reason to make such a switch. However, his successor in Göttingen, Henke, settled mainly on *Drosophila* in pursuing his work in gene physiology.

189. Kühn 1938.

190. Kühn, Letter of December 27, 1946, to Caspari. AGM-UH.

191. See chapter nine in the present book; see also Creager 2004.

192. See Rheinberger 2004.

193. Kühn 1937, 419.

194. Kohler 1994.

195. I have borrowed this term from Terry Shinn, who uses it to characterize multipurpose research instruments. See, for example, Shinn 1997.

196. See, for example, Kühn 1955.

197. Kühn 1960.

198. MPG Archive, Section 3, Folder 5, No. 4, envelope 13.

199. Jacob and Monod 1961.

7. Tobacco Mosaic Virus

1. For greater detail, see Christina Brandt 2004.

2. Butenandt 1977a and 1977b, 4. See also Friedrich-Freksa 1961; Karlson 1990, 114–18. On Stanley's TMV research, see Kay 1986; Creager 2002.

3. ARF, Record Group 1.1, Series 700 D, Box 21, Folder 150.

4. "Tätigkeitsbericht des Institutsdirektors zur Kuratoriumssitzung am 27. April 1937." MPG Archive, Section 1, Folder 1A, No. 2038.

5. Memorandum of a conversation between Adolf Butenandt, Rudolf Mentzel, Ernst Telschow, and Friedrich Glum, MPG Archive, Section 1, Folder 1A, No. 2041.

6. Nordwig 1983; Schüring 2004; Lohff and Conrads 2007.

7. Neuberg, Letter of March 7, 1947, to Butenandt. LAPS, Carl Neuberg Papers.

8. Georg Melchers, Interview of June 28, 1997, in Tübingen. See also Macrakis 1993a, 521.

9. Wettstein, Letter of September 15, 1936, to *Hochverehrter Herr Geheimrat* [?]. AGM-UH, Kühn Correspondence.

10. Wettstein 1938, 153 f. It should be mentioned here that von Wettstein had no reservations about including the head of the Agency of the NSDAP for Racial Policies, Walter Groß, on the reorganized Board of Trustees of the KWI for Biology. Wettstein, Letter of May 17, 1938, to the President of the KWG. MPG Archive, Section 1, Folder 1A, No. 1555.

11. ARF, Record Group 1.1., Series 717, Box 13, Folder 123. For a more detailed discussion of Kühn's research, see chapter six in the present book.

12. Macrakis 1993a and 1993b; Deichmann 1996; Munk 1995, 28–49.

13. Macrakis 1993a. Deichmann 1996, especially 324–25, calls Macrakis's view of things into question.

14. See, for recent discussions, Brandt 2004, chapter three; Gausemeier 2005, chapter four; Lewis 2002.

15. A request for access to the BASF's archive for a search on Bergold was unfortunately unsuccessful. Documents concerning Butenandt in the MPG Archive had

not yet been made accessible when the present text was drafted. Examination of these documents (see Gausemeier 2005, chapter four) has added a few facets to the story that were left out of account here, but has not yielded a fundamentally new picture.

16. MPG Archive, Section 1, Folder 1A, No. 1563.

17. In 1938 the budget of the KWI for Biology was 286,700 RM for the three sections headed by von Wettstein, Kühn, and Max Hartmann. The budget of the KWI for Biochemistry amounted to 101,477 RM for Butenandt alone. MPG Archive, Section 1, Folder 1A, Nos. 1563 and 2042.

18. KWI Institute for Biochemistry, "Tätigkeitsbericht des Direktors zur Kuratoriumssitzung am 10. Mai 1939," 2. Ibid., No. 2038.

19. "Zusätzliche Mittel für die Forschungen des KWI-Instituts für Biochemie für das Rechnungsjahr 1938." Ibid., No. 2041.

20. Butenandt, Letters of March 17, 1939, and October 8, 1940, to the General Administration of the KWG. Ibid., No. 1463a.

21. T. R. Hogness, U. of Chicago Fellowship Report of August 16, 1937. ARF, Record Group 1.1, Series 700 D, Box 21, Folder 150.

22. KWI for Biochemistry, "Tätigkeitsbericht des Direktors zur Kuratoriumssitzung am 10. Mai 1939." MPG Archive, Section 1, Folder 1A, No. 2038.

23. Kühn, Letter to the Reichsminister for Food and Agriculture. Ibid., No. 1550.

24. Melchers, Interview of June 28, 1997, in Tübingen. See also Melchers 1936, 1938a, 1938b, and 1939.

25. Melchers, Interview of June 28, 1997, in Tübingen.

26. Melchers, Letter of March 23, 1938, to Stanley. BLUC, Wendell Stanley Collection, Call Number 78/18c, Carton 10.

27. Melchers and Lang 1941; Melchers and Claes 1943.

28. Tisdale, Letter of August 28, 1934, to Weaver. ARF, Record Group 1.1, Series 717 D, Box 13, Folder 109.

29. Tisdale, Letter of June 26, 1934, to Rector Pohlhausen of the Technical University of Danzig; H. M Miller, Letter of October 9, 1934, to Butenandt; ARF, Record Group 1.1, Series 700 D, Box 21, Folder 150. Between February and May 1935 Butenandt visited Princeton, Philadelphia, Baltimore, Washington, Boston, Montreal, Toronto, Buffalo, Rochester, Detroit, Chicago, St. Louis, Madison, Berkeley, and several universities in and around New York City.

30. ARF, Record Group 10, NS Germany, Hans Bauer.

31. Board of Trustees of the KWI for Biology, "Tätigkeitsbericht des 1. Direktors, 23. Februar 1937." MPG Archive, Section 1, Folder 1A, No. 1556.

32. Ibid., No. 1538.

33. ARF, Record Group 10, NS Germany, Hans Gaffron.

34. ARF, Record Group 10, NS Germany, Ulrich Westphal.

35. Melchers, Schramm, Trurnit, and Friedrich-Freksa 1940, 530n.

36. Butenandt, Letter of August 22, 1939, to Stanley. BLUC, Wendell Stanley Collection, Call Number 78/18c, Carton 6. Stanley, Letter of October 11, 1939, to Butenandt. Ibid., Carton C. 1–5.

37. Letter of Wettstein to the Minister of Science, Education, and Popular Instruction [*Volksbildung*]. MPG Archive, Section 1, Folder 1A, No. 1538.

38. Danneel and Lubnow 1934 and 1936; Danneel 1938; Danneel and Schaumann 1938; Danneel and Paul 1940.

39. Schramm 1941a, 106.

40. Beams and Pickels 1935.

41. See Schramm and Müller 1940a, 55. The company produced electric medical devices and in summer 1943 was to commence production of ultracentrifuges in Hamburg for the German uranium project. See Stumpf 1995.

42. Melchers, Schramm, Trurnit, and Friedrich-Freksa 1940; Melchers and Schramm 1940; Schramm and Müller 1940a and 1940b; Friedrich-Freksa 1940.

43. Schramm and Müller 1940a, 43.

44. The original strain had been left to the group by Stanley. Ibid.

45. MPG Archive, Section 1, Folder 1A, No. 2906 (1).

46. Melchers 1987, 387.

47. MPG Archive, Section 1, Folder 1A, no. 2906 (2).

48. Ibid.

49. Telschow, Letter of July 15, 1940, to Backe. MPG Archive, Section 1, Folder 1A, No. 2906 (2).

50. Telschow, Letter of November 15, 1940, to Bötzkes. Ibid., No. 2906 (3). Bosch had been president of the KWG from 1937–40.

51. Bötzkes, Letter of January 15, 1941, to Telschow. Ibid.

52. Wettstein, "Baupläne und Voranschlag für die Zweigstelle für Virusforschung, 21.3.1941." Ibid., No. 2906 (3).

53. Butenandt 1941.

54. "Richtlinien über die Einrichtung und Ausgestaltung der Arbeitsstätte für Virusforschung der Kaiser-Wilhelm-Institute für Biochemie und Biologie, 17.6.1941," MPG Archive, Section 1, Folder 1A, No. 2906 (4).

55. "Protokoll der 70. Sitzung des Senats der KWG am 31.7.1941 in Berlin-Dahlem," 12 f. MPG Archive, Section 1, Folder 1A.

56. Bergold, personal communication of June 19, 1999.

57. Bergold, personal communication of March 8, 1999.

58. Bergold, personal communication of June 19, 1999.

59. MPG Archive, Section 1, Folder 1A, No. 1564.

60. Telschow, Memorandum of May 29, 1940. Ibid., No. 2605 (1).

61. The Reichsminister for Food and Agriculture, Letter of November 3, 1942. Ibid., No. 2908 (1).

62. Telschow, Memorandum, "Besprechung am 8.5.1944 im Reichsernährungs-ministerium mit Min.-Direktor Schuster." MPG Archive, Section 1, Folder 1A, No. 2906 (6).

63. Butenandt, Letter of August 4, 1944, to the General Administration. Ibid., No. 2906 (6).

64. Ibid., No. 2906 (3).

65. Bergold, Letter of April 24, 1944, to Telschow. Ibid., No. 2906 (6).

66. Josef Hengstenberg, who was born in 1904, studied physics and joined BASF in 1928 where he worked on synthetic materials in the main laboratory. In 1939 he became managing director of the physics department. See *BASF Werkszeitung* 1958/3.

67. Bergold and Hengstenberg 1942.

68. Bergold and Schramm 1942; Bergold and Brill 1942.

69. Bergold 1943a, 292 f.

70. Bergold 1941.

71. Bergold, Interview of April 25, 1998, in Caracas. Butenandt and Schramm had made Bergold's acquaintance upon his return from Africa in the early weeks of the war. Bergold, personal communication of June 19, 1999.

72. Bergold, "Tätigkeitsbericht der Entomologischen Abteilung der Arbeitsstätte für Virusforschung der Kaiser Wilhelm-Institute für Biochemie und Biologie für die Gastzeit in der I.G. Farbenindustrie, Ludwigshafen/Rh[einland]., 21.6.1944." MPG Archive, Section 1, Folder 1A, No. 2906 (6).

73. Ibid.

74. Polyvinylpyrrolidon was developed as a blood substitute under the direction of Oberstabsarzt Weese in the I.G. Farben Pharmacological Institute in Elberfeld. See Hecht and Weese 1943.

75. Bergold, Interview of April 25, 1998, in Caracas.

76. Bergold, personal communication of September 21, 1998.

77. Danneel 1941b and 1941c.

78. Danneel, Letter of January 7, 1946, to Telschow. MPG Archive, Section 1, Folder 46, No. 8.

79. Danneel 1943a.

80. Danneel 1943b.

81. Danneel, Letter of August 12, 1944, to Kühn. Kühn Papers, AGM-UH. Georg Melchers recalled in this connection that "he [Danneel] always wanted to do everything himself." Melchers, Interview of June 28, 1997, in Tübingen.

82. Melchers 1942.

83. These results could be published only after the war. See Friedrich-Freksa, Melchers, and Schramm 1946.

84. Melchers 1943.

85. Born, Lang, Schramm, and Zimmer 1941; Schramm, Born, and Lang 1942; Born, Lang, and Schramm 1943.

86. Melchers 1987, 387.

87. Butenandt, Friedrich-Freksa, Hartwig, and Scheibe 1942, 280 f.; see also Butenandt 1944.

88. Schramm and Friedrich-Freksa 1941; Ardenne, Friedrich-Freksa, and Schramm 1941.

89. Kausche, Pfankuch, and Ruska 1939; Stanley and Anderson 1941.

90. Schramm 1941b.

91. Schramm 1941b, 536.

92. Butenandt 1944, 5.
93. Friedrich-Freksa 1940, 379.
94. Schramm 1941b, 534 f.
95. Stanley, Letter of October 13, 1941, to Schramm. BLUC, Wendell Stanley Collection, Call Number 78/18c, Carton 12.
96. Schramm 1944, 111.
97. Schramm, Letter of October 13, 1941, to Stanley. BLUC, Wendell Stanley Collection, Call Number 78/18c, Carton 12.
98. Schramm 1941b, 536.
99. Kausche and Stubbe 1939 and 1940; Pfankuch, Kausche, and Stubbe 1940; Pfankuch 1940.
100. Schramm 1942a, 793. See also Schramm and Rebensburg 1942.
101. Schramm and Müller 1942, 268.
102. Overview for the Allied Field Information Agency, Technical.
103. Melchers 1948.
104. Schramm 1944, 111.
105. Schramm and Dannenberg 1944.
106. Schramm 1943.
107. Bergold, Interview of April 25, 1998, in Caracas.
108. Schramm 1942a and 1942b; Danneel 1942 and 1944.
109. Karlson, Interview of October 17, 1998, in Marburg.
110. In a manuscript that Friedrich-Freksa, Melchers, and Schramm finished writing in February 1945 the following English-language journals were quoted: *Annual Review of Biochemistry* 1942, *Genetics* 1942, *Journal of General Physiology* 1941, *Nature* 1944, *Journal of Immunology* 1942 (Friedrich-Freksa, Melchers, and Schramm 1946). In an article written in 1943 Butenandt also cites *Science* 1941, *Journal of Biological Chemistry* 1941 and 1942 (Butenandt 1944). In an essay completed in 1942, Bergold also cites *British Journal of Experimental Pathology* 1940 and *Journal of Experimental Medicine* 1941 (Bergold 1943b).
111. *Die Naturwissenschaften* (Butenandt, von Wettstein, Kühn [from 1941 on]); *Berichte der Deutschen Chemischen Gesellschaft* (Butenandt); *Biologisches Zentralblatt* (Butenandt, Kühn, von Wettstein); *Hoppe-Seyler's Zeitschrift für physiologische Chemie* (Butenandt [from 1943 on]).
112. See Gausemeier 2004.
113. *Jahrbuch der Max-Planck-Gesellschaft* 1961, 789–809.
114. For 1945 a sum totaling 147,300 RM had been planned. MPG Archive, Section 1, Folder 1A, Nos. 2907 and 2908.
115. The official budget of the KWI for Biochemistry was 97,535 RM in 1942, 133,387 RM in 1943, and 141,350 RM in 1944. Ibid., Nos. 2041 and 2042.
116. Deichmann 1996, tables, 106–109 and 112–15. Bergold, Danneel, Friedrich-Freksa, Melchers, and Schramm received a total of 91,943 RM for research in the areas of applied zoology, genetics, mutation, and cancer.
117. MPG Archive, Section 1, Folder 8, No. 14. See also Macrakis 1993a, 537 ff.

118. Butenandt, Letter of September 10, 1943, to Kühn. Kühn Papers, AGM-UH.

119. Circular letter of September 13, 1943. MPG Archive, Section 1, Folder 1A, No. 2906 (6).

120. Butenandt, Letter of September 10, 1943, to Kühn. Kühn Papers, AGM-UH.

121. Kühn, Letter of July 29, 1944, to von Wettstein. Kühn Papers, AGM-UH. Karlson recalls the text of the telegram sent from Berlin to Tübingen: "Fungi and insects move immediately." Peter Karlson, Interview of October 17, 1998, in Marburg.

122. Bergold 1946.

123. Melchers, Interview of June 28, 1997, in Tübingen.

124. Schramm, Bergold, and Flammersfeld 1946; Schramm and Bergold 1947.

125. Schramm and Müller 1940a.

126. See Bergold 1947.

127. Bergold 1946.

128. Bergold, personal communication of October 24, 1997.

129. *Reichsgesetzblatt*, No. 33 (1943), Pt. 1, 165.

130. Munk 1995, 10.

131. Bergold, Interview of April 25, 1998, in Caracas.

132. Butenandt, Letter of January 13, 1945, to Kühn. Kühn Papers, AGM-UH. What "the rest of [their] work" involved is to the extent that it can be reconstructed described at length in Gausemeier 2004. Bar was a sex-linked insect eye mutation.

133. Kühn, Letter of November 27, 1945, to Glum. Kühn Papers, AGM-UH.

134. Kühn, Letter of September 3, 1945, to Danneel. Ibid.

135. Bergold, Interview of April 25, 1998, in Caracas.

136. Butenandt joined the party in May 1936, a few months before moving to Dahlem. See Macrakis 1993a, 528.

137. Deichmann 1996, 313.

138. Danneel joined the party in 1937. Danneel, Letter of December 3, 1946, to Kühn. Kühn Papers, AGM-UH.

139. Butenandt, Letter of May 16, 1945, to Kühn. Ibid.

140. Bergold, personal communication of June 19, 1999.

141. Butenandt, Letter of May 16, 1945, to Kühn. Kühn Papers, AGM-UH.

142. Melchers 1987.

143. Kühn, Letter of September 15, 1945, to Hartmann. AGM-UH, Kühn correspondence.

144. Schramm and Bergold 1947; Bergold and Friedrich-Freksa 1947.

145. See Rheinberger 2004.

146. As late as June 1944 Butenandt reported to Kühn, "Yesterday, Wettstein was here for a day for the reception for Minister Backe, who visited my Institute and Physics yesterday. . . . It was a very satisfactory visit. He devoted three hours to biochemistry alone, was extraordinarily interested, and fully shared our views on all questions involving science policy." Butenandt, Letter of June 21, 1944, to Kühn. AGM-UH, Kühn correspondence.

147. Neuberg, Letter of March 7, 1947, to Butenandt. LAPS, Carl Neuberg Papers.

8. The Concept of the Gene

1. Muller 1951, 95 f.
2. A more comprehensive presentation may be found in Rheinberger and Müller-Wille 2004. See also Beurton, Falk, and Rheinberger 2000; Keller 2000.
3. Jacob 1993 (1970).
4. Star and Griesemer 1988.
5. See chapter four.
6. Elkana 1970.
7. Portin 1993, 174.
8. Sarkar 1996.
9. Bachelard 2002 (1938).
10. Kay 2000.
11. Benzer 1955. See also Holmes 2004.
12. Moles 1995.
13. Jacob 1982 (1981).
14. Gros 1991 (1986).
15. Dunn 1991 (1965), 234.
16. Rheinberger 1997, chapter thirteen.
17. Kitcher 1982.
18. The section heading is derived from Paul Zamecnik's essay "Protein Synthesis—Early Waves and Recent Ripples," Zamecnik 1976.
19. For more detailed accounts, see Olby 1994 (1974); Morange 1998 (1994).
20. Muller 1927; Stadler 1928.
21. Timofeev-Ressovskii, Zimmer, and Delbrück 1935.
22. Delbrück 1942.
23. Timofeev-Ressovskii 1939.
24. Beadle 1945.
25. Kühn 1941. See also chapter six in the present book.
26. Avery, MacLeod, and McCarty 1944.
27. Vischer, Zamenhoff, and Chargaff 1949.
28. Portugal and Cohen 1977.
29. Olby 1994 (1974).
30. Kay 2000.
31. Crick 1958.
32. Yanofsky, Carlton, Guest, Helinski, and Henning 1964.
33. Nirenberg and Matthaei 1961.
34. Jacob and Monod 1961.
35. Fischer 1995.
36. For an overview, see Adler and Hajduk 1994.
37. Cooper and Stevens 1995.
38. Gros 1991 (1986), 492–99.
39. Atkins and Gesteland 1996.
40. Gros 1991 (1986), 297.

41. Kitcher 1982, 357.

42. Jacob 1998 (1997), 83 ff.

43. Keller 2000. See also Rheinberger and Müller-Wille 2004.

44. Burian 1985; Falk 1984; Falk 1986; Carlson 1991; and Fischer 1995.

45. Brosius and Gould 1993. Brosius and Gould 1992.

46. Gros 1991, chapter seven.

47. Crick 1958, 153.

48. For an overview see Wright 2004.

49. Jablonka and Lamb 1995; Russo, Martienssen, and Riggs 1996.

50. Keller 1995, 117.

51. Kay 2000. See also chapter ten in the present book.

52. Jacob 1993 (1970).

53. Eigen and Schuster 1979.

54. Polanyi 1969, 236.

55. Pattee 1969, 8.

56. Griesemer 2005.

57. Eigen and Winkler-Oswatitsch 1992 (1987), 33.

58. Jacob 1993 (1970), 297.

59. Atlan and Koppel 1990.

60. Eder and Rembold 1992.

61. Austin entitled his legendary William James Lectures (1955) "How to Do Things with Words."

62. André Lwoff concluded his Marjorie Stephenson Memorial Lecture in April 1957 with as he put it the "prosy, coarse, and vulgar" statement: "Viruses should be considered as viruses because viruses are viruses." Lwoff 1957, 252.

63. Kitcher 1992.

64. Zadeh 1987, 27.

65. Ibid., 23.

66. Coveney and Highfield 1995, 39.

9. The Liquid Scintillation Counter

1. Kay 1988; Elzen 1986; Rasmussen 1997; Creager 2002; Lenoir and Lécuyer 1997; Rabinow 1996; Reinhardt 2006.

2. Kremer, Interview of April 2, 1996, in Zurich.

3. I have discussed this second aspect at length in Rheinberger 1997.

4. For an early overview, see Hevesy 1948. On the production of biologically and medically relevant isotopes in the first particle accelerators see, for example, Heilbron and Seidel 1989, especially chapter eight. On the revolutionary transformation of biomedicine in the wake of the Manhattan Project, see Lenoir and Hays 2000.

5. Alvarez and Cornog 1939.

6. Cited in Heilbron and Seidel 1989, 272.

7. Kamen and Ruben 1940; Ruben and Kamen 1940. See also Kamen 1963. Kamen's

retrospective essay appeared in *The Journal of Chemical Education* (1963), as an introduction to Rothchild 1965, and elsewhere.

8. Whitehouse and Putman 1953, especially chapter four on the production of radioactive isotopes.

9. Hewlett and Anderson Jr. 1962. Vol. 3 appeared later: Hewlett and Holl 1989. The sentence cited is drawn from vol. 2, 109.

10. Hewlett and Anderson Jr. 1962, vol. 2, 96. On the changing image of nuclear energy, see Weart 1988.

11. Lenoir and Hays 2000.

12. Creager 2004.

13. Hewlett and Anderson 1962, vol. 2.

14. Heilbron and Seidel 1989, chapter eight; Lenoir and Hays 2000. The increased use of nuclear medicine in diagnostics and therapy often hovered on the border of experimentation with human beings.

15. Broda 1960, 2. In the first quarter of 1959 the number was 39 percent, in 1961 43 percent, and in 1963 33 percent.

16. Cohn 1968.

17. Hewlett and Anderson 1962, vol. 2, 247.

18. I take these examples from the first 1947 issues of the journal *Nucleonics*.

19. Balogh 1991.

20. Crookes 1903.

21. Elster and Geitel 1903.

22. Regener 1908.

23. Krebs 1953, 362.

24. Further details on the early history of radioactivity research and measurement may be found in Trenn 1976, and Trenn 1986; Rheingans 1988; Hughes 1993.

25. Galison 1997, 454.

26. Krebs 1941. See also Krebs 1955.

27. Curran and Baker 1944, declassified September 23, 1947.

28. Coltman and Marshall 1947.

29. Broser and Kallmann 1947; Hofstadter 1948; Hofstadter 1949.

30. Mandeville and Scherb 1950.

31. Morton and Mitchell 1949, 16.

32. Pringle 1950.

33. Kallmann had written his doctoral dissertation under Max Planck and in 1920 became a staff member of the Kaiser Wilhelm Institute for Physical Chemistry and Electrochemistry in Berlin-Dahlem. He was dismissed when the Nazis took power in 1933. During the war he was forced to work at I.G. Farben (Oster 1966). In 1947 as a professor at the Technical University of Berlin (1945–48) he first published his version of the scintillation counter (Broser and Kallmann 1947). A year later he joined the U.S. Army Signal Corps Laboratories in Belmar, New Jersey, as a researcher. In 1949 he was made a professor in New York University's Physics Department and the Director of its Radiation and Solid State Laboratory.

34. Kallmann and Furst 1950; Kallmann and Furst 1951; Furst and Kallmann 1952.
35. Reynolds, Harrison, and Salvini 1950.
36. Raben and Bloembergen 1951.
37. Hayes, Hiebert, and Schuch 1952.
38. Petersen 1995.
39. Hayes, Hiebert, and Schuch 1952; Hiebert and Watts 1953.
40. Galison 1997, 438–54.
41. Morton and Robinson 1949.
42. Kallmann and Accardo 1950; Reynolds, Harrison, and Salvini 1950; Raben and Bloembergen 1951.
43. Hiebert and Hayes 1958; Langham 1958, 136 ff.
44. Anderson 1958.
45. Reines 1958; Galison 1997, 460–63.
46. Davidson and Feigelson 1957, 3.
47. Holl 1997, especially chapters one and two; Argonne National Laboratory 1986. See also Hewlett and Anderson 1962.
48. Packard, Interview of November 5, 1996.
49. McNeill 1991, 123 f., 158.
50. University of Chicago Publication Office 1991, 105.
51. Feld and Weiss Szilard 1972, especially part four: Published Papers in Biology (1949–64), with an introduction by Aaron Novick, 389–524.
52. Grandy 1996, especially chapters five and six; Lanouette 1992.
53. Packard, Interview of November 5, 1996.
54. Feld and Weiss Szilard 1972, 389.
55. Packard, Interview of November 5, 1996.
56. Ibid.
57. Libby 1952.
58. Packard, Interview of November 5, 1996.
59. "A method for converting samples to aliphatic hydrocarbon is being worked out, so that solutions of good efficiency can be prepared that are 80% sample." Arnold 1954.
60. Packard, Interview of November 5, 1996; *idem*, personal communication of August 26, 1998.
61. Holl 1997, 75.
62. Kabara, Okita, and LeRoy 1958.
63. Packard, Interview of November 5, 1996.
64. Goldwasser, Interview of November 5, 1996, in Chicago; Goldwasser, personal communication of November 4, 1998.
65. Goldwasser 1953.
66. Goldwasser 1955.
67. Packard, personal communication of February 25, 1998.
68. Packard, Interview of November 5, 1996.
69. Davidson and Feigelson 1957.

70. Ibid., 17.
71. Loftfield, personal communication of September 24, 1998.
72. Rapkin 1970, 47.
73. On this instrument, see Utting 1958.
74. Packard, Interview of November 5, 1996.
75. Rapkin 1970, 48.
76. Packard Instrument Company, *Annual Report 1965, Ten Year Financial Highlights*.
77. Davidson and Feigelson 1957, 3.
78. Packard 1958.
79. Packard, Interview of November 5, 1996.
80. Loftfield, personal communication of September 24, 1998.
81. Packard, Interview of November 5, 1996.
82. Langham 1958, 136 ff.
83. Rapkin, cited in Packard, Interview of November 5, 1996.
84. Packard Instrument Company, *Liquid Scintillation Counting Systems, Advertisement*, September 1961.
85. Rapkin 1970, 50.
86. Packard, personal communication of August 26, 1998.
87. Ibid.
88. The patent was for a "Method and Apparatus for Automatic Standardization in Liquid Scintillation Spectrometry" and covered both automatic internal and automatic external standardization. The upshot was that the biggest producers could no longer produce first-class machines without a license.
89. Packard had its own Absolute Activity Analyzer (AAA) that after counting each sample printed out not only the counting events per minute (cpm), but also the disintegrations per minute (dpm). Beckman offered an Automatic Quench Calibration (AQC) with which the counting efficiency of each quenched sample could be projected on the basis of a previously measured reference sample.
90. See Gaudillière 1998.
91. Loftfield, personal communication of September 24, 1998.
92. Okita, Kabara, Richardson, and LeRoy 1957. The authors here thank Slattery for "suggesting the discriminator-ratio method." The engineer George Utting of the Technical Measurement Corporation had a similarly productive relationship with his scientific customers. See Blau 1957. Blau thanks Utting for "help and advice."
93. Packard, Interview of November 5, 1996.
94. Rapkin, cited in ibid.
95. Kremer, interview of April 2, 1996, in Zurich.
96. Rapkin, cited in Lyle E. Packard, Interview of November 5, 1996.
97. Rapkin and Packard 1961.
98. Goldwasser, Interview of November 5, 1996, in Chicago.
99. Loftfield, personal communication of September 24, 1998.
100. Loftfield and Eigner 1960.

101. Packard Instrument Company, Inc., *Common Shares Offer*, 1961, 6 f.

102. See, for example, the advertisement in the October 1947 issue of *Nucleonics*, 85.

103. Packard, Interview of November 5, 1996.

104. Rapkin 1961.

105. Eighteen of these Technical Bulletins appeared between 1961 and 1969, one-third of them with Rapkin's signature.

106. Between 1957 and 1966 a series of conferences were held in New York, Chicago, Washington, Los Angeles, San Francisco, Zurich, and Boston. See, among others, *Proceedings of the Symposium on Tritium in Tracer Applications*, sponsored by New England Nuclear, Atomic Associates, Packard Instrument, New York, November 22, 1957; *Proceedings of the Symposium on Tritium in Tracer Applications*, sponsored by New England Nuclear, Atomic Associates, Packard Instrument, New York, October 31, 1958; Rothchild 1963–68.

107. Packard, Interview of November 5, 1996; Rapkin and Packard 1961.

108. See also Schram and Lombaert 1963. In their introduction the authors dryly observe that "the field of organic scintillation detectors may be said to extend to several sciences, physics, electronics, organic and biological chemistry."

109. Arnold 1958, 129.

110. Rapkin, cited in Packard, Interview of November 5, 1996.

111. Luntz 1957, 79.

112. What is Nucleonics?—The Magazine. *Nucleonics*, 1 (1) (September 1947): 2.

113. Rapkin contributed a valuable historical overview to this conference. See Rapkin 1970 and Rapkin, cited in Packard, Interview of November 5, 1996.

114. Horrocks and Peng 1971.

115. Schram and Lombaert 1963; Kobayashi and Maudsley 1974.

116. Packard, Interview of November 5, 1996.

117. Kaartinen 1969; Everett, Kaartinen, and Kreveld 1974.

118. Packard, Interview of November 5, 1996.

119. Packard, personal communication of February 25, 1998.

10. The Concept of Information

1. Jacob 1972 (1966), 148.

2. Nadine Peyrieras's and Michel Morange's edition of François Jacob's scientific writings has greatly facilitated this task: *Travaux scientifiques de François Jacob* (2002). The book includes a comprehensive bibliography of Jacob's publications.

3. Gunther Stent, a representative of the "information school" in molecular biology, assumes that its importance was already on the wane by the late 1970s. See Stent 1977, 137.

4. See, for example, Gaudillière 2002, 279.

5. Not least by Jacob himself. See his autobiography *The Statue Within* (1988 [1987]), chapter seven. See also Judson 1979, chapter seven; Pardee 1985; Morange 1998 (1994), chapters thirteen and fourteen; and Burian and Gayon 1999. Lily E. Kay

offers a very detailed account in her *Who Wrote the Book of Life?* (2000), chap. five. See further Gaudillière 2002, chapter seven.

6. Pardee, Jacob, and Monod 1958; Pardee, Jacob, and Monod 1959; Riley, Pardee, Jacob, and Monod 1960.
7. Jacob 1988 (1987), 290.
8. Pardee, Jacob, and Monod 1958.
9. Pardee, Jacob, and Monod 1959, 165.
10. Ibid., 175.
11. Ibid., 174.
12. Ibid., 177.
13. Jacob and Monod 1959. See also Morange 1990.
14. Riley, Pardee, Jacob, and Monod 1960.
15. Ibid., 216.
16. Ibid., 224.
17. Ibid.
18. Ibid., 225.
19. Thieffry 1997; Kay 2000, chap. five.
20. Chantrenne, personal communication of March 19, 1996.
21. Jacob, Perrin, Sanchez, and Monod 1960.
22. Brenner, Jacob, and Meselson 1961.
23. For a more detailed discussion see Judson 1979 and Jacob 1988 (1987). See also Grmek and Fantini 1982.
24. Brenner, Jacob, and Meselson 1961, 576.
25. Ibid., 580.
26. Jacob and Monod 1961.
27. Jacob 1988 (1987), 318.
28. Jacob, Perrin, Sanchez, and Monod 1960.
29. Ibid., 1729.
30. Jacob and Monod 1961, 318 f.
31. Ibid., 340 f., 349.
32. Ibid., 353.
33. Ibid., 354.
34. Jacob 1960, 33.
35. Jacob and Monod 1961, 354.
36. See also chapter eight in the present book.
37. Monod and Jacob 1961, 400.
38. Ibid., 393.
39. Jacob 1993 (1970). See also Rheinberger 2002.
40. Jacob 1965, 28.
41. Ibid., 24.
42. Ibid., 27 f.
43. Ibid., 25.
44. Ibid., 22.
45. Ibid., 23.

46. Ibid., 24 f.

47. Ibid., 31.

48. Jacob 1977 (1974).

49. Jacob 1993 (1970), 2.

50. Ibid., 297.

51. Ibid., 8, translation modified.

52. Ibid., 9.

53. Ibid., 299–324 ("Conclusion: The Integron"). See also the last section of chapter eight in the present book.

54. Ibid., 278.

55. See, for example, Keller 2000.

56. Jacob 1993 (1970), 1.

57. Ibid., 324.

58. See Brandt 2004.

11. Intersections

1. See chapter two in the present book.

2. A good example is given in chapter nine in the present book.

3. Schleiden 1849 (1846), vol. 1, 415.

4. See also Chadarevian 1993a.

5. See Schickore 2002a.

6. Schickore 2002b and 2007.

7. Dierig 2006. On this and similar instruments, see the Virtual Physiology Laboratory, http://vlp.mpiwg-berlin.mpg.de.

8. Mayer 2005; Mayer forthcoming.

9. Geimer 2002.

10. Kohler 1994.

11. On the historical place of this concept, also see chapter three in the present book.

12. This was even true of Alfred Kühn, for example, and down to the 1930s and 1940s at that. See chapter six in the present book.

13. See Rheinberger 1993; Creager 2002.

14. Haraway 1976.

15. Chadarevian 2002.

16. On electron microscopy see also Rasmussen 1993 and 1997; Strasser 2002 and 2006.

17. See also chapter nine in the present book.

18. Krige 2006.

19. See the special issue of the *Journal of the History of Biology* 34 (4) (2006) on radioactivity and biomedical research in the latter half of the twentieth century with contributions by Soraya de Chadarevian, Angela Creager, Karen Rader, and Maria Jesus Santesmases.

20. Rheinberger 2000a.

12. Preparations

1. For a detailed discussion, see chapter two in the present book.
2. *Duden, vol. 5: Das Fremdwörterbuch* (1974). The 1997 edition of *Duden* mentions only the "didactic purposes." On that definition preparations would be nothing more than passive objects used for purposes of demonstration, not objects of research, that is, things around which and by means of which scientific work is carried out. A better definition would read, "an organism, or parts thereof, prepared for research or didactic purposes."
3. Heidegger 2002, 71.
4. On the subject of the following section see also Rheinberger 2003.
5. François Jacob has described its critical role in the organization of the life sciences of the seventeenth and early eighteenth centuries in Jacob 1993 (1970).
6. See Daston 2004.
7. See chapter eleven in the present book.

13. The Economy of the Scribble

1. See, for example, Shapin and Schaffer 1985; Myers 1990; Lenoir 1998.
2. From a science studies perspective see Latour and Woolgar 1986 (1979); for an overview from the perspective of history of science see Holmes, Renn, and Rheinberger 2003.
3. See, for example, Berz and Hoffmann 2001.
4. For a detailed account see Rheinberger 1997.
5. Jacob 1988 (1987), 9.
6. Rheinberger 1998.
7. Derrida 1982 (1972).
8. Chadarevian 1993b; Geimer 2001.
9. See Latour 1988. See also Berz and Hoffmann, "Machs Notizbuch," in Berz and Hoffmann 2001, 19–41.
10. For a nice example of such reversibility see Latour 1995 (1993).
11. Jacob 1998 (1997), 125.
12. Ibid., 126.
13. See, among the studies of this subject, Steinle 2005.
14. See Macho and Wunschel 2004.
15. See also chapter four in the present book.
16. MPG Archive, Section 3, Folder 17, no. 115.
17. Mendel 1901 (1866), trans. revised by Roger Blumberg, http://www.mendelweb.org/.
18. Vries 1900a; Correns 1900a.
19. Foucault 1972 (1969), 140.
20. Specialized versions of such technologies may be found in different scientific disciplines. On their use in chemistry see Klein 2003.
21. For a good overview see Geison and Holmes 1993.

Adler, Brian K., and Stephen L. Hajduk. 1994. Mechanism and origins of RNA editing. *Current Opinion in Genetics and Development* 4 (2): 316–22.

Allen, Garland E. 1974. Opposition to the Mendelian-chromosome theory: The physiological and developmental genetics of Richard Goldschmidt. *Journal of the History of Biology* 7:49–92.

Alvarez, Luis W., and Robert Cornog. 1939. Helium and hydrogen of mass 3. *The Physical Review* 56:613.

Anderson, Ernest C. 1958. The Los Alamos human counter. In *Liquid Scintillation Counting*, ed. Carlos G. Bell and Francis Newton Hayes, 211–19. New York: Pergamon.

Ardenne, Manfred von, Hans Friedrich-Freksa, and Gerhard Schramm. 1941. Elektronenmikroskopische Untersuchung der Präcipitinreaktion von Tabakmosaikvirus mit Kaninchenantiserum. *Archiv für die gesamte Virusforschung* 2:80–86.

Argonne National Laboratory. 1986. *Argonne News* 30 (5): 3–15.

Arnold, James R. 1954. Scintillation counting of natural radiocarbon. I: The counting method. *Science* 119:155–57.

———. 1958. Archaeology and chemistry. In *Liquid Scintillation Counting*, ed. Carlos G. Bell and Francis Newton Hayes, 129–34. New York: Pergamon.

Atkins, John F., and Raymond F. Gesteland. 1996. A case for *trans* translation. *Nature* 379:769–71.

Atlan, Henri, and Moshe Koppel. 1990. The cellular computer DNA: Program or data? *Bulletin of Mathematical Biology* 52:335–48.

Austin, John L. 1962. *How to Do Things with Words*, ed. James O. Urmson and Marina Sbisà. Cambridge, Mass.: Harvard University Press.

Avery, Oswald T., Colin M. MacLeod, and Maclyn McCarty. 1944. Studies on the chemical transformation of pneumococcal types. *Journal of Experimental Medicine* 79:137–58.

Bachelard, Gaston. 1928. *Essai sur la connaissance approchée*. Paris: Vrin.

———. 1932. *Le pluralisme cohérent de la chimie moderne*. Paris: Vrin.

———. 1933. *Les intuitions atomistiques*. Paris: Boivin.

———. 1949. *Le rationalisme appliqué*. Paris: Presses universitaires de France.

———. 1951. *L'activité rationaliste de la physique contemporaine*. Paris: Presses universitaires de France.

———. 1953. *Le matérialisme rationnel*. Paris: Presses universitaires de France.

———. 1964 (1938). *The Psychoanalysis of Fire*. Trans. Alan C. M. Ross. Boston: Beacon.

———. 1968 (1940). *The Philosophy of No: A Philosophy of the New Scientific Mind*. Trans. G. C. Waterston. New York: Orion.

———. 1970 (1931–32). Noumène et microphysique. In *Etudes*, by Gaston Bachelard, 11–24. Paris: Vrin.

———. 1972a (1951). Le problème philosophique des méthodes scientifiques. In *L'engagement rationaliste*, by Gaston Bachelard, 35–44. Paris: Presses universitaires de France.

———. 1972b (1951). L'actualité de l'histoire des sciences. In *L'engagement rationaliste*, by Gaston Bachelard, 137–52. Paris: Presses universitaires de France.

———. 1984 (1934). *The New Scientific Spirit*. Trans. Arthur Goldhammer. Boston: Beacon.

———. 1987 (1928). *Essai sur la connaissance approchée*. Paris: Vrin.

———. 2002 (1938). *The Formation of the Scientific Mind: A Contribution to a Psychoanalysis of Objective Knowledge*. Trans. Mary McAllester Jones. Manchester: Clinamen.

Balibar, Etienne. 1993. Science et vérité dans la philosophie de Georges Canguilhem. In *Georges Canguilhem, philosophe, historien des sciences. Actes du Colloque du Collège International de Philosophie, 6–8/12/1990*, 58–76. Paris: Albin Michel.

Balogh, Brian. 1991. *Chain Reaction: Expert Debate and Public Participation in American Commercial Nuclear Power, 1945–1975*. Cambridge: Cambridge University Press.

Beadle, George W. 1945. The genetic control of biochemical reactions. *The Harvey Lectures* 40:179–94.

Beadle, George W., and Boris Ephrussi. 1935. Transplantation in Drosophila. *Proceedings of the National Academy of Sciences of the United States of America* 21:642–46.

———. 1936. The differentiation of eye pigments in Drosophila as studied by transplantation. *Genetics* 21:225–47.

Beadle, George W., and Edward L. Tatum. 1941. Genetic control of biochemical reactions in Neurospora. *Proceedings of the National Academy of the Sciences of the United States of America* 27:499–506.

Beams, Jesse W., and Edward G. Pickels. 1935. The production of high rotational speeds. *Review of Scientific Instruments* 6:299–308.

Becker, Erich. 1937. Extraktion des bei der Mehlmotte Ephestia kühniella die dunkle Ausfärbung der Augen auslösenden Gen-A-Hormons. *Die Naturwissenschaften* 25:507.

———. 1938. Die Gen-Wirkstoff-Systeme der Augenausfärbung bei Insekten. *Die Naturwissenschaften* 26:433–41.

———. 1939. Über die Natur des Augenpigments von Ephestia kühniella und seinen Vergleich mit den Augenpigmenten anderer Insekten. *Biologisches Zentralblatt* 59:597–627.

———. 1941. Über Versuche zur Anreicherung und physiologischen Charakterisierung des Wirkstoffs der Puparisierung. *Biologisches Zentralblatt* 61:360–88.

Becker, Erich, and Ernst Plagge. 1937. Vergleich der die Augenausfärbung bedingenden Gen-Wirkstoffe von Ephestia und Drosophila. *Die Naturwissenschaften* 25 (50): 809.

Behrends, Johannes. 1935. Über die Entwicklung des Lakunen-, Ader- und Tracheensystems während der Puppenruhe im Flügel der Mehlmotte Ephestia kühniella Z. *Zeitschrift für Morphologie und Ökologie der Tiere* 30:573–96.

Behrens, Wilhelm J. 1883. *Hilfsbuch zur Ausführung mikroskopischer Untersuchungen im Botanischen Laboratorium.* Braunschweig: C. A. Schwetschke und Sohn.

Benzer, Seymour. 1955. Fine structure of a genetic region in bacteriophage. *Proceedings of the National Academy of Sciences of the United States of America* 41:344–54.

Bergold, Gernot. 1941. Eine Mikroinjektionsspritze und Mikrobürette bis zu 0,1 cmm. *Biologisches Zentralblatt* 61:158–62.

———. 1943a. Röhrchen und Messzelle zur Ultrazentrifuge aus Superpolyamid. *Kolloid-Zeitschrift* 102:292–93.

———. 1943b. Über Polyederkrankheiten bei Insekten. *Biologisches Zentralblatt* 63:1–55.

———. 1946. Diffusions- und Sedimentationsmessungen zur Bestimmung des Molekulargewichts von Proteinen und hochpolymeren Kunststoffen. *Zeitschrift für Naturforschung* 1:100–108.

———. 1947. Die Isolierung des Polyeder-Virus und die Natur der Polyeder. *Zeitschrift für Naturforschung* 2b:122–43.

Bergold, Gernot, and Rudolf Brill. 1942. Spreitungsversuche mit Insektenviren. *Kolloid-Zeitschrift* 99:1–6.

Bergold, Gernot, and Hans Friedrich-Freksa. 1947. Zur Größe und Serologie des Bombyx-mori-Polyedervirus. *Zeitschrift für Naturforschung* 2b:410–14.

Bergold, Gernot, and Josef Hengstenberg. 1942. Ultrazentrifugenversuche mit Insektenviren. *Kolloid-Zeitschrift* 98:304–11.

Bergold, Gernot, and Gerhard Schramm. 1942. Biochemische Charakterisierung von Insektenviren. *Biologisches Zentralblatt* 62:105–18.

Bernard, Claude. 1949 (1865). *An Introduction to the Study of Experimental Medicine.* Trans. Henry Copley Greene. New York: Schuman.

———. 1954. *Philosophie: Manuscrit inédit*, ed. Jacques Chevalier. Paris: Editions Hatier-Boivin.

Berz, Peter, and Christoph Hoffmann, eds. 2001. *Über Schall: Ernst Machs und Peter Salchers Geschoßfotografien.* Göttingen: Wallstein.

Beurton, Peter, Raphael Falk, and Hans-Jörg Rheinberger, eds. 2000. *The Concept of the Gene in Development and Evolution.* Cambridge: Cambridge University Press.

Bing, François, and Jean-François Braunstein. 1998. Entretien avec Georges Canguilhem. In *Actualité de Georges Canguilhem: Le normal et le pathologique*, 121–35. Paris: Synthélabo.

Blau, Monte. 1957. Separated channels improve liquid scintillation counting. *Nucleonics* 15 (4).

Blaustein, Werner. 1935. Histologische Untersuchungen über die Metamorphose der Mehlmotte Ephestia kühniella Z. *Zeitschrift für Morphologie und Ökologie der Tiere* 30:333–54.

Boltzmann, Ludwig. 1905. Der zweite Hauptsatz der mechanischen Wärmetheorie. In *Populäre Schriften*, by Ludwig Boltzmann, 25–50. Leipzig: Barth.

Born, Hans-Joachim, Anton Lang, and Gerhard Schramm. 1943. Markierung von Tabakmosaikvirus mit Radiophosphor. *Archiv für die gesamte Virusforschung* 2:461–79.

Born, Hans Joachim, Anton Lang, Gerhard Schramm, and Karl Günter Zimmer. 1941. Versuche zur Markierung von Tabakmosaivirus mit Radiophosphor. *Die Naturwissenschaften* 29 (14/15): 222–23.

Brachet, Jean. 1960. *The Biological Role of Ribonucleic Acids (Sixth Weizmann Memorial Lecture Series)*. Amsterdam: Elsevier.

Brandt, Christina. 2004. *Metapher und Experiment*. Göttingen: Wallstein.

Brandt, Herbert. 1934. Die Lichtorientierung der Mehlmotte Ephestia kühniella Z. *Zeitschrift für vergleichende Physiologie* 20:645–73.

Braunstein, Jean-François, et al. 1998. *Actualité de Georges Canguilhem: Le normal et le pathologique*. Paris: Synthélabo.

Brenner, Sidney, François Jacob, and Matthew Meselson. 1961. An unstable intermediate carrying information from genes to ribosomes for protein synthesis. *Nature* 190:576–81.

Broda, Engelbert. 1960. *Radioactive Isotopes in Biochemistry*. Amsterdam: Elsevier.

Broglie, Louis de. 1947. *Physique et microphysique*. Paris: Albin Michel.

Broser, Immanuel, and Hartmut Kallmann. 1947. Über die Anregung von Leuchtstoffen durch schnelle Korpuskularteilchen I, and Über den Elementarprozess der Lichtanregung in Leuchtstoffen durch alpha-Teilchen, schnelle Elektronen und gamma-Quanten II. *Zeitschrift für Naturforschung* 2a:439–40, 642–50.

Brosius, Jürgen, and Stephen Jay Gould. 1992. On "genomenclature." A comprehensive (and respectful) taxonomy for pseudogenes and other junk DNA. *Proceedings of the National Academy of Sciences of the United States of America*. 89:10706–10.

———. 1993. Molecular constructivity. *Nature* 365:102.

Burian, Richard M. 1985. On conceptual change in biology: The case of the gene. In *Evolution at a Crossroads: The New Biology and the New Philosophy of Science*, ed. David J. Depew and Bruce H. Weber. Cambridge, Mass.: MIT Press.

Burian, Richard M., and Jean Gayon. 1999. The French school of genetics: From physiological and population genetics to regulatory molecular genetics. *Annual Review of Genetics* 33:313–49.

Burian, Richard M., Jean Gayon, and Doris T. Zallen. 1988. The singular fate of genetics in the history of French biology, 1900–1940. *Journal of the History of Biology* 21:357–402.

———. 1991. Boris Ephrussi and the synthesis of genetics and embryology. In *Developmental Biology: A Comprehensive Synthesis, Vol. VII: A Conceptual History of Modern Embryology*, ed. Scott H. Gilbert, 207–27. New York: Plenum.

Busselmann, Antonette. 1934. Bau und Entwicklung der Raupenocellen der Mehlmotte Ephestia kühniella Z. *Zeitschrift für Morphologie und Ökologie der Tiere* 29:218–28.

Butenandt, Adolf. 1941. Die biologische Chemie im Dienste der Volksgesundheit. *Preußische Akademie der Wissenschaften, Vorträge und Schriften, Heft 8*. Berlin: De Gruyter.

———. 1944. Zur Feinstruktur des Tabakmosaik-Virus. In *Abhandlungen der Preußischen Akademie der Wissenschaften, Mathematisch-naturwissenschaftliche Klasse, Heft 10*, 1–11.

———. 1977a. Die Entwicklung der modernen Virusforschung. Werner Schäfer zum 65. Geburtstag. *MPG-Spiegel* 6:39–45.

———. 1977b. The historical development of modern virus research in Germany, especially in the Kaiser-Wilhelm-/Max-Planck-Society, 1936–1954. *Medical Microbiology and Immunology* 164:3–14.

Butenandt, Adolf, Hans Friedrich-Freksa, St. Hartwig, and G. Scheibe. 1942. Beitrag zur Feinstruktur des Tabakmosaikvirus. *Hoppe Seyler's Zeitschrift für physiologische Chemie* 274:276–84.

Butenandt, Adolf, Wolfhard Weidel, and Erich Becker. 1940a. Kynurenin als Augenpigmentbildung auslösendes Agens bei Insekten. *Die Naturwissenschaften* 28:63–64.

———. 1940b. Alpha-Oxytryptophan als "Prokynurenin" in der zur Augenpigmentbildung führenden Reaktionskette bei Insekten. *Die Naturwissenschaften* 28:447–48.

Bütschli, Otto. 1876. *Studien über die ersten Entwicklungsvorgänge der Eizelle, die Zelltheilung und die Conjugation der Infusorien*. Frankfurt am Main: Christian Winter.

Butterfield, Herbert. 1951. *The Whig Interpretation of History*. New York: Charles Scribner's Sons.

Canguilhem, Georges. 1952. *La connaissance de la vie*. Paris: Hachette.

———. 1955. *La formation du concept de réflexe aux XVIIe et XVIIIe siècles*. Paris: Presses universitaires de France.

———. 1963. L'histoire des sciences dans l'œuvre épistémologique de Gaston Bachelard. *Annales de l'Université de Paris* 33:24–39.

———. 1965. Philosophie et science. (Interviewed by Alain Badiou). *Revue de l'enseignement philosophique*, February 15, 1965, 10–17.

———. 1968a. L'objet de l'histoire des sciences / Galilée: La signification de l'œuvre et la leçon de l'homme / L'idée de médecine expérimentale selon Claude Bernard / Théorie et technique de l'expérimentation chez Claude Bernard / La constitution de la physiologie comme science / Modèles et analogies dans la découverte en biologie. In *Etudes d'histoire et de philosophie des sciences*, by Georges Canguilhem, 9–23, 37–50, 127–42, 143–55, 226–73, 305–15. Paris: Vrin.

———. 1968b. [Remarks on] Objectivité et historicité de la pensée scientifique. *Raison présente* 8:39–41, 46 f., 51 f.

———. 1975 (1952). La théorie cellulaire. In *La connaissance de la vie*, by Georges Canguilhem, 43–80. Paris: Vrin.

———. 1979. *Wissenschaftsgeschichte und Epistemologie: Gesammelte Aufsätze*. Trans. Michael Bischoff and Walter Seitter, ed. Wolf Lepenies. Frankfurt am Main: Suhrkamp.

———. 1981. Qu'est-ce qu'une idéologie scientifique? In *Idéologie et rationalité dans l'histoire des sciences de la vie*, by Georges Canguilhem. Paris: Vrin.

———. 1988 (1971, 1977, 1981). Introduction: The role of epistemology in contemporary history of science (1977) / What is a scientific ideology? (1981) / The development of the concept of biological regulation in the eighteenth and nineteenth centuries (1977) / On the history of the Life sciences since Darwin. Opening lecture at the twelfth international conference on the history of the sciences in Moscow, 18–24 August 1971. In *Ideology and Rationality in the History of the Life Sciences*, by Georges Canguilhem, trans. Arthur Goldhammer. 1–23, 27–40, 81–102, 103–23. Cambridge, Mass.: MIT Press.

———. 1991 (1966). *The Normal and the Pathological*. Trans. Carolyn R. Fawcett with Robert S. Cohen. New York: Zone Books.

———. 1994. *A Vital Rationalist*, trans. Arthur Goldhammer, ed. François Delaporte. New York: Zone Books.

Carlson, Elof Axel. 1991. Defining the gene: An evolving concept. *American Journal of Human Genetics* 49:475–87.

Caspari, Ernst. 1933. Über die Wirkung eines pleiotropen Gens bei der Mehlmotte Ephestia kühniella Zeller. *Wilhelm Roux' Archiv für Entwicklungsmechanik der Organismen* 130:353–81.

Cassirer, Ernst. 1950. *The Problem of Knowledge: Philosophy, Science, and History Since Hegel*. Trans. William H. Woglom and Charles W. Hendel. New Haven: Yale University Press.

———. 1961 (1942). *The Logic of the Humanities*. Trans. Clarence Smith Howe. New Haven: Yale University Press.

Castelão-Lawless, Teresa. 1995. Phenomenotechnique in historical perspective: Its origins and implications for philosophy of science. *Philosophy of Science* 62:44–59.

Chadarevian, Soraya de. 1993a. Instruments, illustrations, skills, and laboratories in 19th-century German botany. In *Non-Verbal Communication in Science Prior to 1900*, ed. Renato G. Mazzolini, 529–62. Florence: Leo S. Olschki.

———. 1993b. Graphical method and discipline: Self-recording instruments in nineteenth-century physiology. *Studies in History and Philosophy of Science* 24:267–91.

———. 2002. *Designs for Life: Molecular Biology after World War II*. Cambridge: Cambridge University Press.

Chen, Heng-an. 2003. *Die Sexualitätstheorie und "Theoretische Biologie" von Max Hartmann in der ersten Hälfte des zwanzigsten Jahrhunderts*. Wiesbaden: Franz Steiner.

Chimisso, Cristina. 2001. *Gaston Bachelard: Critic of Science and the Imagination*. London: Routledge.

Claude, Albert. 1947–48, 1950. Studies on cells: morphology, chemical constitution, and distribution of biochemical function. *The Harvey Lectures* 43:121–64.

Cohn, Waldo E. 1968. Introductory remarks by chairman. In *Advances in Tracer Methodology, Volume 4*, ed. Seymour Rothchild. New York: Plenum.

Coltman, John W., and Fitz-Hugh Marshall. 1947. Photomultiplier radiation detector. *Nucleonics* 1 (3): 58–64.

Cooper, Antony A., and Tom H. Stevens. 1995. Protein splicing: Self-splicing of genetically mobile elements at the protein level. *TIBS* 20:351–56.

Correns, Carl. 1889. Über Dickenwachstum durch Intussusception bei einigen Algenmembranen. Diss. University of Munich. *Flora* 72:298–347.

———. 1892. Über die Abhängigkeit der Reizerscheinungen höherer Pflanzen von der Gegenwart freien Sauerstoffs [Habilitationsschrift]. Tübingen.

———. 1893. Zur Kenntnis der inneren Struktur einiger Algenmembranen. In *Beiträge zur Morphologie und Physiologie der Pflanzenzelle, Volume 1*, ed. Albrecht Zimmermann, 260–305. Tübingen: Verlag der Laupp'schen Buchhandlung.

———. 1895. Floristische Bemerkungen über das Ursernthal. *Berichte der Schweizerischen Botanischen Gesellschaft* 5:86–93.

———. 1896. Zur Physiologie der Ranken. *Botanische Zeitung* 54:1–20.

———. 1897. Über die Membran und die Bewegung der Oscillarien: Vorläufige Mittheilung. *Berichte der Deutschen Botanischen Gesellschaft* 15:139–48.

———. 1899a. *Untersuchungen über die Vermehrung der Laubmoose durch Brutorgane und Stecklinge*. Jena: G. Fischer.

———. 1899b. Untersuchungen über die Xenien bei Zea Mays. *Berichte der Deutschen Botanischen Gesellschaft* 17:410–17.

———. 1900a. G. Mendel's Regel über das Verhalten der Nachkommenschaft der Rassenbastarde. *Berichte der Deutschen Botanischen Gesellschaft* 18:158–68.

———. 1900b. Über den Einfluss, welchen die Zahl der zur Bestäubung verwendeten Pollenkörner auf die Nachkommenschaft hat. *Berichte der Deutschen Botanischen Gesellschaft* 18:422–35.

———. 1900c. Gregor Mendel's 'Versuche über Pflanzen-Hybriden' und die Bestätigung ihrer Ergebnisse durch die neuesten Untersuchungen. *Botanische Zeitung* 58:229–35.

———. 1900d. Über Levkojenbastarde. *Botanisches Centralblatt* 84:97–113.

———. 1901. *Bastarde zwischen Maisrassen: Mit besonderer Berücksichtigung der Xenien*. Stuttgart: Erwin Nägele.

———. 1902. Über den Modus und den Zeitpunkt der Spaltung der Anlagen bei den Bastarden vom Erbsen-Typus. *Botanische Zeitung* 60:65–82.

Coveney, Peter, and Roger Highfield. 1995. *Frontiers of Complexity: The Search for Order in a Chaotic World*. London: Faber and Faber.

Creager, Angela N. H. 2002. *The Life of a Virus: Tobacco Mosaic Virus as an Experimental Model, 1930–1965*. Chicago: University of Chicago Press.

———. 2004. The industrialization of radioisotopes by the U.S. Atomic Energy Commission. In *The Science-Industry Nexus: History, Policy, Implications*, ed. Karl Grandin, Nina Wormbs, and Sven Widmalm, 141–67. Sagamore Beach, Mass.: Science History Publications/USA and The Nobel Foundation.

Crick, Francis. 1958. On protein synthesis. *Symposia of the Society for Experimental Biology London* 12:138–63.

Crookes, William. 1903. The emanation of radium. *Proceedings of the Royal Society of London* 71:405–8.

Curran, Samuel Crowe, and W. R. Baker. 1947 (1944). A photoelectric alpha particle detector. *U.S. Atomic Energy Commission Rpt. MDDC 1296*, 17 November 1944, declassified 23 September 1947.

da Cunha, Alberto Xavier. 1935. Estudo da acção dum gene pleiotropo na "Ephestia kühniella" Zeller. *Revista da Faculdade de Ciências, Universidade de Coimbra* 5:177–278.

Dagognet, François. 2003. Sur une seconde rupture. In *Bachelard et l'épistémologie française*, ed. Jean-Jacques Wunenburger, 13–17. Paris: Presses universitaires de France.

Danneel, Rolf. 1938. Die Wirkungsweise der Grundfaktoren für Haarfärbung beim Kaninchen. *Die Naturwissenschaften* 26:505–9.

———. 1941a. Die Ausfärbung überlebender c- und cn-Drosophila-Augen mit Produkten des Tryptophanstoffwechsels. *Biologisches Zentralblatt* 61:388–99.

———. 1941b. Ein Papillom-Virus aus Kaninchenhaut. *Die Naturwissenschaften* 29:364–65.

———. 1941c. Untersuchungen über virusbedingte Kaninchenpapillome I. *Biologisches Zentralblatt* 61:441–52.

———. 1942. Virus-Krankheiten. *Westermanns Monatshefte* 86 (1027): 357–60.

———. 1943a. Bemerkungen zu meiner Mitteilung: Ein Papillom-Virus aus Kaninchenhaut. *Die Naturwissenschaften* 31:551.

———. 1943b. Melaninbildende Fermente bei Drosophila melanogaster. *Biologisches Zentralblatt* 63:377–94.

———. 1944. Wege und Ziele der Virusforschung. *Archiv für Kinderheilkunde* 131:42–53.

Danneel, Rolf, and Ernst Lubnow. 1934. Zur Physiologie der Kälteschwärzung beim Russenkaninchen I. *Biologisches Zentralblatt* 54:287–91.

———. 1936. Zur Physiologie der Kälteschwärzung beim Russenkaninchen II. *Biologisches Zentralblatt* 56:572–84.

Danneel, Rolf, and Hildegard Paul. 1940. Zur Physiologie der Kälteschwärzung beim Russenkaninchen IV. *Biologisches Zentralblatt* 60:79–85.

Danneel, Rolf, and Kurt Schaumann. 1938. Zur Physiologie der Kälteschwärzung beim Russenkaninchen III. *Biologisches Zentralblatt* 58:242–60.

Daston, Lorraine, ed. 2000. *Biographies of Scientific Objects*. Chicago: University of Chicago Press.

———. 2004. Type specimens and scientific memory. *Critical Inquiry* 31 (1): 153–82.

Davidson, Jack D., and Philip Feigelson. 1957. Practical aspects of internal-sample liquid-scintillation counting. *International Journal of Applied Radiation and Isotopes* 2:1–18.

Deichmann, Ute. 1996. *Biologists under Hitler*. Trans. Thomas Dunlap. Cambridge, Mass.: Harvard University Press.

Delbrück, Max. 1942. Bacterial viruses (bacteriophages). *Advances in Enzymology* 2:1–32.

Derrida, Jacques. 1982 (1972). Signature event context. In *Margins of Philosophy* by Jacques Derrida, trans. Alan Bass, 307–30. Chicago: University of Chicago Press.

———. 1999. *Sur parole*. Paris: Editions de l'Aube.

Dierig, Sven. 2006. *Wissenschaft in der Maschinenstadt: Emil Du Bois Reymond und seine Laboratorien in Berlin*. Göttingen: Wallstein.

Dietrich, Michael. 1995. Richard Goldschmidt's "heresis" and the evolutionary synthesis. *Journal of the History of Biology* 28:431–61.

Dingler, Max. 1925. *Die Hausinsekten und ihre Bekämpfung*. Berlin.

Driesch, Hans. 1901. *Die organischen Regulationen: Vorbereitungen zu einer Theorie des Lebens*. Leipzig: W. Engelmann.

Du Bois-Reymond, Emil. 1912. Über Geschichte der Wissenschaft. A lecture delivered at the Leibniz-Session of the Academy of Sciences in Berlin on 4 July 1872. In *Reden von Emil Du Bois-Reymond, mit einer Gedächtnisrede von Julius Rosenthal, Volume 1*, ed. Estelle Du Bois-Reymond, 431–41. Leipzig: Veit.

Duden, vol. 5: Das Fremdwörterbuch. Mannheim: Bibliographisches Institut & F. A. Brockhaus AG (1974).

Dunn, Leslie C. 1991 (1965). *A Short History of Genetics: The Development of Some of the Main Lines of Thought, 1864-1939*. Ames: Iowa State University Press.

Eder, Jörg, and Heinz Rembold. 1992. Biosemiotics—a paradigm of biology. *Die Naturwissenschaften* 79:60–67.

Egelhaaf, Albrecht. 1996. Alfred Kühn, his work and his contribution to molecular biology. *International Journal of Developmental Biology* 40:69–75.

Eicher, Eva M. 1987. Ernst W. Caspari: Geneticist, teacher, and mentor. *Advances in Genetics* 24:xv–xxix.

Eigen, Manfred, and Peter Schuster. 1979. *The Hypercycle: A Principle of Natural Self-Organization*. Heidelberg: Springer.

Eigen, Manfred, with Ruthild Winkler-Oswatitsch. 1992 (1987). *Steps towards Life*. Trans. Paul Woolley. Oxford: Oxford University Press.

Elkana, Yehuda. 1970. Helmholtz' 'Kraft': An illustration of concepts in flux. *Historical Studies in the Physical Sciences* 2:263–98.

Elster, Julius, and Hans Geitel. 1903. Über die durch radioaktive Emanation erregte scintillierende Phosphoreszenz der Sidot-Blende. *Physikalische Zeitschrift* 4:439–40.

Elzen, Boelie. 1986. Two ultracentrifuges: A comparative study of the social construction of artefacts. *Social Studies of Science* 16:621–62.

Ephrussi, Boris. 1938. Aspects of the physiology of gene action. *The American Naturalist* 72:5–23.

Ephrussi, Boris, and George W. Beadle. 1935. La transplantation des disques imaginaux chez la Drosophile. *Comptes rendus de l'Académie des Sciences Paris* 201:98–100.

Ephrussi, Boris, C. W. Clancy, and George W. Beadle. 1936. Influence de la lymphe sur la couleur des yeux vermilion chez la Drosophile (Drosophila melanogaster). *Comptes rendus de l'Académie des Sciences Paris* 203:545–46.

Everett, Leroy J., Niilo Kaartinen, and P. Kreveld. 1974. An advanced automatic sample oxidizer—new horizons in liquid scintillation sample preparation. In *Liquid Scintillation Counting*, ed. Philip E. Stanley and Bruce A. Scoggins, 139–52. New York: Academic.

Falk, Raphael. 1984. The gene in search of an identity. *Human Genetics* 68:195–204.

——. 1986. What is a gene? *Studies in the History and Philosophy of Science* 17:133–73.

Feld, Bernard T., and Gertrud Weiss Szilard, ed. 1972. *The Collected Works of Leo Szilard: Scientific Papers, Volume 1*. Cambridge, Mass.: MIT Press.

Feldotto, Wolfgang. 1933. Sensible Perioden des Flügelmusters bei Ephestia kühniella Z. *Wilhelm Roux' Archiv für Entwicklungsmechanik der Organismen* 128:299–341.

Fischer, Ernst Peter. 1995. How many genes has a human being? The analytical limits of a complex concept. In *The Human Genome*, ed. Ernst Peter Fischer and Sigmar Klose, 223–56. Munich: Piper.

Fleck, Ludwik. 1929. Zur Krise der 'Wirklichkeit.' *Die Naturwissenschaften* 17:425–30.

——. 1979 (1935). *Genesis and Development of a Scientific Fact*. Trans. Fred Bradley and Thaddeus J. Trenn. Chicago: University of Chicago Press.

Focke, Wilhelm Olbers. 1881. *Die Pflanzen-Mischlinge: Ein Beitrag zur Biologie der Gewächse*. Berlin: Borntraeger.

Forman, Paul. 1997. Recent science: Late-modern and post-modern. In *The Historiography of Contemporary Science and Technology*, ed. Thomas Söderqvist, 179–213. Amsterdam: Harwood.

Foucault, Michel. 1972 (1971). The discourse on language. In *The Archaeology of Knowledge and The Discourse on Language*, by Michel Foucault, trans. A. M. Sheridan Smith, 215–37. New York: Pantheon.

——. 1972 (1969). *The Archaeology of Knowledge and the Discourse on Language*. Trans. A. M. Sheridan Smith. New York: Pantheon.

——. 1973 (1963). *The Birth of the Clinic: An Archaeology of Medical Perception*. Trans. A. M. Sheridan Smith. London: Tavistock.

——. 1998 (1994). Life: Experience and science. In *Essential Works of Foucault, 1954-1984, Volume 2, Aesthetics, Method and Epistemology*, by Michel Foucault, trans. Robert Hurley et al., ed. James Faubion, 465–78. London: Penguin.

——. 2006 (1961). *History of Madness*. Trans. Jonathan Murphy and Jean Khalfa, ed. Jean Khalfa. London: Routledge.

Friedman, Michael. 2000. *A Parting of the Ways: Carnap, Cassirer, Heidegger*. La Salle: Open Court.

Friedrich-Freksa, Hans. 1940. Bei der Chromosomenkonjugation wirksame Kräfte und ihre Bedeutung für die identische Verdopplung von Nucleoproteinen. *Die Naturwissenschaften* 28:376–79.

——. 1961. Genetik und biochemische Genetik in den Instituten der Kaiser-Wilhelm-Gesellschaft und der Max-Planck-Gesellschaft. *Die Naturwissenschaften* 48:10–22.

Friedrich-Freksa, Hans, Georg Melchers, and Gerhard Schramm. 1946. Biologischer, chemischer und serologischer Vergleich zweier Parallelmutanten phytopathogener Viren mit ihren Ausgangsformen. *Biologisches Zentralblatt* 65:187–222.

Furst, Milton, and Hartmut Kallmann. 1952. Fluorescence of solutions bombarded with high energy radiation (Energy transport in liquids), Pt. III. *The Physical Review* 85:816–25.

Galison, Peter. 1997. *Image and Logic: A Material Culture of Microphysics.* Chicago: University of Chicago Press.

Gärtner, Carl Friedrich. 1849. *Versuche und Beobachtungen über die Bastarderzeugung im Pflanzenreich: Mit Hinweisung auf die ähnlicheren Erscheinungen im Thierreiche.* Stuttgart: K. F. Hering.

Gaudillière, Jean-Paul. 1998. The molecularization of cancer etiology in the postwar United States: Instruments, politics, and management. In *Molecularizing Biology and Medicine: New Practices and Alliances 1910s–1970s,* ed. Soraya de Chadarevian and Harmke Kamminga, 139–70. Amsterdam: Harwood.

———. 2002. *Inventer la biomédicine: La France, l'Amérique et la production des savoirs du vivant (1945–1965).* Paris: La Découverte.

Gaukroger, Stephen W. 1976. Bachelard and the problem of epistemological analysis. *Studies in History and Philosophy of Science* 7:189–244.

Gausemeier, Bernd. 2004. An der Heimatfront, 'Kriegswichtige' Forschungen am Kaiser-Wilhelm-Institut für Biochemie. In *Adolf Butenandt und die Kaiser-Wilhelm-Gesellschaft: Wissenschaft, Industrie und Politik im "Dritten Reich,"* ed. Wolfgang Schieder and Achim Trunk, 134–68. Göttingen: Wallstein.

———. 2005. *Natürliche Ordnungen und politische Allianzen: Biologische und biochemische Forschungen an den Kaiser-Wilhelm-Instituten 1933–1945.* Göttingen: Wallstein.

Gayon, Jean. 1994. Génétique de la pigmentation de l'oeil de la drosophile: La contribution spécifique de Boris Ephrussi. In *Les sciences biologiques et médicales en France,* ed. Claude Debru, Jean Gayon, and J. F. Picard, 187–206. Paris: Editions du CNRS.

———. 1995. *Bachelard: Le rationalisme appliqué* (manuscript). Centre de Vanves, Ministère de l'Education nationale, Centre national d'enseignement à distance.

———. 1998. The concept of individuality in Canguilhem's philosophy of biology. *Journal of the History of Biology* 31:305–25.

———. 2003. Bachelard et l'histoire des sciences. In *Bachelard et l'epistémologie française,* ed. Jean-Jacques Wunenburger, 51–113. Paris: Presses universitaires de France.

Geimer, Peter, ed. 2001. *Ordnungen der Sichtbarkeit: Fotografie in Wissenschaft, Kunst und Technologie.* Frankfurt: Suhrkamp.

———, ed. 2002. *Untot.* Berlin: Max Planck Institute for the History of Science, Preprint 250.

Geison, Gerald L., and Frederic L. Holmes, eds. 1993. *Research Schools: Historical Reappraisals* (Osiris 8). Chicago: University of Chicago Press.

Gierke, Else von. 1932. Über die Häutungen und die Entwicklungsgeschwindigkeit der Larven der Mehlmotte Ephestia kühniella Z. *Wilhelm Roux' Archiv für Entwicklungsmechanik der Organismen* 127:387–410.

Goette, Alexander. 1883. *Über den Ursprung des Todes*. Hamburg and Leipzig: Leopold Voss.

Goldschmidt, Richard B. 1934. Lymantria. *Bibliographia Genetica* 11:1–185.

———. 1935. Gen und Außeneigenschaft (Untersuchungen an Drosophila) I u. II. *Zeitschrift für induktive Abstammungs- und Vererbungslehre* 69:38–131.

———. 1960. *In and Out of the Ivory Tower: The Autobiography of Richard B. Goldschmidt*. Seattle: University of Washington Press.

Goldwasser, Eugene. 1953. The incorporation of adenine into ribonucleic acid in vitro. *Journal of Biological Chemistry* 202:751–55.

———. 1955. Incorporation of adenosine-5'-phosphate into ribonucleic acid. *Journal of the American Chemical Society* 77:6083.

Grandy, David A. 1996. *Leo Szilard: Science as a Mode of Being*. Lanham, Md.: University Press of America.

Grasse, G., ed. 1972. *Alfred Kühn zum Gedächtnis*. Verband Deutscher Biologen, 5. Iserlohn: Biologisches Jahresheft.

Grene, Marjorie. 2000. The philosophy of science of Georges Canguilhem: A transatlantic view. *Revue d'histoire des sciences* 53:47–63.

Griesemer, James R. 2005. The informational gene and the substantial body: On the generalization of evolutionary theory by abstraction. In *Idealization XII: Correcting the Model, Idealization and Abstraction in the Sciences (Poznan Studies in the Philosophy of the Sciences and the Humanities, Volume 86)*, ed. Martin R. Jones and Nancy Cartwright, 59–115. Amsterdam: Rodopi.

Grmek, Mirko D., and Bernardino Fantini. 1982. Le rôle du hasard dans la naissance du modèle de l'opéron. *Revue d'histoire des sciences* 35:193–215.

Gros, François. 1991 (1986). *Les secrets du gène*. Paris: Odile Jacob.

Grossbach, Ulrich. 1996. Genes and development: An early chapter in German developmental biology. *International Journal of Developmental Biology* 40:83–87.

Guignard, Léon. 1899a. Sur les anthérozoïdes et la double copulation sexuelle chez les végétaux angiospermes. *Comptes rendus de l'Académie des Sciences Paris* 128:864–71.

———. 1899b. Les découvertes récentes sur la fécondation chez les végétaux angiospermes. *Cinquantenaire de la Société de Biologie*, 189–98.

Gutting, Gary. 2001. *French Philosophy in the Twentieth Century*. Cambridge: Cambridge University Press.

Haraway, Donna J. 1976. *Crystals, Fabrics, and Fields: Metaphors of Organicism in Twentieth-Century Developmental Biology*. New Haven: Yale University Press.

Harrington, Anne. 1996. *Reenchanted Science: Holism in German Culture from Wilhelm II to Hitler*. Princeton: Princeton University Press.

Hartmann, Max. 1917. Untersuchungen über die Morphologie und Physiologie des Formwechsels (Entwicklung, Fortpflanzung, Befruchtung und Vererbung) der Phytomonadinen (Volvocales), II. Mitteilung: Über die dauernde, rein agame Züchtigung von Eudorina elegans und ihre Bedeutung für das Befruchtungs- und Todesproblem. In *Sitzungsberichte der Preußischen Akademie der Wissenschaften zu Berlin*, 760–76.

————. 1919. Untersuchungen über die Morphologie und Physiologie des Formwechsels (Entwicklung, Fortpflanzung, Befruchtung und Vererbung) der Phytomonadinen (Volvocales), Programm der Untersuchungen und I. Mitt.: Über die Kern- und Zellteilung von Chlorogonium elongatum Dangeard. *Archiv für Protistenkunde* 39:1–33.

————. 1920. Otto Bütschli und das Befruchtungs- und Todproblem. *Die Naturwissenschaften* 8:555–58.

————. 1921. Untersuchungen über die Morphologie und Physiologie des Formwechsels (Entwicklung, Fortpflanzung, Befruchtung und Vererbung) der Phytomonadinen (Volvocales). Programm der Untersuchungen und III. Mitteilung. Die dauernd agame Zucht von Eudorina elegans, experimentelle Beiträge zum Befruchtungs- und Todesproblem. *Archiv für Protistenkunde* 43:223–86.

————. 1927. *Allgemeine Biologie.* Jena: Fischer.

————. 1956 (1937). Die Kausalität in Physik und Biologie. In *Gesammelte Vorträge und Aufsätze, Volume 2, Naturphilosophie,* 144–56. Stuttgart: Fischer.

Harwood, Jonathan. 1984. The reception of Morgan's chromosome theory in Germany: Inter-war debate over cytoplasmic inheritance. *Medizinhistorisches Journal* 19:3–32.

————. 1985. The reaction against specialization in 20th century biology: A study of Alfred Kühn. *Freiburger Universitätsblätter* 87/88:193–203.

————. 1993. *Styles of Scientific Thought: The German Genetics Community 1900-1933.* Chicago: University of Chicago Press.

Hase, Albrecht. 1928. Insekten. In *Methodik der wissenschaftlichen Biologie,* Vol. 2, ed. Tibor Péterfi, 265–89. Berlin: Springer.

Hayes, Francis Newton, Richard D. Hiebert, and Robert L. Schuch. 1952. Low energy counting with a new liquid scintillation solute. *Science* 116:140.

Hecht, G., and H. Weese. 1943. Periston, ein neuer Blutflüssigkeitsersatz. *Münchner Medizinische Wochenschrift* 90:11–15.

Heidegger, Martin. 2002 (1977). The age of the world picture. In *Off the Beaten Track* by Martin Heidegger, ed. and trans. Julian Young and Kenneth Haynes, 57–85. Cambridge: Cambridge University Press.

Heilbron, John L., and Robert W. Seidel. 1989. *Lawrence and His Laboratory: A History of the Lawrence Berkley Laboratory,* Pt. 1. Berkeley: University of California Press.

Henke, Karl. 1924. Die Färbung und Zeichnung der Feuerwanze (Pyrrhocoris apterus L.) und ihre experimentelle Beeinflußbarkeit. *Zeitschrift für vergleichende Physiologie* 1:297–499.

Hertwig, Richard. 1889. Über die Konjugation der Infusorien. *Abhandlungen der Bayerischen Akademie der Wissenschaften, II. Klasse,* Vol. 17, Pt. 1.

————. 1899. Mit welchem Recht unterscheidet man geschlechtliche und ungeschlechtliche Fortpflanzung? *Sitzungsberichte der Gesellschaft für Morphologie und Physiologie,* 142–53. Munich.

————. 1902/1903. Über das Wechselverhältnis von Kern und Protoplasma. *Sitzungsberichte der Gesellschaft für Morphologie und Physiologie,* 77–100. Munich.

————. 1903. Über Korrelation von Zell- und Kerngröße und ihre Bedeutung für die geschlechtliche Differenzierung und die Teilung der Zelle. *Biologisches Central-blatt* 23:49–62, 108–19.

Hesse, Hans. 2001. *Augen aus Auschwitz: Ein Lehrstück über nationalsozialistischen Rassenwahn und medizinische Folgen: Der Fall Dr. Karin Magnussen*. Essen: Klartext Verlag.

Hevesy, Georg von. 1948. Historical sketch of the biological application of tracer elements. *Cold Spring Harbor Symposia on Quantitative Biology* 13:129–50.

Hewlett, Richard G., and Oscar E. Anderson Jr. 1962. *A History of the United States Atomic Energy Commission, Volume 1: The New World 1939/1946; Volume 2: Atomic Shield 1947/1952*. University Park: Pennsylvania State University Press.

Hewlett, Richard G., and Jack M. Holl. 1989. *Atoms for Peace and War, 1953–1961*. Berkeley: University of California Press.

Hiebert, Richard D., and F. Newton Hayes. 1958. Instrumentation for liquid scintillation counting at Los Alamos. In *Liquid Scintillation Counting*, ed. Carlos G. Bell and F. Newton Hayes, 41–49. New York: Pergamon.

Hiebert, Richard D., and Richard J. Watts. 1953. Fast-coincidence circuit for H^3 and C^{14} measurements. *Nucleonics* 11 (12): 38–41.

Hofstadter, Robert. 1948. Alkali halide scintillation counters. *The Physical Review* 74:100–101.

———. 1949. The detection of gamma-rays with thallium-activated sodium iodide crystals. *The Physical Review* 75:796–810.

Holl, Jack M. 1997. *Argonne National Laboratory, 1946–96*. Urbana and Chicago: University of Illinois Press.

Holmes, Frederic L. 2004. Seymour Benzer and the convergence of molecular biology with classical genetics. In *From Molecular Genetics to Genomics: The Mapping Cultures of Twentieth Century Genetics*, ed. Jean-Paul Gaudillière and Hans-Jörg Rheinberger, 42–62. London: Routledge.

Holmes, Frederic L., Jürgen Renn, and Hans-Jörg Rheinberger, eds. 2003. *Reworking the Bench: Research Notebooks in the History of Science*. Dordrecht: Kluwer.

Horrocks, Donald L., and Chin-Tzu Peng, eds. 1971. *Organic Scintillators and Liquid Scintillation Counting*. New York and London: Academic.

Hügel, Eckhardt. 1933. Über das genetische Verhalten der weißen Distalbinde und ihre genetischen Korrelationen zu anderen Merkmalen auf dem Vorderflügel der Mehlmotte Ephestia kühniella Z. *Wilhelm Roux' Archiv für Entwicklungsmechanik der Organismen* 130:202–42.

Hughes, Jeff A. 1993. *The Radioactivists: Community, Controversy, and the Rise of Nuclear Physics*. Ph.D. diss., University of Cambridge.

Husserl, Edmund. 1962. L'origine de la géométrie. In *Edmund Husserl, l'origine de la géométrie*, trans. and ed. Jacques Derrida. Paris: Presses universitaires de France.

———. 1970 (1954). *The Crisis of the European Sciences and Transcendental Phenomenology*. Trans. David Carr. Evanston, Ill.: Northwestern University Press.

Hyder, David. 2003. Foucault, Cavaillès, and Husserl on the historical epistemology of the sciences. *Perspectives on Science* 10 (4): 107–29.

Jablonka, Eva, and Marion J. Lamb. 1995. *Epigenetic Inheritance and Evolution*. Oxford: Oxford University Press.

Jacob, François. 1960. Genetic control of viral functions. In *The Harvey Lectures Series* 54, 1–39. New York: Academic Press.

———. 1965. *Leçon inaugurale au Collège de France* (May 7).

———. 1972 (1966). Genetics of the bacterial cell. In *Nobel Lectures: Physiology or Medicine, 1963–1970*, trans. anon., 148–71. Amsterdam: Elsevier.

———. 1977 (1974). The linguistic model in biology. In *Roman Jakobson*, ed. Daniel Armstrong and C. H. van Schooneveld, 185–92. Lisse: Peter De Ridder.

———. 1982 (1981). *The Possible and the Actual*. Seattle: University of Washington Press.

———. 1988 (1987). *The Statue Within: An Autobiography*. Trans. Franklin Philip. New York: Basic Books.

———. 1993 (1970). *The Logic of Life: A History of Heredity*. Trans. Betty E. Spillmann. Princeton: Princeton University Press.

———. 1998 (1997). *Of Flies, Mice, and Men: On the Revolution in Modern Biology*. Trans. Giselle Weiss. Cambridge, Mass.: Harvard University Press.

Jacob, François, and Jacques Monod. 1959. Gènes de structure et gènes de régulation dans la biosynthèse des protéines. *Comptes rendus hebdomadaires des séances de l'Académie des Sciences de Paris* 249:1282–84.

———. 1961. Genetic regulatory mechanisms in the synthesis of proteins. *Journal of Molecular Biology* 3:318–56.

Jacob, François, David Perrin, Carmen Sanchez, and Jacques Monod. 1960. L'opéron: Groupe de gènes à expression coordonnée par un opérateur. *Comptes rendus hebdomadaires des séances de l'Académie des Sciences de Paris* 250:1727–29.

Jahn, Ilse. 1957–58. Zur Geschichte der Wiederentdeckung der Mendelschen Gesetze. *Wissenschaftliche Zeitschrift der Friedrich-Schiller Universität Jena, mathematisch-naturwissenschaftliche Reihe* 7:215–27.

Jahrbuch der Max-Planck-Gesellschaft. 1961. Max-Planck-Institut für Virusforschung in Tübingen, Pt. 2.

Jansen, Sarah. 2003. 'Schädlinge': Geschichte eines wissenschaftlichen und politischen Konstrukts: 1840–1920. Frankfurt am Main: Campus.

Joerges, Bernward, and Terry Shinn, eds. 2001. *Instrumentation between Science, State, and Industry*. Dordrecht: Reidel.

Judson, Horace F. 1979. *The Eighth Day of Creation: Makers of the Revolution in Biology*. New York: Simon & Schuster.

Kaartinen, Niilo. 1969. *Packard Technical Bulletin* No. 18. Downers Grove, Ill.: Packard Instrument Co., Inc.

Kabara, Jon J., George T. Okita, and George V. LeRoy. 1958. Simultaneous use of H^3 and C^{14} compounds to study cholesterol metabolism. In *Liquid Scintillation Counting*, ed. Carlos G. Bell and Francis Newton Hayes, 191–97. New York: Pergamon.

Kaestner, Hans. 1931. Die Wirkung von Temperaturreizen auf die Pigmentierung und ihre Nachwirkung in den folgenden Generationen bei Habrobracon juglandis Ash. *Wilhelm Roux' Archiv für Entwicklungsmechanik der Organismen* 124:1–16.

Kallmann, Hartmut, and Carl A. Accardo. 1950. Coincidence experiments for noise reduction in scintillation counting. *Review of Scientific Instruments* 21:48–51.

Kallmann, Hartmut, and Milton Furst. 1950. Fluorescence of solutions bombarded with high energy radiation (Energy transport in liquids), Pt. I. *The Physical Review* 79:857–70.

———. 1951. Fluorescence of solutions bombarded with high energy radiation (Energy transport in liquids), Pt. II. *The Physical Review* 81:853–64.

Kamen, Martin D. 1963. Early history of carbon-14. *Science* 140:584–90.

Kamen, Martin D., and Samuel Ruben. 1940. Production and properties of carbon 14. *The Physical Review* 58:194.

Karlson, Peter. 1972. Das Häutungshormon der Insekten. In *Alfred Kühn zum Gedächtnis*, ed. G. Grasse, 183–92. Verband Deutscher Biologen, 5. Iserlohn: Biologisches Jahresheft.

———. 1990. *Adolf Butenandt: Biochemiker, Hormonforscher, Wissenschaftspolitiker.* Stuttgart: Wissenschaftliche Verlagsgesellschaft.

Kausche, Gustav A., Edgar Pfankuch, and Helmut Ruska. 1939. Die Sichtbarmachung von pflanzlichem Virus im Übermikroskop. *Die Naturwissenschaften* 27:292–99.

Kausche, Gustav A., and Hans Stubbe. 1939. Über die Entstehung einer mit Röntgenstrahlen induzierten 'Mutation' des Tabakmosaikvirus. *Die Naturwissenschaften* 27:501 f.

———. 1940. Zur Frage der Entstehung röntgenstrahleninduzierter Mutationen beim Tabakmosaikvirusprotein. *Die Naturwissenschaften* 28:824.

Kay, Lily E. 1986. W. M. Stanley's crystallization of the tobacco mosaic virus, 1930–1940. *Isis* 77:1–34.

———. 1988. The Tiselius electrophoresis apparatus and the life sciences, 1930–1945. *History and Philosophy of the Life Sciences* 10:51–72.

———. 1989. Selling pure science in wartime: The biochemical genetics of G. W. Beadle. *Journal of the History of Biology* 22:73–101.

———. 1993. *The Molecular Vision of Life: Caltech, the Rockefeller Foundation, and the Rise of the New Biology.* Oxford: Oxford University Press.

———. 2000. *Who Wrote the Book of Life? A History of the Genetic Code.* Stanford: Stanford University Press.

Keller, Evelyn Fox. 1995. *Refiguring Life: Metaphors of Twentieth-Century Biology.* New York: Columbia University Press.

———. 2000. *The Century of the Gene.* Cambridge, Mass.: Harvard University Press.

Khouvine, Yvonne, and Boris Ephrussi. 1937. Fractionnement des substances qui interviennent dans la pigmentation des yeux de Drosophila melanogaster. *Comptes rendus des séances de la Société de Biologie* 124:885–87.

Khouvine, Yvonne, Boris Ephrussi, and Simon Chevais. 1938. Development of eye colors in Drosophila: Nature of the diffusible substances; effects of yeast, peptones and starvation on their production. *The Biological Bulletin* 75:425–45.

Khouvine, Yvonne, Boris Ephrussi, and Morris Henry Harnly. 1936. Extraction et solubilité des substances intervenant dans la pigmentation des yeux de Drosophila melanogaster. *Comptes rendus hebdomadaires des séances de l'Académie des Sciences de Paris* 203:1542–44.

Kitcher, Philip. 1982. Genes. *British Journal for the Philosophy of Science* 33:337–59.

———. 1992. Gene: Current usages. In *Keywords in Evolutionary Biology*, ed. Evelyn Fox Keller and Elisabeth A. Lloyd, 128–31. Cambridge, Mass.: Harvard University Press.

Klebs, Georg. 1896. *Die Bedingungen der Fortpflanzung bei einigen Algen und Pilzen*. Jena: Fischer.

———. 1899. Ueber den Generationswechsel der Thallophyten. *Biologisches Centralblatt* 19:209–26.

Klein, Ursula. 2003. *Experiments, Models, Paper Tools—Cultures of Organic Chemistry in the Nineteenth Century*. Stanford: Stanford University Press.

Knorr Cetina, Karin. 1999. *Epistemic Cultures: How the Sciences Make Knowledge*. Cambridge, Mass.: Harvard University Press.

Kobayashi, Yutaka, and David V. Maudsley. 1974. *Biological Applications of Liquid Scintillation Counting*. New York: Academic.

Kohler, Robert. 1991. Systems of production: Drosophila, Neurospora and biochemical genetics. *Historical Studies in the Physical and Biological Sciences* 22:87–130.

———. 1994. *Lords of the Fly: Drosophila Genetics and the Experimental Life*. Chicago: University of Chicago Press.

Köhler, Wilhelm. 1932. Die Entwicklung der Flügel bei der Mehlmotte Ephestia kühniella Z. mit besonderer Berücksichtigung des Zeichnungsmusters. *Zeitschrift für Morphologie und Ökologie der Tiere* 24:582–681.

Kragh, Helge. 1987. *An Introduction to the Historiography of Science*. Cambridge: Cambridge University Press.

Krebs, Adolf. 1941. Ein Demonstrationsversuch zur Emanationsdiffusion. *Annalen der Physik* 39 (5): 330–32.

———. 1953. Szintillationszähler. *Ergebnisse der exakten Naturwissenschaften* 27:361–409.

———. 1955. Early history of the scintillation counter. *Science* 122:17–18.

Krige, John. 2006. Atoms for peace, scientific internationalism, and scientific intelligence. In *Global Power Knowledge: Science and Technology in International Affairs*, ed. John Krige and Kai Henrik Bork, *Osiris* 21:161–81.

Kühn, Alfred. 1908. Die Entwicklung der Keimzellen in den parthenogenetischen Generationen der Cladoceren Daphnia pulex De Geer und Polyphemus pediculus De Geer. *Archiv für Zellforschung* 1:2–50.

———. 1909. Sproßwachstum und Polypenknospung bei den Thecaphoren. Studien zur Ontogenie und Phylogenie der Hydroiden I. *Zoologische Jahrbücher (Abteilung für Anatomie und Ontogenie der Tiere)* 28:387–476.

———. 1910. Die Entwicklung der Geschlechtsindividuen der Hydromedusen. Studien zur Ontogenie und Phylogenie der Hydroiden II. *Zoologische Jahrbücher (Abteilung für Anatomie und Ontogenie der Tiere)* 30:43–174.

———. 1917. *Anleitung zu Tierphysiologischen Grundversuchen.* Leipzig: Quelle & Meyer.

———. 1919. *Die Orientierung der Tiere im Raum.* Jena: Fischer.

———. 1920. Untersuchungen zur kausalen Analyse der Zellteilung I. Zur Morphologie und Physiologie der Kernteilung von Vahlkampfia bistadialis. *Wilhelm Roux' Archiv für Entwicklungsmechanik der Organismen* 46:259–327.

———. 1926. Über die Änderung des Zeichnungsmusters von Schmetterlingen durch Temperaturreize und das Grundschema der Nymphalidenzeichnung. *Nachrichten von der Gesellschaft der Wissenschaften zu Göttingen, Mathematisch-Physikalische Klasse, Heft 2*, 120–41.

———. 1927. Die Pigmentierung von Habrobracon juglandis Ash., ihre Prädetermination und ihre Vererbung durch Gene und Plasmon. *Nachrichten von der Gesellschaft der Wissenschaften zu Göttingen, Mathematisch-Physikalische Klasse, Heft 4*, 407–21.

———. 1932. Entwicklungsphysiologische Wirkungen einiger Gene von Ephestia kühniella. *Die Naturwissenschaften* 20:974–77.

———. 1934. Erbkunde. In *Erbkunde, Rassenpflege, Bevölkerungspolitik*, ed. Heinz Woltereck, 1–96. Leipzig: Quelle & Meyer.

———. 1936a. Versuche über die Wirkungsweise der Erbanlagen. *Die Naturwissenschaften* 24:1–10.

———. 1936b. Weitere Untersuchungen über den Gen-A-Wirkstoff bei der Mehlmotte Ephestia kühniella Z. *Nachrichten von der Gesellschaft der Wissenschaften zu Göttingen, Mathematisch-Physikalische Klasse. Neue Folge. Fachgruppe VI. Nachrichten aus der Biologie* 2:239–49.

———. 1937. Entwicklungsphysiologisch-genetische Ergebnisse an Ephestia kühniella Z. *Zeitschrift für induktive Abstammungs- und Vererbungslehre* 73:419–55.

———. 1938. Grenzprobleme zwischen Vererbungsforschung und Chemie. *Berichte der Deutschen Chemischen Gesellschaft* 71:107–14.

———. 1939, 1950, 1961, and 1965. *Grundriss der Vererbungslehre.* Berlin: Quelle & Meyer.

———. 1941. Über eine Gen-Wirkkette der Pigmentbildung bei Insekten. *Nachrichten der Akademie der Wissenschaften in Göttingen, Mathematisch-Physikalische Klasse, Heft 6*, 231–61.

———. 1944a. Über eine die Schuppenformbildung und Schuppenpigmentierung beeinflussende Mutation (vd) von Ephestia kühniella. *Biologisches Zentralblatt* 64:81–97.

———. 1944b. Eine neue Mutation der Schuppenpigmentierung (ca) bei Ptychopoda seriata. *Biologisches Zentralblatt* 64:154–57.

———. 1948a. XIV. Neue Mutationen und Phänogenetik bei Tieren. In *FIAT Report on Science and Medicine: Naturforschung und Medizin in Deutschland 1939–1946, Volume 53, Part 2*, 77–93. Wiesbaden: Dieterich.

———. 1948b. XIX. Entwicklungsphysiologie der Wirbellosen. In *FIAT Report on Sci-*

ence and Medicine: Naturforschung und Medizin in Deutschland 1939-1946, Volume 53, Part 2, 157-80. Wiesbaden: Dieterich.

———. 1955. Versuche zur Entwicklung eines Modells der Genwirkungen. *Die Naturwissenschaften* 43:25-28.

———. 1959. *Nova Acta Leopoldina* 21:274-80.

———. 1960. Genetisch bedingte Mosaikbildungen bei Ephestia kühniella. *Zeitschrift für Vererbungslehre* 91:1-26.

———. 1964 (1922). *Grundriß der allgemeinen Zoologie.* 15th ed. Leipzig: Georg Thieme.

Kühn, Alfred, and Erich Becker. 1942. Quantitative Beziehungen zwischen zugeführtem Kynurenin und Augenpigment bei Ephestia kühniella Z. *Biologisches Zentralblatt* 62:303-17.

Kühn, Alfred, Ernst Caspari, and Ernst Plagge. 1935. Über hormonale Genwirkungen bei Ephestia kühniella Z. *Nachrichten von der Gesellschaft der Wissenschaften zu Göttingen, Mathematisch-Physikalische Klasse* 2:1-29.

Kühn, Alfred, and Melitta von Engelhardt. 1933. Über die Determination des Symmetriesystems auf dem Vorderflügel von Ephestia kühniella. *Wilhelm Roux' Archiv für Entwicklungsmechanik der Tiere* 130:660-703.

———. 1937. Über eine melanistische Mutation von Ptychopoda seriata Schrk. (at⁺ → At) und die Abhängigkeit der at⁺ und At zugeordneten Merkmale von Außenbedingungen. *Biologisches Zentralblatt* 57:329-47.

———. 1940. Ein das Flügelmuster beeinflussender Letalfaktor bei Ptychopoda seriata. *Biologisches Zentralblatt* 60:561-66.

———. 1943a. Über zwei melanistische Mutationen (At und ni) bei Ptychopoda seriata. *Biologisches Zentralblatt* 63:251-67.

———. 1943b. Über zwei Entschuppung bewirkende Mutationen (Vi und cal) bei Ptychopoda seriata. *Biologisches Zentralblatt* 63:470-78.

———. 1944. Mutationen und Hitzemodifikationen des Zeichnungsmusters von Ptychopoda seriata. *Biologisches Zentralblatt* 64:24-73.

———. 1946. Konstruktionsprinzipien von Schmetterlingsschuppen nach elektronenmikroskopischen Aufnahmen. *Zeitschrift für Naturforschung* 1:348-51.

Kühn, Alfred, and Karl Henke. 1929 (I-VII), 1932 (VIII-XII), 1936 (XIII-XIV). Genetische und entwicklungsphysiologische Untersuchungen an der Mehlmotte Ephestia kühniella Zeller, I-XIV. *Abhandlungen der Gesellschaft der Wissenschaften zu Göttingen, Mathematisch-Physikalische Klasse, n.s., Heft 15.* Berlin: Weidmannsche Buchhandlung.

———. 1930. Eine Mutation der Augenfarbe und der Entwicklungsgeschwindigkeit bei der Mehlmotte Ephestia Kühniella Z. *Wilhelm Roux' Archiv für Entwicklungsmechanik der Organismen* 122:204-12.

Kühn, Alfred, and Hans Piepho. 1936. Über hormonale Wirkungen bei der Verpuppung der Schmetterlinge. *Nachrichten von der Gesellschaft der Wissenschaften zu Göttingen, Mathematisch-Physikalische Klasse. Nachrichten aus der Biologie* 2:141-54.

Kühn, Alfred, and W. von Schuckmann. 1912. Cytologische Studien über Trypanosomen. *Zoologische Jahrbücher*, Supplement 15, vol. 2: 329-82.

Kühn, Alfred, and Victor Schwartz. 1942. Über eine weißäugige Mutante (wa) von Ephestia kühniella. *Biologisches Zentralblatt* 62:226–30.

Kühn, Alfred, and Wilhelm Trendelenburg. 1911. Die exogenen und endogenen Bahnen des Rückenmarks der Taube mit der Degenerationsmethode untersucht. *Archiv für Anatomie und Physiologie (Anatomische Abteilung)*: 35–48.

Kühn, Alfred, and Th. von Wasielewski. 1914. Untersuchungen über Bau und Teilung des Amöbenkernes. *Zoologische Jahrbücher (Abteilung für Anatomie und Ontogenie der Tiere)* 38:253–326.

Langendorff, Oskar. 1891. *Physiologische Graphik: Ein Leitfaden der in der Physiologie gebräuchlichen Registrirmethoden*. Leipzig, Wien: Deuticke.

Langham, Wright H. 1958. Application of liquid scintillation counting to biology and medicine. In *Liquid Scintillation Counting*, ed. Carlos G. Bell and Francis Newton Hayes, 135–49. New York: Pergamon.

Lanouette, William. 1992. *Genius in the Shadows: A Biography of Leo Szilard; The Man behind the Bomb*. New York: Scribner's.

Latour, Bruno. 1988. Drawing things together. In *Representation in Scientific Practice*, ed. Michael Lynch and Steve Woolgar, 19–68. Cambridge, Mass.: MIT Press.

———. 1990. The force and the reason of experiment. In *Experimental Inquiries*, ed. Homer E. Le Grand, 49–80. Dordrecht: Kluwer.

———. 1993 (1991). *We Have Never Been Modern*. Trans. Catherine Porter. Cambridge, Mass.: Harvard University Press.

———. 1995 (1993). 'The Pédofil' of Boa Vista: A photo-philosophical montage. Trans. Bart Simon and Katia Verresen. *Common Knowledge* 4 (1): 144–87.

Latour, Bruno, and Steve Woolgar. 1986 (1979). *Laboratory Life: The Construction of Scientific Facts*. Princeton: Princeton University Press.

Laubichler, Manfred. 2000. The organism is dead. Long live the organism! *Perspectives on Science* 8:286–315.

Lecourt, Dominique. 1975 (1969). Gaston Bacherlard's historical epistemology. In *Marxism and Epistemology: Bachelard, Canguilhem, Foucault*, by Dominique Lecourt, trans. Ben Brewster, 23–116. London: Verso.

———. 1975 (1972). Georges Canguilhem's historical epistemology. In *Marxism and Epistemology: Bachelard, Canguilhem, Foucault*, by Dominique Lecourt, trans. Ben Brewster, 162–86. London: Verso.

———, ed. 2001 (1971). *Bachelard: Epistémologie*. Paris: Presses universitaires de France.

———. 2002. *L'épistémologie historique de Gaston Bachelard*. 11th ed., with an afterword by Dominique Lecourt. Paris: Vrin.

Lenoir, Timothy, ed. 1998. *Inscribing Science: Scientific Texts and the Materiality of Communication*. Stanford: Stanford University Press.

Lenoir, Timothy, and Marguerite Hays. 2000. The Manhattan project for biomedicine. In *Controlling Our Destinies: Historical, Philosophical, Social and Ethical Perspectives on the Human Genome Project*, ed. Phillip R. Sloan. South Bend, Ind.: University of Notre Dame Press.

Lenoir, Timothy, and Christophe Lécuyer. 1997. Instrument makers and discipline builders: The case of nuclear magnetic resonance. In *Instituting Science: The Cultural Production of Scientific Disciplines*, ed. Timothy Lenoir, 239-92. Stanford: Stanford University Press.

Lévi-Strauss, Claude. 1966 (1962). *The Savage Mind*. Chicago: University of Chicago Press.

Lewis, Jeffrey. 2002. *Continuity in German Science, 1937-1972*. Ph.D. diss., Ohio State University, Columbus.

Libby, Willard Frank. 1952. *Radiocarbon Dating*. Chicago: University of Chicago Press.

Lindsley, Dan L., and Ed H. Grell. 1967. *Genetic Variants of Drosophila melanogaster*. Carnegie Institution of Washington, Publication No. 627.

Loftfield, Robert B., and Elizabeth A. Eigner. 1960. Scintillation counting of paper chromatograms. *Biochemical and Biophysical Research Communications* 2:72-75.

Lohff, Brigitte, and Hinderk Conrads. 2007. *From Berlin to New York: Life and Work of the almost Forgotten German-Jewish Biochemist Carl Neuberg (1877-1956)*. Stuttgart: Steiner.

Lösch, Niels C. 1997. *Rasse als Konstrukt: Leben und Werk Eugen Fischers*. Frankfurt am Main: Peter Lang.

Luntz, Jerome D. 1957. The story of a magazine and an industry. *Nucleonics* 15 (9): 78-83.

Lwoff, André. 1957. The concept of virus. *Journal of General Microbiology* 17:239-53.

Macho, Thomas, and Annette Wunschel, eds. 2004. *Science & Fiction: Über Gedankenexperimente in Wissenschaft, Philosophie und Literatur*. Frankfurt: Fischer.

Macrakis, Kristie. 1993a. The survival of basic biological research in National Socialist Germany. *Journal of the History of Biology* 26:519-43.

———. 1993b. *Surviving the Swastika: Scientific Research in Nazi Germany*. New York: Oxford University Press.

Mandeville, Charles E., and Morris V. Scherb. 1950. Photosensitive Geiger counters: Their applications. *Nucleonics* 7 (5): 34-38.

Mamoli, Luigi. 1939. Über biochemische Dehydrierungen in der Cortingruppe. *Berichte der deutschen chemischen Gesellschaft* 72:1863-65.

Maupas, Émile. 1888. Recherches expérimentales sur la multiplication des infusoires ciliés. *Archives de zoologie expérimentale et générale*, Deuxième Série, Tome Sixième, 165-227. Paris: Librairie de C. Reinwald.

———. 1889. Le rajeunissement karyogamique chez les ciliés. *Archives de zoologie expérimentale et générale*. Deuxième Série, Tome Septième, 149-517. Paris: Librairie de C. Reinwald.

Mayer, Andreas. 2005. Autographien des Ganges: Repräsentation und Redressement bewegter Körper im 19. Jahrhundert. In *Kunstmaschinen: Spielräume des Sehens zwischen Wissenschaft und Kunst*, ed. Andreas Mayer and Alexandre Métaux, 101-38. Frankfurt: Fischer.

———. Forthcoming. Redressment by representation. Locomotion physiology, graphic methods, and equestrian science in nineteenth-century France. *Representation*.

McNeill, William H. 1991. *Hutchins' University: A Memoir of the University of Chicago 1929–1950*. Chicago: University of Chicago Press.

Meijer, Onno G. 1985. Hugo de Vries no Mendelian? *Annals of Science* 42:189–232.

Melchers, Georg. 1936. Versuche zur Genetik und Entwicklungsphysiologie der Blühreife. *Biologisches Zentralblatt* 56:567–70.

———. 1938a. Die Wirkung von Genen, tiefen Temperaturen und blühenden Pfropfpartnern auf die Blühreife von Hyoscyamus niger L. *Biologisches Zentralblatt* 57:568–614.

———. 1938b. Die Auslösung von Blütenbildung an zweijährigen Pflanzen im ersten Sommer durch implantierte Reiser selbst nicht blühfähiger Kurztagspflanzen in Langtagbedingungen. *Die Naturwissenschaften* 26:496.

———. 1939. Die Blühhormone. *Berichte der Deutschen Botanischen Gesellschaft* 57:29–48.

———. 1942. Über einige Mutationen des Tabakmosaikvirus und eine 'Parallelmutation' des Tomatenmosaikvirus. *Die Naturwissenschaften* 30:48 f.

———. 1943. Die Darstellung des Nucleoproteids des Kartoffel-Y-Virus. *Berichte der Deutschen Botanischen Gesellschaft* 61:89 f.

———. 1948. Phytopathogene Viren: Naturforschung und Medizin in Deutschland 1939–1946. In FIAT *Review of German Science, Volume 52: Biologie*, Pt. 1: 111–29.

———. 1987. Ein Botaniker auf dem Wege in die Allgemeine Biologie auch in Zeiten moralischer und materieller Zerstörung und Fritz von Wettstein 1895–1945 mit Liste der Veröffentlichungen und Dissertationen (Persönliche Erinnerungen). *Berichte der Deutschen botanischen Gesellschaft* 100:373–405.

Melchers, Georg, and Hedwig Claes. 1943. Auslösung von Blütenbildung bei der Langtagpflanze Hyoscyamus niger in Kurztagsbedingungen durch Hemmung der Atmung in den Dunkelphasen. *Die Naturwissenschaften* 31:249.

Melchers, Georg, and Anton Lang. 1941. Weitere Untersuchungen zur Frage der Blühhormone. *Biologisches Zentralblatt* 61:16–39.

Melchers, Georg, and Gerhard Schramm. 1940. Über den Verlauf der Viruskrankheit in anfälligen und resistenten Rassen von Nicotiana tabacum. *Die Naturwissenschaften* 28:476–78.

Melchers, Georg, Gerhard Schramm, Hans Joachim Trurnit, and Hans Friedrich-Freksa. 1940. Die biologische, chemische und elektronenmikroskopische Untersuchung eines Mosaikvirus aus Tomaten. *Biologisches Zentralblatt* 60:524–56.

Mendel, Gregor. 1901 (1866). Experiments in plant hybridization. *Journal of the Royal Horticultural Society* 26:1–32. Trans. William Bateson.

Moles, Abraham A. 1995. *Les sciences de l'imprécis*. Paris: Seuil.

Monod, Jacques, and François Jacob. 1961. General conclusions: Teleonomic mechanisms in cellular metabolism, growth, and differentiation. *Cold Spring Harbor Symposia on Quantitative Biology* 26:389–401.

Morange, Michel. 1990. Le concept de gène régulateur. In *Histoire de la génétique, pratiques, techniques et théories*, ed. Jean-Louis Fischer and William H. Scheider, 271–91. Paris: A.R.P.E.M.

———. 1998 (1994). *A History of Molecular Biology*. Trans. Matthew Cobb. Cambridge, Mass.: Harvard University Press.

Morgan, Thomas Hunt, Calvin B. Bridges, and Alfred H. Sturtevant. 1925. The genetics of Drosophila, chap. 6: Modifying factors and selection. In *Biobliographia Genetia, Volume 2*, ed. Johannes Paulus Lotsy and Havik Nicolaas Kooiman, 1–262. 's Gravenhage: Martinus Nijhoff.

Morton, George A., and Jane A. Mitchell. 1949. Performance of 931-A type multiplier as a scintillation counter. *Nucleonics* 4 (1): 16–23.

Morton, George A., and Kenneth W. Robinson. 1949. A coincidence scintillation counter. *Nucleonics* 4 (2): 25–29.

Muller, Hermann J. 1927. Artificial transmutation of the gene. *Science* 66:84–87.

———. 1951. The development of the gene theory. In *Genetics in the Twentieth Century*, ed. Leslie C. Dunn. New York: Macmillan.

Munk, Klaus. 1995. *Virologie in Deutschland: Die Entwicklung eines Fachgebietes*. Basel: Karger.

Myers, Greg. 1990. *Writing Biology*. Madison: University of Wisconsin Press.

Navashin, Sergej G. 1898. Resultate einer Revision der Befruchtungsvorgänge bei Lilium Martagon und Fritillaria tenella. *Bulletin de l'Académie Impériale des Sciences de Saint-Pétersbourg* 9:377–82.

———. 1899. Neue Beobachtungen über Befruchtung bei Fritillaria tenella und Lilium Martagon. *Botanisches Centralblatt* 77:62.

Nirenberg, Marshall W., and Heinrich J. Matthaei. 1961. The dependence of cell-free protein synthesis in *E. coli* upon naturally occurring or synthetic polyribonucleotides. *Proceedings of the National Academy of Sciences of the United States of America* 47:1588–1602.

Nordwig, Arnold. 1983. Vor fünfzig Jahren: Der Fall Neuberg. Aus der Geschichte des Kaiser-Wilhelm-Instituts für Biochemie zur Zeit des Nationalsozialismus. *MPG-Spiegel* 6:49–53.

Okita, George T., Jon J. Kabara, Florence Richardson, and George V. LeRoy. 1957. Assaying compounds containing H^3 and C^{14}. *Nucleonics* 15 (6): 111–14.

Olby, Robert C. 1966. *Origins of Mendelism*. New York: Schocken.

———. 1971. Correns. *Dictionary of Scientific Biography*, 421–23. New York: Scribners.

Orel, Vítezslaw. 1996. *Gregor Mendel: The First Geneticist*. Oxford: Oxford University Press.

Oster, Gerald. 1966. A young physicist at seventy: Hartmut Kallmann. *Physics Today* (April): 51–54.

Packard Instrument Company. *Annual Report 1965: Ten Year Financial Highlights*.

Packard, Lyle E. 1958. Instrumentation for internal sample liquid scintillation counting. In *Liquid Scintillation Counting*, ed. Carlos G. Bell and Francis Newton Hayes, 50–66. New York: Pergamon.

Pardee, Arthur B. 1985. Molecular basis of gene expression: Origins from the Pajama experiment. *BioEssays* 2:86–89.

Pardee, Arthur B., François Jacob, and Jacques Monod. 1958. Sur l'expression et le rôle des allèles 'inductible' et 'constitutif' dans la synthèse de la β-galactosidase chez des zygotes d'Escherichia coli. *Comptes rendus de l'Académie des Sciences Paris* 246:3125–28.

———. 1959. The genetic control and cytoplasmic expression of 'inducibility' in the synthesis of β—galactosidase by *E. coli*. *Journal of Molecular Biology* 1:165–78.

Pattee, Howard. 1969. How does a molecule become a message? *Developmental Biology Supplement* 3:1–16.

Petersen, Donald. 1995. Los Alamos radiation detectors. *Los Alamos Science* 23:274–79.

Peyrieras, Nadine, and Michel Morange. 2002. *Travaux scientifiques de François Jacob*. Paris: Editions Odile Jacob.

Pfankuch, Edgar. 1940. Über die Spaltung von Virusproteinen der Tabakmosaik-Gruppe. *Biochemische Zeitschrift* 306:125–29.

Pfankuch, Edgar, Gustav A. Kausche, and Hans Stubbe. 1940. Über die Entstehung, die biologische und physikalisch-chemische Charakterisierung von Röntgen- und (gamma)-Strahlen induzierten 'Mutationen' des Tabakmosaikvirus-proteins. *Biochemische Zeitschrift* 304:238–58.

Picard, Jean-François. 1999. *La Fondation Rockefeller et la recherche médicale*. Paris: Presses universitaires de France.

Pickering, Andrew, ed. 1992. *Science as Practice and Culture*. Chicago: University of Chicago Press.

Piepho, Hans. 1942. Untersuchungen zur Entwicklungsphysiologie der Insekten-metamorphose. Über die Puppenhäutung der Wachsmotte Galleria mellonella L. *Wilhelm Roux' Archiv für Entwicklungsmechanik der Organismen* 141:500–583.

———. 1972. Hormonphysiologie der Schmetterlingsmetamorphose. In *Alfred Kühn zum Gedächtnis*, ed. G. Grasse, 177–82. Verband Deutscher Biologen, 5. Iserlohn: Biologisches Jahresheft.

Plagge, Ernst. 1935. Die Pigmentierung der Imaginal- und Raupenaugen der Mehl-motte Ephestia kühniella Z. bei verschiedenen Rassen, Transplantatträgern und Rassenkreuzungen. *Wilhelm Roux' Archiv für Entwicklungsmechanik der Organismen* 132:648–70.

———. 1936a. Bewirkung der Augenausfärbung der rotäugigen Rasse von Ephestia kühniella durch Implantation artfremder Hoden. *Nachrichten von der Gesellschaft der Wissenschaften zu Göttingen, Mathematisch-Physikalische Klasse. Neue Folge. Fachgruppe VI. Nachrichten aus der Biologie* 2:251–56.

———. 1936b. Transplantationen von Augenimaginalscheiben zwischen der schwarz- und der rotäugigen Rasse von Ephestia kühniella Z. *Biologisches Zentralblatt* 56:406–9.

———. 1936c. Der zeitliche Verlauf der Auslösbarkeit von Hoden- und Imaginal-augenfärbung durch den A-Wirkstoff bei Ephestia und die zur Ausscheidung einer wirksamen Menge nötige Zeitdauer. *Zeitschrift für induktive Abstammungs- und Vererbungslehre* 72:127–37.

Polanyi, Michael. 1969. Life's irreducible structure. In *Knowing and Being: Essays by*

Michael Polanyi, ed. Marjorie Grene, 225–39. Chicago: University of Chicago Press.

Portin, Petter. 1993. The concept of the gene: Short history and present status. *The Quarterly Review of Biology* 68:173–223.

Portugal, Franklin H., and Jack S. Cohen. 1977. *A Century of DNA*. Cambridge, Mass.: MIT Press.

Prigogine, Ilya, and Isabelle Stengers. 1979. *La nouvelle alliance*. Paris: Gallimard.

Pringle, Robert W. 1950. The scintillation counter. *Nature* 166:11–14.

Proceedings of the Symposium on Tritium in Tracer Applications, sponsored by New England Nuclear, Atomic Associates, Packard Instrument, New York, November 22, 1957, and October 31, 1958.

Raben, Maury S., and Nicolaas Bloembergen. 1951. Determination of radioactivity by solution in a liquid scintillator. *Science* 114:363–64.

Rabinow, Paul. 1996. *Making PCR: A Story of Biotechnology*. Chicago: University of Chicago Press.

———. 2000. Epochs, presents, events. In *Living and Working with the New Medical Technologies: Intersections of Inquiry*, ed. Margaret Lock, Alan Young, and Alberto Cambrosio, 31–46. Cambridge Studies in Medical Anthropology 8. Cambridge: Cambridge University Press.

———. 2004. Anthropologie des Zeitgenössischen. Ein Gespräch mit Paul Rabinow. In *Anthropologie der Vernunft*, ed. and trans. Carlo Caduff and Tobias Rees, 56–65. Frankfurt am Main: Suhrkamp.

Rapkin, Edward. 1961. Hydroxide of hyamine 10-X. *Technical Bulletin Number 3*, revised June 1961.

———. 1970. Development of the modern liquid scintillation counter. In *The Current Status of Liquid Scintillation Counting*, ed. Edwin D. Bransome, 45–68. New York: Grune & Stratton.

Rapkin, Edward, and Lyle E. Packard. 1961. New accessories for liquid scintillation counting. In *Proceedings of the University of New Mexico Conference on Organic Scintillation Detectors, August 15–17, 1960*, ed. Guido H. Daub, F. Newton Hayes, and Elizabeth Sullivan, 216–31. Washington, D.C.: U.S. Government Printing Office.

Rasmussen, Nicolas. 1993. Facts, artifacts, and mesosomes: Practicing epistemology with the electron microscope. *Studies in History and Philosophy of Science* 24:227–65.

———. 1997. *Picture Control: The Electron Microscope and the Transformation of Biology in America, 1940–1960*. Stanford: Stanford University Press.

Regener, Erich. 1908. Über Zählung der α-Teilchen durch die Szintillation und die Größe des elektrischen Elementarquantums. *Verhandlungen der deutschen physikalischen Gesellschaft* 10:78–83.

Reines, Frederick. 1958. Giant liquid scintillation detectors and their applications. In *Liquid Scintillation Counting*, ed. Carlos G. Bell and Francis Newton Hayes, 246–57. New York: Pergamon.

Reinhardt, Carsten. 2006. *Shifting and Rearranging: Physical Methods and the Trans-*

formation of Modern Chemistry. Sagamore Beach, Mass.: Science History Publications.

Renard, Gilles. 1996. *L'épistémologie chez Georges Canguilhem*. Paris: Editions Nathan.

Revue d'histoire des sciences 53 (1) (2000), special issue, Canguilhem en son temps: Etudes et témoignages.

Reynolds, George T., Frank B. Harrison, and Giorgio Salvini. 1950. Liquid scintillation counters. *The Physical Review* 78:488.

Rheinberger, Hans-Jörg. 1993. Vom Mikrosom zum Ribosom: 'Strategien' der 'Repräsentation' 1935–1955. In *Die Experimentalisierung des Lebens*, ed. Hans-Jörg Rheinberger and Michael Hagner, 162–87. Berlin: Akademie Verlag.

———. 1995. When did Carl Correns read Gregor Mendel's paper? A research note. *Isis* 86:612–16.

———. 1997. *Toward a History of Epistemic Things: Synthesizing Proteins in the Test Tube*. Stanford: Stanford University Press.

———. 1998. Experimental systems—graphematic spaces. In *Inscribing Science*, ed. Timothy Lenoir and Hans Ulrich Gumbrecht, 285–303. Stanford: Stanford University Press.

———. 2000a. Kurze Geschichte der Molekularbiologie. In *Geschichte der Biologie*, ed. Ilse Jahn, 642–63. Heidelberg: Spektrum Akademischer Verlag.

———. 2000b. Mendelian inheritance in Germany between 1900–1910. The case of Carl Correns (1864–1933). *Comptes rendus de l'Académie des Sciences, Série III, Sciences de la vie* 323:1089–96.

———. 2002. Afterword to François Jacob, *Die Logik des Lebenden*, 345–54. Frankfurt: S. Fischer.

———. 2003. 'Präparate'—'Bilder' ihrer selbst: Eine bildtheoretische Glosse. *Bildwelten des Wissens, Kunsthistorisches Jahrbuch für Bildkritik* 1/2:9–19.

———. 2004. Die Zusammenarbeit zwischen Adolf Butenandt und Alfred Kühn. In *Adolf Butenandt und die Kaiser-Wilhelm-Gesellschaft: Wissenschaft, Industrie und Politik im "Dritten Reich,"* ed. Wolfgang Schieder and Achim Trunk, 169–97. Göttingen: Wallstein.

———. 2010. *On Historicizing Epistemology: An Essay*. Stanford: Stanford University Press.

Rheinberger, Hans-Jörg, and Staffan Müller-Wille. 2004. Gene. Stanford Encyclopedia of Philosophy, http://plato.stanford.edu.

Rheingans, Friedrich G. 1988. *Hans Geiger und die elektrischen Zählmethoden, 1908–1928*. Berlin: D.A.V.I.D. Verlagsgesellschaft.

Riezler, Kurt. 1928. Die Krise der 'Wirklichkeit.' *Die Naturwissenschaften* 16:705–12.

Riley, Monica, Arthur B. Pardee, François Jacob, and Jacques Monod. 1960. On the expression of a structural gene. *Journal of Molecular Biology* 2:216–25.

Rimpau, Wilhelm. 1884. Die Kreuzung als Mittel zur Erzeugung neuer Varietäten von landwirthschaftlichen Culturpflanzen. *Tageblatt der 57. Versammlung Deutscher Naturforscher und Ärzte*, 179–86. Magdeburg: Faber.

Risler, Helmut. 1972. Alfred Kühn zum Gedächtnis. In *Alfred Kühn zum Gedächtnis*.

5. *Biologisches Jahresheft*, ed. G. Grasse, 47–51. Iserlohn: Verband Deutscher Biologen.

Roberts, Herbert F. 1929. *Plant Hybridization before Mendel*. Princeton: Princeton University Press.

Rothchild, Seymour, ed. 1963–1968. *Advances in Tracer Methodology*, 4 vols. New York: Plenum.

Ruben, Samuel, and Martin D. Kamen. 1940. Radioactive carbon of long half-life. *The Physical Review* 57:549.

Russo, Vincenzo E. A., Robert A. Martienssen, and Arthur D. Riggs, eds. 1996. *Epigenetic Mechanisms of Gene Regulation*. New York: Cold Spring Harbor Laboratory Press.

Sapp, Jan. 1987. *Beyond the Gene: Cytoplasmic Inheritance and the Struggle for Authority in Genetics*. Oxford: Oxford University Press.

Sarkar, Sahotra. 1996. Biological information: A skeptical look at some central dogmas of molecular biology. In *The Philosophy and History of Molecular Biology*, ed. Sahorta Sarkar, 187–231. Dordrecht: Kluwer.

Schaudinn, Fritz. 1905. Neuere Forschungen über die Befruchtung bei Protozoen. *Verhandlungen der Deutschen Zoologischen Gesellschaft* 21:16–35.

Schickore, Jutta. 2002a. Fixierung mikroskopischer Beobachtungen: Zeichnung, Dauerpräparat, Mikrofotografie. In *Ordnungen der Sichtbarkeit: Fotografie in Wissenschaft, Kunst und Technologie*, ed. Peter Geimer, 285–310. Frankfurt: Suhrkamp.

———. 2002b. (Ab)using the past for present purposes: Exposing contextual and trans-contextual features of error. *Perspectives on Science* 10:443–56.

———. 2007. *The Microscope and the Eye: A History of Reflections, 1740–1870*. Chicago: University of Chicago Press.

Schiemann, Gregor. 1997. Geschichte und Natur zwischen Differenz und Konvergenz. In *Geschichtsdiskurs, Volume 4, Krisenbewußtsein, Katastrophenerfahrungen und Innovationen, 1880–1945*, ed. Wolfgang Küttler, Jörn Rüsen, and Ernst Schulin, 153–61. Frankfurt am Main: Fischer.

Schleiden, Matthias Jacob. 1849 (1846). *Principles of Scientific Botany, or, Botany as an Inductive Science*, vol. 1. London: Longman, Brown, Green & Longmans.

Schlottke, Egon. 1925. Über die Variabilität der schwarzen Pigmentierung und ihre Beeinflußbarkeit durch Temperaturen bei Habrobracon juglandis Ashmead. *Zeitschrift für vergleichende Physiologie* 3:692–736.

Scholthof, Karen-Beth G., John G. Shaw, and Milton Zaitlin, eds. 1999. *Tobacco Mosaic Virus*. St. Paul, Minn.: APS.

Schram, Eric, and Robert Lombaert. 1963. *Organic Scintillation Detectors: Counting of Low-Energy Beta Emitters*. Amsterdam: Elsevier.

Schramm, Gerhard. 1941a. Die luftgetriebene Ultrazentrifuge. *Kolloid-Zeitschrift* 97:106–15.

———. 1941b. Enzymatische Abspaltung der Nucleinsäure aus dem Tabakmosaikvirus. *Berichte der Deutschen Chemischen Gesellschaft* 74:532–36.

———. 1942a. Neuere Ergebnisse und Probleme in der Untersuchung der Virus-arten. *Deutsche Medizinische Wochenschrift* 68:791–94.

———. 1942b. Viruskranke Pflanzen. *Das Reich* 16 (April 16).

———. 1943. Über die Spaltung des Tabakmosaikvirus in niedermolekulare Pro-teine und die Rückbildung hochmolekularen Proteins aus den Spaltstücken. *Die Naturwissenschaften* 31:94–96.

———. 1944. Über die Konstitution des Tabakmosaikvirus. *Die Chemie* 57:109–13.

Schramm, Gerhard, and Gernot Bergold. 1947. Über das Molekulargewicht des Tabakmosaikvirus. *Zeitschrift für Naturforschung* 2b:108–12.

Schramm, Gerhard, Gernot Bergold, and Heinrich Flammersfeld. 1946. Zur Konsti-tution der Hefenucleinsäure. *Zeitschrift für Naturforschung* 1:328–36.

Schramm, Gerhard, Hans-Joachim Born, and Anton Lang. 1942. Versuch über den Phosphoraustausch zwischen phosphorhaltigem Tabakmosaikvirus und Natriumphosphat. *Die Naturwissenschaften* 30:170 f.

Schramm, Gerhard, and Heinz Dannenberg. 1944. Über die Ultraviolettabsorption des Tabakmosaikvirus. *Berichte der Deutschen Chemischen Gesellschaft* 77:53–60.

Schramm, Gerhard, and Hans Friedrich-Freksa. 1941. Die Präcipitinreaktion des Tabakmosaikvirus mit Kaninchen- und Schweineantiserum. *Hoppe Seyler's Zeit-schrift für physiologische Chemie* 270:233–46.

Schramm, Gerhard, and Hans Müller. 1940a. Zur Chemie des Tabakmosaikvirus. Über die Einwirkung von Keten und Phenylisocyanat auf das Virusprotein. *Hoppe-Seyler's Zeitschrift für physiologische Chemie* 266:43–55.

———. 1940b. Über die Konfiguration der im Tabakmosaikvirus enthaltenen Aminosäuren. *Die Naturwissenschaften* 28:223 f.

———. 1942. Über die Bedeutung der Aminogruppen für die Vermehrungsfähig-keit des Tabakmosaikvirus. *Hoppe-Seyler's Zeitschrift für physiologische Chemie* 274:267–75.

Schramm, Gerhard, and L. Rebensburg. 1942. Zur vergleichenden Charakterisier-ung einiger Mutanten des Tabakmosaikvirus. *Die Naturwissenschaften* 30:48–51.

Schüring, Michael. 2004. Der Vorgänger: Carl Neubergs Verhältnis zu Adolf Buten-andt. In *Adolf Butenandt und die Kaiser-Wilhelm-Gesellschaft: Wissenschaft, Indus-trie und Politik im "Dritten Reich,"* ed. Wolfgang Schieder and Achim Trunk, 346–68. Göttingen: Wallstein.

Seidel, Friedrich. 1924. Die Geschlechtsorgane in der embryonalen Entwicklung von Pyrrhocoris apterus L. *Zeitschrift für Morphologie und Ökologie der Tiere* 1:429–506.

Shapin, Steven, and Simon Schaffer. 1985. *Leviathan and the Air-Pump: Hobbes, Boyle, and the Experimental Life.* Princeton: Princeton University Press.

Shinn, Terry. 1997. Crossing boundaries: The emergence of research-technology communities. In *Universities and the Global Knowledge Economy,* ed. Henry Etzko-witz and Loet Leydesdorff, 85–96. London: Pinter.

Singer, Charles. 1950. *History of Biology: A General Introduction to the Study of Living Things.* New York: Schuman.

Stadler, Lewis J. 1928. Genetic effects of x-rays in maize. *Proceedings of the National Academy of Sciences of the United States of America* 14:69–75.

Stanley, Wendell M., and Thomas F. Anderson. 1941. A study of purified viruses with the electron microscope. *Journal of Biological Chemistry* 139:325–38.

Star, Susan Leigh, and James R. Griesemer. 1988. Institutional ecology, 'translations,' and boundary objects: Amateurs and professionals in Berkeley's Museum of Vertebrate Zoology, 1907–39. *Social Studies of Science* 19:387–420.

Stein, Emmy. 1950. Dem Gedächtnis von Carl Erich Correns nach einem halben Jahrhundert der Vererbungswissenschaft. *Die Naturwissenschaften* 37:457–63.

Steinle, Friedrich. 2005. *Explorative Experimente*. Stuttgart: Franz Steiner.

Stent, Gunther. 1977. Explicit and implicit semantic content of the genetic information. In *Foundational Problems in the Special Sciences*, ed. Robert E. Butts and Jaakko Hintikka, 131–49. Dordrecht: Reidel.

Stöffler, Georg, Rolf Bald, Berthold Kastner, Reinhard Lührmann, Marina Stöffler-Meilicke, and Gilbert W. Tischendorf. 1980. Structural organization of the Escherichia coli ribosome and localization of funcional domains. In *Ribosomes, Structure, and Genetics*, ed. Glenn Chambliss, Gary R. Craven, Julian Davies, K. Davis, Lawrence Kahan, and Masayasu Nomura, 171–205. Baltimore: University Park Press.

Strasser, Bruno. 2002. Microscopes électroniques, totems de laboratoires et réseaux transnationaux: L'émergence de la biologie moléculaire à Genève (1945–1960). *Revue d'histoire des sciences* 55:5–53.

———. 2006. *La fabrique d'une nouvelle science: La biologie moléculaire à l'âge atomique*. Florence: Olschki.

Stumpf, Hans-Friedrich. 1995. Kernenergieforschung in Celle 1944/45. In *Celler Beiträge zur Landes- und Kulturgeschichte*, 30–32. Celle.

Sturtevant, Alfred H. 1920. The vermilion gene and gynandromorphism. *Proceedings of the Society of Experimental Biology and Medicine* 17:70–71.

———. 1927. The effect of the bar gene of Drosophila in mosaic eyes. *Journal of Experimental Zoology* 46:493–98.

Szöllösi-Janze, Margit. 1994. Von der Mehlmotte zum Holocaust. Fritz Haber und die chemische Schädlingsbekämpfung während und nach dem Ersten Weltkrieg. In *Von der Arbeiterbewegung zum modernen Sozialstaat*, ed. Jürgen Kocka, Hans-Jürgen Puhle, and Klaus Tenfelde, 658–82. Munich: K. G. Saur.

Tatum, Edward L. 1939. Development of eye-colors in Drosophila: Bacterial synthesis of v[+] hormone. *Proceedings of the National Academy of the Sciences of the United States of America* 25:486–90.

Tatum, Edward L., and George W. Beadle. 1938. Development of eye colors in Drosophila: Some properties of the hormones concerned. *Journal of General Physiology* 22:239–53.

Tatum, Edward L., and Arie J. Haagen-Smit. 1941. Identification of Drosophila v[+] hormone of bacterial origin. *Journal of Biological Chemistry* 140:575–80.

Thieffry, Denis. 1997. Contributions of the 'Rouge-Cloître Group' to the notion of 'messenger RNA.' *History and Philosophy of the Life Sciences* 19:89–113.

Thimann, Kenneth V., and George W. Beadle. 1937. Development of eye colors in

Drosophila: Extraction of the diffusible substances concerned. *Proceedings of the National Academy of Sciences of the United States of America* 23:143–46.

Timofeev-Ressovskii, Nikolai. 1939. Genetik und Evolution. *Zeitschrift für induktive Abstammungs- und Vererbungslehre* 76:158–219.

Timofeev-Ressovskii, Nikolai, Karl Zimmer, and Max Delbrück. 1935. Über die Natur der Genmutation und der Genstruktur. *Nachrichten von der Gesellschaft der Wissenschaften zu Göttingen, Mathematisch-physikalische Klasse: Nachrichten aus der Biologie, n. s. 1, Heft 13,* 190–245.

Trenn, Thaddeus J. 1976. Die Erfindung des Geiger-Müller-Zählrohres. *Deutsches Museum, Abhandlungen und Berichte* 44:54–64.

———. 1986. The Geiger-Müller counter of 1928. *Annals of Science* 43:111–35.

Troeltsch, Ernst. 1922. *Der Historismus und seine Probleme.* Tübingen: J. C. B. Mohr.

Tschermak-Seysenegg, Erich von. 1900. Über künstliche Kreuzung bei Pisum sativum. *Berichte der Deutschen Botanischen Gesellschaft* 18:232–39.

Umbach, Wilhelm. 1934. Entwicklung und Bau des Komplexauges der Mehlmotte Ephestia kühniella Z. *Zeitschrift für Morphologie und Ökologie der Tiere* 28:561–94.

University of Chicago Publication Office. 1991. *One in Spirit: A Retrospective View of the University of Chicago on the Occasion of its Centennial.* Chicago: University of Chicago Publication Office.

Utting, George R. 1958. The design of a commercial liquid scintillation coincidence counter. In *Liquid Scintillation Counting,* ed. Carlos G. Bell and Francis Newton Hayes, 67–73. New York: Pergamon.

Vischer, Ernst, Stephen Zamenhoff, and Erwin Chargaff. 1949. Microbial nucleic acids: The desoxypentose nucleic acids of avian tubercle bacilli and yeast. *Journal of Biological Chemistry* 177:429–38.

Vries, Hugo de. 1899. Sur la fécondation hybride de l'albumen. *Comptes rendus hebdomadaires des séances de l'Académie des Sciences de Paris* 973–75.

———. 1900a. Sur la loi de disjonction des hybrides. *Comptes rendus de l'Académie des Sciences Paris* 130:845–47.

———. 1900b. Sur la fécondation hybride de l'endosperme chez le Maïs. *Revue générale de Botanique* 12:129–37.

Wagner, Helmuth O. 1931. Samen- und Eireifung der Mehlmotte Ephestia kühniella Z. *Zeitschrift für Zellforschung und mikroskopische Anatomie* 12:749–71.

Weart, Spencer R. 1988. *Nuclear Fear: A History of Images.* Cambridge, Mass.: Harvard University Press.

Weidel, Wolfhard. 1940. *Chemische Untersuchungen über die zur Augenpigmentierung führende Reaktionskette bei Insekten,* Inaugural-Dissertation. Düsseldorf: Dissertations-Verlag G. H. Nolte.

Weindling, Paul. 2003. Genetik und Menschenversuche in Deutschland 1940–1950. In *Rassenforschung an den Kaiser-Wilhelm-Instituten vor und nach 1933,* ed. Hans-Walter Schmuhl, 245–74. Göttingen: Wallstein.

Weismann, August. 1882. *Ueber die Dauer des Lebens.* Jena: Fischer.

———. 1892 (1884). *Über Leben und Tod,* 2nd ed. Jena: Fischer.

Wettstein, Fritz von. 1938. Carl Erich Correns. *Berichte der Deutschen Botanischen Gesellschaft* 56:140–60.

What is Nucleonics? *Nucleonics: Techniques and Applications of Nuclear Science and Engineering* 1 (1) (September 1947): 2.

Whitehouse, W. J., and J. L. Putman. 1953. *Radioactive Isotopes: An Introduction to their Preparation, Measurement and Use.* Oxford: Clarendon.

Whiting, Phineas W. 1919. Genetic studies on the mediterranean flour-moth, Ephestia kühniella Zeller. *Journal of Experimental Zoology* 28:413–45.

———. 1921. Rearing meal moths and parasitic wasps for experimental purposes. *The Journal of Heredity* 12:255–61.

Woodruff, Lorande Loss. 1911. Two thousand generations of paramaecium. *Archiv für Protistenkunde* 21:263–66.

Woodruff, Lorande Loss, and Rhoda Erdmann. 1914. A normal periodic reorganization process without cell fusion in paramaecium. *Journal of Experimental Zoology* 17:425–502.

Wright, Barbara E. 2004. Stress-directed adaptive mutations and evolution. *Molecular Microbiology* 52:643–50.

Yanofsky, Charles, Bruce C. Carlton, John R. Guest, Don R. Helinski, and Ulf Henning. 1964. On the colinearity of gene structure and protein structure. *Proceedings of the National Academy of Sciences of the United States of America* 51:266–72.

Yonath, Ada, William S. Bennet, Sulamith Weinstein, and Heinz Günther Wittmann. 1990. Crystallography and image reconstructions of ribosomes. In *The Ribosome, Structure, Function and Evolution*, ed. Walter E. Hill, Albert Dahlberg, Roger A. Garrett, Peter B. Moore, David Schlessinger, and Jonathan R. Warner, 134–47. Washington, D.C.: American Society for Microbiology.

Zadeh, Lofti A. 1987. Coping with the imprecision of the real world. In *Fuzzy Sets and Applications: Selected Papers by Lofti A. Zadeh*, ed. R. R. Yager, S. Ovchinnikov, R. M. Tong, and H. T. Nguyen, 9–28. New York: John Wiley.

Zamecnik, Paul. 1976. Protein synthesis—early waves and recent ripples. In *Reflections in Biochemistry*, ed. Arthur Kornberg, Bernard L. Horecker, Lluís Cornudella, and John Oro, 303–8. New York: Pergamon.

Zeller, Philipp Christian. 1879. Lepidopterologische Bemerkungen. *Stettiner entomologische Zeitung* 40:462–73.

Zirkle, Conway. 1964. Some oddities in the delayed discovery of Mendelism. *Journal of Heredity* 55:65–72.

Page numbers in italics indicate illustrations.

and, 24; objective, 31; science as progress of, 25–26; societies, 4

Kohler, Robert E., 127, 224

Köhler, Wilhelm, 102

Köhler, Wolfgang, 17

Kölreuter, Joseph G., 60

Körnicke, Friedrich A., 238, 239

Koyré, Alexandre, 40, 44

Kragh, Helge, 41

Krauch, Carl, 136, 137, 138, 148

Krebs, Adolf T., 175

Kremer, Gerhard, 171, 194

Kries, Johannes von, 94

Kriner, James, 200

Kühn, Alfred, 1, 7, 49, 94–127, 129–34, 137, 144–48, 160; on biochemistry, 113–15; biographical note, 94–96; international competition and, 116–18; Rockefeller Foundation support for, 104–9, 113–15; work during war years of, 121–26

Kühn, Julius, 97

Kuhn, Thomas, 20, 42

KWI. See Kaiser Wilhelm Institutes for Biology and Biochemistry (KWI)

Kymographion, 222–23, 223

Laboratory writing, 244–47, 251–52

Lamarck, Jean B., 44

Lang, Anton, 132, 133, 141

Langham, Wright H., 188, 189

Latour, Bruno, xv, 3, 25

Laubichler, Manfred, 17

Lawrence, Ernest, 172

Lecourt, Dominique, 37

LeRoy, George V., 182, 184, 195

Lévi-Strauss, Claude, 8

Libby, Frank W., 181

Life sciences: fragmentation in disciplines, 16, 28; historicism and, 13; history of, 42, 44–45, 49, 220, 222, 238; induction and, 92; model organisms in, 1, 6, 223; physics and, 16;

research technologies in, 215. *See also specific disciplines*

Light microscope, preparations, 219–22, 238–40

Liquid scintillation counter (LSC), 2, 5, 151, *171*, *183*, *232*; automation and, 186–94; coincidence counters, 177–78, *178*; conferences, 197–98, 199–200; cooperative networks, 196–201; development of, 170–72, 176–79; eclipse of, 201–2; industry and, 194–96; market shares and, 189–91; patent for, 283 n. 88; production of, 181–84; Tri-Carb systems and, 170, 182, 184–86, *186*, *187*, 188–90; ultracentrifugation and, 187

Literal techniques, 251

Loftfield, Robert B., 192, 195, 196

Lovejoy, Arthur, 44

LSC. *See* Liquid scintillation counter (LSC)

Ludwig, Carl, 222

Lwoff, André, 147, 168, 202

Magnussen, Karin, 102

Maize (*Zea mays*), 51, 53–54, 57, 71–79, 81, 247

Mamoli, Luigi, 118

Manhattan Project, 172–73, 229

Marey, Etienne-Jules, 223

Marmont, George, 180

Matthaei, Heinrich J., 157, 162

Maupas, Émile, 83–84

Max Planck Institute for Biology, 128

Max Planck Institute for Virus Research, 128, 136, 148

McClintock, Barbara, 164

Met Lab, 199

Meijer, Onno G., 57, 58

Melchers, Georg, 1, 49, 129, 132–36, 141–48

Mendel, Gregor, 42, 51–52, 54–60, 68, 79–80, 100, 153–54, 247–49

Mentzel, Rudolf, 129

Merleau-Ponty, Maurice, 25

Meselson, Matthew, 208

Meyen, Franz F., 219

Meyerson, Emile, 40

Microscopy. *See* Electron microscopy; Light microscope

Miller, Harry M., 105–6

Model organisms, 223–24; conditioning of, 124; defined, 6–7; epistemic objects and, 224; as instrument, 224; molecular genetics and, 7, 127, 157; use of, 5, 7–8, 49, 92, 125, 154, 223–24. *See also specific model organisms*

Molecular biology: assemblages in, 5, 6, 10; defined, 4–5, 157; on gene, 153; history of, 9–10, 49, 128; preparations in, 241–43; specificity in, 5–6; time-frames in, 10. *See also* Classical genetics

Molecular genetics: emergence of, 6, 157–58; gene as central to, 1, 151, 156, 160; genetic manipulation and, 154–55, 159; humanists and, 4; linguistics and, 212, 213; microbial systems and, 210; model organisms and, 7, 127; oversimplification in, 162; terminology in, 203, 204, 205–6, 211, 212–14. *See also* Classical genetics

Monod, Jacques, 127, 162, 203–11, 213; PaJaMo experiments and, 204–8

Morange, Michel, 4, 9, 10

Morgan, Thomas Hunt, 7, 96, 132, 224

Müller, Hans, 133, 143, 147

Muller, Herman J., 154

Müller-Cunradi, Martin, 138

Mutations: as analytical tools, 120, 124, 134; characters and, 101; gene-action chain and, 122; genes and, 153, 156, 159, 164–67, 209; genetic factors and, 100; long-term effects, 180; in tobacco mosaic virus, 142–44; triggering of, 107

Nägeli, Carl W. von, 52, 57

National Socialism, effect on research, 130, 148

Natural sciences: cultural sciences and, 9; defined, 19; epistemology and, 14, 22, 233; history and, 13, 23; humanities and, 3, 4; image of rigidity and, 19–20; as mechanistic, 19–20, 21; naturalism in, 23; phenomeno-technical aspects of, 21; positivism in, 2; research objects and, 2, 15, 247; self-conception of, 22

Naudin, Charles, 60

Navashin, Sergej G., 54

Neuberg, Carl, 116, 128–29, 132, 149

New England Nuclear Corporation, 192, 197

Nirenberg, Marshall W., 157, 162

North American Philips Company, 173

Northwestern University (Chicago, Ill.), 197–98

Noumenology, 27, 33, 34, 35–36

Novick, Aaron, 180

Nuclear Chicago, 185, 189–90

Nuclear technology journals, 199

Olby, Robert, 52

Operon, use of term, 209

Organisms. *See* Model organisms

Packard Bioscience Company, 202

Packard Instrument Company, 171, 182, 184, 196, 197, 199, 200–201, *201*, 202

Packard, Lyle E., 170, 179–202

PaJaMo experiments, 204–8

Papilloma virus, 140

Pardee, Arthur, 204, 205

Pasteur Institute, 151, 203, 204, 205, 207, 210

Patat, Franz, 139

Pattee, Howard, 167

Pea. *See* Garden pea (*Pisum sativum*)

Pfankuch, Edgar, 142, 143, 144

Pfeffer, Wilhelm, 52

Phage research. *See* Bacteriophage research

Phenomenology: concept of, 25, 26–27; empiricism and, 19, 27; phenomeno-technique as extension of, 31, 215; subjectivism/conceptualism and, 25

Phenomenotechnique: as extension of phenomenology, 31, 215; in natural sciences, 21; rational application and, 34; scientific objectivity and, 31

Phenomes: genomes and, 154, 166

Phenotypes. *See* Genotype-phenotype distinction

"Philosophy and the Crisis of European Humanity" (Husserl), 22

Philosophy of science: as distributed, 28; horizontal/vertical mobility in, 28–29; justification and discovery in, 2; positioning in dichotomies of, 27

Philosophy of the As If (Vaihinger), 35

Phytopathology, 8

Pickels, Edward G., 133

Picker Nuclear Company, 190

Piepho, Hans, 94, 108, 112, 115

Pisum sativum. *See* Garden pea (*Pisum sativum*)

Plagge, Ernst, 94, 109–15

Pohl, Robert, 95

Polanyi, Michael, 167

Polyhedral virus, 139

Portin, Petter, 155

Positivism: crisis in, 2; historical, 39; ideological, 22; knowledge of nature and, 23

Preparation of objects: anatomical, 235–37, *235–36*; epistemologica and, 233–35, 242–43; herbaria and, 238, 239; microscopic, 219–22, 228, 238–40, *240*; in molecular biology, 241–43

Propagation, 83, 88, 91

Protein synthesis, 49; genome and, 210;

homopolymer translation in, 162; models of, 207; regulation of, 206; RNA and, 157, 207, 208

Protozoa: biological regulation and, 82; death and, 85; fertilization and, 86; rejuvenation and, 83–87, 89–90

Ptychopoda seriata, 108–9, 124–25

Raben, Maury S., 176–77

Rabinow, Paul, 5

Radiation Counter Laboratories, 173

Radiation research, 174–76

Radioactive tracing, 2, 141, 144, 151, 171, 172–74, 202, 229–31. *See also* Liquid scintillation counter (LSC)

Radiochemical Center, 192

Rapkin, Edward, 185–86, 189–91, 194–95, 197, 199

Rationalism: applied, 28, 34; industry and, 33

Realities, three, 17–18, 21

Realization and objects, 31

Recursion, 29, 32–33, 36, 40, 42

Regener, Erich, 174

Reines, Frederick, 178

Rejuvenation and protozoa, 83–86, 89–90

Renard, Gilles, 41, 44, 47

Renner, Otto, 123

Reynolds, George T., 176, 177, 182

Ribonucleic acid. *See* RNA (ribonucleic acid)

Rickert, Heinrich, 14, 94

Riezler, Kurt, 16–18, 21–22

Rimpau, Wilhelm, 54, 58

RNA (ribonucleic acid): information-bearing, 207–11; protein synthesis and, 157, 161, 207, 208, 209, 211; splicing, 163; structure analysis of, 146; transcripts, 162, 163; viruses and, 143

Roberts, Herbert F., 55, 58

Rockefeller Foundation: on molecular

Hans-Jörg Rheinberger is professor and director at
the Max Planck Institute for the History of Science,
Berlin. He is the author of *On Historicizing Epistemology:
An Essay* (2010) and *Toward a History of Epistemic Things:
Synthesizing Proteins in the Test Tube* (1997), and the co-
editor of *Heredity Produced* (2007); *Classical Genetic
Research and Its Legacy* (2004); *From Molecular Genetics to
Genomics* (2004); *Reworking the Bench: Research Notebooks
in the History of Science* (2003); and *The Concept of the Gene
in Development and Evolution* (2000).

∴

Library of Congress Cataloging-in-Publication Data
Rheinberger, Hans-Jörg.
[Epistemologie des Konkreten. English]
An epistemology of the concrete : twentieth-century
histories of life / Hans-Jörg Rheinberger ; foreword
by Tim Lenoir.
p. cm. — (Experimental futures : technological lives,
scientific arts, anthropological voices)
Includes bibliographical references and index.
ISBN 978-0-8223-4560-2 (cloth : alk. paper)
ISBN 978-0-8223-4575-6 (pbk. : alk. paper)
1. Biology—Philosophy. 2. Biology—History—20th
century. 3. Biology—Methodology. 4. Knowledge,
Theory of. I. Title. II. Series: Experimental futures.
QH331.R4513 2010
570.1—dc22   2010014581